A Textbook of Partial Differential Equation

Classical and Mathematical Software-based Solution Approach

A Textbook of Partial Differential Equation

Classical and Mathematical Software-based Solution Approach

Dr Yogeshwari F. Patel
Dr. Jayesh M. Dhodiya
Dr. Feni Chevli

CWP

This edition has been published by Central West Publishing PTY LTD, (ABN 13 683 898 722) Australia

For more information about the books published by Central West Publishing PTY LTD, please visit https://centralwestpublishing.com

Disclaimer
Every effort has been made by the publisher, editors and authors while preparing this book, however, no warranties are made regarding the accuracy and completeness of the content. The publisher, editors and authors disclaim without any limitation all warranties as well as any implied warranties about sales, along with fitness of the content for a particular purpose. Citation of any website and other information sources does not mean any endorsement from the publisher, editors and authors. For ascertaining the suitability of the contents contained herein for a particular lab or commercial use, consultation with the subject expert is needed. In addition, while using the information and methods contained herein, the practitioners and researchers need to be mindful for their own safety, along with the safety of others, including the professional parties and premises for whom they have professional responsibility. To the fullest extent of law, the publisher, editors and authors are not liable in all circumstances (special, incidental, and consequential) for any injury and/or damage to persons and property, along with any potential loss of profit and other commercial damages due to the use of any methods, products, guidelines, procedures contained in the material herein.

NATIONAL LIBRARY OF AUSTRALIA

A catalogue record for this book is available from the National Library of Australia

ISBN (print): 978-1-922617-69-9

Preface

Partial differential equations provide a generic description of the majority of physical events, whether they are related to heat flow, fluid dynamics, electricity, magnetism, mechanics, or optics. Such equations are, in fact, essential to mathematical physics. Even if these equations can be simplified to become ordinary differential equations, the full description of physical systems can be found in the general field of partial differential equations. This extremely helpful literature demonstrates to the reader how to create a mathematical model of the physical problem, how to generate a partial differential equation from it, and how to solve the equation (along with initial and boundary conditions).

This succinct guide is devoted to assisting engineering students in solving problems in their profession that include partial differential equations and only requires a basic understanding of ordinary differential equations. Questions of existence, uniqueness, and convergence are not emphasised; instead, the treatment concentrates solely on the useful solution of partial differential equations. The more prevalent partial differential equations related to phenomena including vibration, heat transport, electricity, and elasticity are derived. The book is organized in the following manner:

Chapter 1 entitled Partial Differential Equation describes the definitions of partial differential equation, formulation of PDE, classification of PDE, solution linear and nonlinear PDE. It chapter also deals with the general method to solve the PDE equation of linear and nonlinear PDE.

Chapter 2 entitled First Order Partial Differential Equations and its Solution discusses the linear and nonlinear first order PDE and different approaches to solve it.

Chapter 3 entitled Higher Order Partial Differential Equation deals with the solution of linear homogenous and non-homogenous PDE of second order with constant as well as variable coefficients. Method of separation of variables is also described to find the solution of higher order PDE.

Chapter 4 Solution Method for Higher Order Partial Differential Equation discusses the analytical and numerical approaches to find the solution of PDE with MATLAB coding.

Chapter 5 entitled Heat Equation initially describes the derivation of one-dimensional heat equation of parabolic nature. Chapter 3 explains the solution of the heat equation with initial and boundary condition with large illustration in details.

Chapter 6 entitled Wave Equation deals with solution of vibrating string and it is explained using bisection problem, section problem and Trisection problem. It also describes the solution of Wave Equation using D Almbert 's Method.

Chapter 7 Laplace Equation deals with solution of Laplace Equation using Fourier series using large number of illustrations.

An in-depth mathematical treatment of nonlinear partial differential equations, and the method for solving them is key feature of the proposed book. Readers will gain an understanding of the subject and its many applications through these solutions. Every concept and method are explained thoroughly and clearly along with carefully selected examples, with special emphasis on areas in which students have difficulty. The book's organization makes it possible for instructors to select chapters based on their length and emphasis on a different level of courses. Every chapter concludes with a wide selection of exercises that will assist the reader in developing both manipulative skills and greater insight into nonlinear equations. At the end of the exercise, students will find answers and hints to selected exercises with original source material.

Table of Contents

Chapter 1 Partial Differential Equation

1.1 Introduction

In the field of applied science and technology, the analysis of dynamical systems is fundamental to understanding the behavior of various physical, chemical, and biological phenomena. One of the most powerful mathematical tools for studying such systems is the partial differential equation (PDE). As the number of independent variables increases, so does the complexity and diversity of the problems—making PDEs essential in modeling systems that vary with respect to space and time.

Real-world problems often involve functions of several independent variables, such as time, temperature, position, or pressure, which naturally lead to the formulation of PDEs. Applications of PDEs are found in nearly every branch of science and engineering, including fluid mechanics, heat and mass transfer, electromagnetic theory, and even biological systems. For example, in fluid dynamics, PDEs describe the movement of fluids; in biology, they can model population dynamics or the spread of disease; in physics, they are central to the laws of quantum mechanics and electromagnetism.

To solve a PDE that models a real-world phenomenon, we often require additional conditions—known as initial and boundary conditions—that reflect the physical constraints of the problem. Together with the differential equation, these conditions form an initial value problem or a boundary value problem, depending on the context.

Many real-life variables such as temperature, pressure, or concentration depend on multiple parameters—such as time, latitude, longitude, and altitude. PDEs allow us to model how such quantities change with respect to one parameter while holding others constant. For instance, if we are interested in how temperature changes over time at a fixed location, we differentiate with respect to time while holding spatial coordinates constant. Similarly, if we want to see how temperature changes with altitude, we take the derivative with respect to altitude, keeping other variables fixed.

Partial differential equations lie at the heart of multivariable calculus and advanced mathematical modeling. They enable us to describe the evolution of physical systems, from the diffusion of gases and wave propagation to the formation of galaxies and the underlying structure of quantum mechanics.

In this chapter, we begin by introducing the definition of PDEs, followed by their classification into types based on order and linearity. We then explore various solution techniques, including analytical and numerical methods, and finally, we study the formulation of PDEs derived from physical laws. This structured approach will provide a solid foundation for understanding and applying PDEs to solve real-world scientific and engineering problems.

1.2 Partial Differential Equation

Partial differential equations (PDE) is a relation between an unknown function and its partial derivatives. It is defined as an equation involving two (or more) independent variables like *x, y* and a dependent variable *u* and its partial derivatives such as

$$\frac{\partial u}{\partial x}, \frac{\partial u}{\partial y}, \frac{\partial^2 u}{\partial x^2}, \frac{\partial^2 u}{\partial y^2}, \frac{\partial^2 u}{\partial x \partial y} \text{ etc.}$$

In general, mathematically, PDE can be written as

$$F\left(x_1, x_2, x_3, \ldots\ldots\ldots, x_n, u_{x_1}, u_{x_2}, u_{x_3}, \ldots\ldots u_{x_n}\right) = 0 \qquad (1.1)$$

where $x_1, x_2, x_3, \ldots\ldots\ldots, x_n$ are independent variables, *u* is unknown function of these variables and $u\left(x_1, x_2, x_3, \ldots\ldots\ldots, x_n\right)$ are dependent variables , $u_{x_i} = \frac{\partial u}{\partial x_i}$.

The following are some of the well-known as well as other example of PDEs

$$\frac{\partial^2 u}{\partial x^2}+\frac{\partial^2 u}{\partial y^2}=0 \text{ (Two-Dimensional Laplace Equation)} \tag{1.2}$$

$$\frac{\partial^2 u}{\partial t^2}=c^2\frac{\partial^2 u}{\partial x^2} \text{ (One Dimensional Wave Equation)} \tag{1.3}$$

$$\frac{\partial u}{\partial t}=c^2\frac{\partial^2 u}{\partial x^2} \text{ (One Dimensional Heat Equation)} \tag{1.4}$$

$$\frac{\partial u}{\partial t}+c\frac{\partial u}{\partial x}=0 \tag{1.5}$$

$$u_x u_{xx}+\left(u_y\right)^2=0 \tag{1.6}$$

$$\left(u_{xx}\right)^2+\left(u_{yy}\right)^2+f(x,y)u_x+g(x,y)u=0 \tag{1.7}$$

$$\frac{\partial^2 u}{\partial x^2}+\frac{\partial^2 u}{\partial y^2}=f\left(x,y\right) \tag{1.8}$$

Here *c* is a constant, *t* is time, *x, y, z* are Cartesian coordinates, and dimension is the number of these coordinates in the equation.

Notation: In the study of PDE following notation are generally considered:

$$p=\frac{\partial u}{\partial x},q=\frac{\partial u}{\partial y},r=\frac{\partial^2 u}{\partial x^2},t=\frac{\partial^2 u}{\partial y^2},s=\frac{\partial^2 u}{\partial x\partial y}.$$

Order of a PDE: The order of PDE is the order of the highest order derivative present in the given partial differential equation.

Example:

The order of the partial differential equation

$$\left(\frac{\partial^3 z}{\partial x^3}\right)^4+\frac{\partial^2 z}{\partial x \partial y}+\frac{\partial z}{\partial x}=x \text{ is 3.}$$

Example:

The order of the partial differential equation

$$\frac{\partial^2 u}{\partial x^2}+\frac{\partial^2 u}{\partial y^2}=0 \text{ is 2.}$$

Degree of a PDE: The degree of PDE is the greatest power of the highest order partial derivatives involved in partial differential equation provided the equation can be written in some polynomial form.

Example:

The degree of the partial differential equation

$$\left(\frac{\partial^3 z}{\partial x^3}\right)^4+\frac{\partial^2 z}{\partial x \partial y}+\frac{\partial z}{\partial x}=x \text{ is 4.}$$

Example:

The degree of the partial differential equation

$$\frac{\partial^2 u}{\partial x^2}+\frac{\partial^2 u}{\partial y^2}=0 \text{ is 1.}$$

1.3 Classification of PDE and Type of Solution

Understanding the classification of partial differential equations is essential for choosing suitable techniques for solving them and comprehending the behaviour of solutions in diverse physical scenarios. Different PDE types can have different qualitative behaviours and call for different approaches to solving them. PDEs can be classified in several ways based on various criteria. The common classification conditions are as follows:

Linear PDE: A PDE is said to be linear if it is linear in the unknown function and all its derivatives with coefficients depending only on the independent variables. For example, (1.2) - (1.5) are linear PDEs.

Non-Linear PDE: A PDE is said to be nonlinear if it is not linear. For example, (1.6), (1.7) are nonlinear P.D.Es.

Quasi linear PDE: A PDE is said to be quasi linear if it is linear in the highest order derivatives with coefficients depending on the independent variables, the unknown functional and its derivatives of order lower than the order of the equation. These are the special cases of nonlinear P.D.Es. For example, (1.6) is quasi linear second order PDEs.

Homogeneous and Non-Homogeneous PDE: If in the P.D. E each term contains the dependent variable or its derivatives then it is called homogeneous P.D.E. Otherwise it is called non-homogeneous P.D.E. For example, (1.2) -(1.5) are all homogeneous PDEs while (1.8) is a non-homogeneous equation where $f(x,y)\neq 0$.

Solution of PDE: A solution $F\left(x,y,\ldots\ldots,u,u_x,u_y,\ldots.,u_{xx},u_{yy},u_{xy},\ldots\ldots\right)=0$ is function *u* satisfying the given P.D.E. A solution of PDE is generally not unique, an additional condition must generally be specified on the boundary of the region where solution is defined i.e. once initial conditions and the boundary conditions, where appropriate, are specified, then there will be a unique solution to PDE. The various solution associate with the PDEs

Solution or Integral of P.D.E: It is a relation between dependent and independent variable that satisfies the differential equation.

Complete Integral or Complete Solution of PDE: It is a relation between the variables which satisfies the equation and contains as many arbitrary constants as the number of independent variables.

Particular Integral: It is the complete integral with arbitrary constants assigned particular values.

Singular solution or Integral Solution of PDE: It is relation between the variables that involves no arbitrary constant and is not obtainable as a particular integral from the complete integral.

General Solution or Integral of PDE: A general solution is a solution that contains several arbitrary independent functions equal to the order of the equation.

1.4 Formulation of PDE

This section describes the two different procedures for the formation of partial differential equations from multi-dimensional mathematical equation:

I. By eliminating the arbitrary constant.
II. By eliminating arbitrary function.

Method I: By eliminating arbitrary constant

Suppose $F(x, y, z, a, b) = 0$, where *a* and *b* are arbitrary constants. Let *z* be considered as a function of two independent variables *x* and *y*. Differentiating $F(x, y, z, a, b) = 0$ partially with respect to *x* and *y*, we get

$$\frac{\partial F}{\partial x}+p\frac{\partial F}{\partial z}=0 \tag{1.9}$$

$$\frac{\partial F}{\partial x}+q\frac{\partial F}{\partial z}=0 \tag{1.10}$$

Now, eliminating two constant a and b from (1.9) and (1.10), the required partial differential equation of order one is $f(x,y,z,p,q)=0$.

Notes:

1. The obtained partial differential equation may or may not be unique.

2. If number of independent variables is equal to number of arbitrary constants, then first order PDE exit.

3. If the number of independent variables is less than the number of arbitrary constants, then PDE is of second or higher order.
4. If there are less arbitrary constants than the number of independent variables, then the elimination of constants usually give rise to more than one PDE of first order. For example, if $z=x+by$ then the partial differential equations are $q=\frac{z-x}{y}$ and $p=1$.

Example: Derive PDE by eliminating *a* and *b* from $z=(x-a)^2+(y-b)^2$.

Solution: Here $z=(x-a)^2+(y-b)^2$ (1.11)

Since the number of independent variables is equal to number of arbitrary constants therefore first order PDE exist. Differentiating equation (1.11) partially w.r.t x and y, we get

$$\frac{\partial z}{\partial x}=2(x-a)\Rightarrow\frac{1}{2}\frac{\partial z}{\partial x}=(x-a) \quad (1.12)$$

$$\frac{\partial z}{\partial y}=2(y-a)\Rightarrow\frac{1}{2}\frac{\partial z}{\partial y}=(y-a) \quad (1.13)$$

Putting the value of $(x-a)$and $(y-a)$ from equations (1.12) and (1.13) in equation (1.11), we get

$$\left(\frac{1}{2}\frac{\partial z}{\partial x}\right)^2+\left(\frac{1}{2}\frac{\partial z}{\partial y}\right)^2=z$$, which is required PDE.

Example:

Find partial differential equation by eliminating *a* and *b* from $z=ax+by+a^2+b^2$.

Solution:

Here $z=ax+by+a^2+b^2$ (1.14)

Since the number of independent variables is equal to number of arbitrary constants, therefore, first order PDE exist. Differentiating equation (1.14) partially w.r.t x and y, we get

$$\frac{\partial z}{\partial x}=a \quad (1.15)$$

$$\frac{\partial z}{\partial y}=b \quad (1.16)$$

Putting the value of a and b from equations (1.15) and (1.16) in equation (1.14), we get

$$z = x\left(\frac{\partial z}{\partial x}\right) + y\left(\frac{\partial z}{\partial y}\right) + \frac{\partial z}{\partial x}.\frac{\partial z}{\partial y}$$, which is required PDE.

Example: Find PDE by eliminating *a* and *b* from

$z = axe^{y} + \frac{1}{2}a^{2}e^{2y} + b$.

Solution:

Here $z = axe^{y} + \frac{1}{2}a^{2}e^{2y} + b$ (1.17)

Since the number of independent variables is equal to number of arbitrary constants therefore first order PDE exist. Differentiating equation (1.17) partially w.r.t *x* and *y*, we get

$$\frac{\partial z}{\partial x} = ae^{y} \quad (1.18)$$

$$\frac{\partial z}{\partial y} = axe^{y} + a^{2}e^{2y} = a\left(xe^{y}\right) + \left(ae^{y}\right)^{2} \quad (1.19)$$

Putting the value of ae^{y} from equation (1.18) into equation (1.19), we get

$$\frac{\partial z}{\partial y} = x\left(\frac{\partial z}{\partial x}\right) + \left(\frac{\partial z}{\partial x}\right)^{2}$$, which is required PDE.

Example:

Find the differential equation of the set of all right circular cones whose axes coincide with z-axis.

Solution:

The general equation of the set of all right circular cones having semi vertical angle α and centre $(0,0,a)$ whose axes coincide with z-axis is given by

$$x^2+y^2=(z-a)^2\tan^2\alpha \tag{1.20}$$

where a and α are arbitrary constants. Since the number of independent variables is equal to number of arbitrary constants therefore first order PDE exist. Differentiating equation (1.20) partially w.r.t x and y, we get

$$2x=2(z-a)\frac{\partial z}{\partial x}\tan^2\alpha \tag{1.21}$$

$$2y=2(z-a)\frac{\partial z}{\partial y}\tan^2\alpha \tag{1.22}$$

Dividing equation (1.21) by equation (1.22), we get

$$\frac{2x}{2y}=\frac{2(z-c)\frac{\partial z}{\partial x}\tan^2\alpha}{2(z-c)\frac{\partial z}{\partial y}\tan^2\alpha}\Rightarrow\frac{x}{y}=\frac{\frac{\partial z}{\partial x}}{\frac{\partial z}{\partial y}}\Rightarrow x\frac{\partial z}{\partial y}=y\frac{\partial z}{\partial x},$$

which is required PDE.

Example:

Find PDE by eliminating a, b and c from $\frac{x^2}{a^2}+\frac{y^2}{b^2}+\frac{z^2}{c^2}=1.$

Solution: Here $\frac{x^2}{a^2}+\frac{y^2}{b^2}+\frac{z^2}{c^2}=1$ (1.23)

Since the number of independent variables is not equal to number of arbitrary constants therefore second order PDE exist. Differentiating equation (1.23) partially w.r.t x and y, we get

$$\frac{2x}{a^2}+\frac{2z}{c^2}\frac{\partial z}{\partial x}=0 \Rightarrow -\frac{z}{x}\frac{\partial z}{\partial x}=\frac{c^2}{a^2} \tag{1.24}$$

$$\frac{2y}{b^2}+\frac{2z}{c^2}\frac{\partial z}{\partial y}=0 \Rightarrow -\frac{z}{y}\frac{\partial z}{\partial y}=\frac{c^2}{b^2} \tag{1.25}$$

Differentiating Eq. (1.24) partially w.r.t x, we get

$$\frac{z}{x}\frac{\partial^2 z}{\partial x^2}+\frac{\partial z}{\partial x}\left(\frac{x\frac{\partial z}{\partial x}-z}{x^2}\right)=0 ,$$

$$\frac{z}{x}\frac{\partial^2 z}{\partial x^2}+\frac{1}{x}\left(\frac{\partial z}{\partial x}\right)^2-\frac{z}{x^2}\left(\frac{\partial z}{\partial x}\right)=0$$, which is required PDE.

The other possible PDE can be obtained by differentiating equation (1.24) w.r.t to y or equation (1.25) w.r.t to y.

Method: II: Method of eliminating arbitrary functions

Case 1: Let the given equation be $z=f(u)$. (1.26)

where *u* is a function of x, y, z. Differentiating equation (1.26), we get

$$\frac{\partial z}{\partial x}=\frac{\partial f}{\partial u}.\frac{\partial u}{\partial x}+\frac{\partial f}{\partial u}.\frac{\partial u}{\partial z}.\frac{\partial z}{\partial x} \tag{1.27}$$

$$\frac{\partial z}{\partial y}=\frac{\partial f}{\partial u}.\frac{\partial u}{\partial y}+\frac{\partial f}{\partial u}.\frac{\partial u}{\partial z}.\frac{\partial z}{\partial y} \tag{1.28}$$

By eliminating the arbitrary function *f* from equations (1.26), (1.27), and (1.28), a PDE of first order is obtained.

Case 2: Let the given equation be $F(u,v)=0$ (1.29)

where *u* is a function of x, y, z.

Differentiating equation (1.29) partially w.r.t x and y, we get

$$\frac{\partial F}{\partial u}\left(\frac{\partial u}{\partial x}+\frac{\partial u}{\partial z}.\frac{\partial z}{\partial x}\right)+\frac{\partial F}{\partial v}\left(\frac{\partial v}{\partial x}+\frac{\partial v}{\partial z}.\frac{\partial z}{\partial x}\right)=0,$$

$$\frac{\partial F}{\partial u}\left(\frac{\partial u}{\partial y}+\frac{\partial u}{\partial z}.\frac{\partial z}{\partial y}\right)+\frac{\partial F}{\partial v}\left(\frac{\partial v}{\partial y}+\frac{\partial v}{\partial z}.\frac{\partial z}{\partial y}\right)=0 \text{ respectively.}$$

Eliminating $\frac{\partial F}{\partial u}$ and $\frac{\partial F}{\partial v}$, we get

$$\begin{vmatrix} \frac{\partial u}{\partial x}+\frac{\partial u}{\partial z}.\frac{\partial z}{\partial x} & \frac{\partial v}{\partial x}+\frac{\partial v}{\partial z}.\frac{\partial z}{\partial x} \\ \frac{\partial u}{\partial y}+\frac{\partial u}{\partial z}.\frac{\partial z}{\partial y} & \frac{\partial v}{\partial y}+\frac{\partial v}{\partial z}.\frac{\partial z}{\partial y} \end{vmatrix}=0,$$

Expanding this determinant, a PDE of first order is obtained.

Example:

Form a partial differential equation by elimination of the arbitrary functions from $z=f(x+ay)+g(x-ay)$.

Solution:

$$z=f(x+ay)+g(x-ay) \tag{1.30}$$

Differentiating partially w.r.t *x* and *y* we get

$$\frac{\partial z}{\partial x}=f'(x+ay)+g'(x-ay) \tag{1.31}$$

$$\frac{\partial z}{\partial y} = af'(x+ay) - ag'(x-ay) \tag{1.32}$$

Again, differentiating partially equation (1.31) and (1.32) w.r.t *x* and *y*, we get

$$\frac{\partial^2 z}{\partial x^2} = f''(x+ay) + g''(x-ay) \tag{1.33}$$

$$\frac{\partial^2 z}{\partial y^2} = a^2 f''(x+ay) + a^2 g''(x-ay) \tag{1.34}$$

From equations (1.33) and (1.34), we get $a^2 \frac{\partial^2 z}{\partial x^2} = \frac{\partial^2 z}{\partial y^2}$, which is required PDE.

Example:

Form a partial differential equation by elimination of the arbitrary functions from $x+y+z = f\left(x^2+y^2+z^2\right)$.

Solution:

$$x+y+z = f\left(x^2+y^2+z^2\right) \tag{1.35}$$

Differentiating Eq. (1.35) partially w.r.t *x* and *y*, we get

$$1+\frac{\partial z}{\partial x} = 2x\, f'\left(x^2+y^2+z^2\right) \tag{1.36}$$

$$1+\frac{\partial z}{\partial y} = 2y\, f'\left(x^2+y^2+z^2\right) \tag{1.37}$$

Dividing equation (1.36) by equation (1.37), we get

$$\frac{1+\frac{\partial z}{\partial x}}{1+\frac{\partial z}{\partial y}}=\frac{2x f'\left(x^2+y^2+z^2\right)}{2y f'\left(x^2+y^2+z^2\right)} \Rightarrow \frac{1+\frac{\partial z}{\partial x}}{1+\frac{\partial z}{\partial y}}=\frac{x}{y},$$

$y\left(1+\frac{\partial z}{\partial x}\right)=x\left(1+\frac{\partial z}{\partial y}\right)$, which is required PDE.

Example:

Form a partial differential equation by elimination of the arbitrary functions from $z=e^{ax+by} f(ax-by)$

Solution:

Here $z=e^{ax+by} f(ax-by)$ (1.38)

Differentiating partially (1.38) w.r.t *x* and *y*, we get

$$\frac{\partial z}{\partial x}=e^{ax+by} a f'(ax-by)+ae^{ax+by} f(ax-by) \quad (1.39)$$

$$\frac{\partial z}{\partial y}=e^{ax+by}\left(-bf'(ax-by)\right)+be^{ax+by} f(ax-by) \quad (1.40)$$

Multiplying equation (1.39) by *b* and equation (1.40) by *a* and adding, we get

$$b\left(\frac{\partial z}{\partial x}\right)+a\left(\frac{\partial z}{\partial y}\right)=2abe^{ax+by} f(ax-by).$$

From equation (1.38), we get

$b\left(\frac{\partial z}{\partial x}\right)+a\left(\frac{\partial z}{\partial y}\right)=2abz$, which is required PDE.

Example:

Form a partial differential equation by elimination of the arbitrary function from $f\left(xy+z^2,x+y+z\right)=0$.

Solution:

Here $f\left(xy+z^2,x+y+z\right)=0$ (1.41)

Let $u=xy+z^2, v=x+y+z$ then above equation reduces to

$$f\left(u,v\right)=0. \tag{1.42}$$

Differentiating Eq. (1.42) partially w.r.t x and y we get

$$\frac{\partial f}{\partial u}\left(\frac{\partial u}{\partial x}+\frac{\partial u}{\partial z}.\frac{\partial z}{\partial x}\right)+\frac{\partial f}{\partial v}\left(\frac{\partial v}{\partial x}+\frac{\partial v}{\partial z}.\frac{\partial z}{\partial x}\right)=0 \tag{1.43}$$

$$\frac{\partial f}{\partial u}\left(\frac{\partial u}{\partial y}+\frac{\partial u}{\partial z}.\frac{\partial z}{\partial y}\right)+\frac{\partial f}{\partial v}\left(\frac{\partial v}{\partial y}+\frac{\partial v}{\partial z}.\frac{\partial z}{\partial y}\right)=0 \tag{1.44}$$

respectively. Here $\frac{\partial u}{\partial x}=y, \frac{\partial u}{\partial y}=x, \frac{\partial u}{\partial z}=2z, \frac{\partial v}{\partial x}=1, \frac{\partial v}{\partial y}=1, \frac{\partial v}{\partial z}=1.$

Therefore, equations (1.43) and (1.44) reduces to

$$\frac{\partial f}{\partial u}\left(y+2z\frac{\partial z}{\partial x}\right)+\frac{\partial f}{\partial v}\left(1+\frac{\partial z}{\partial x}\right)=0 \tag{1.45}$$

$$\frac{\partial f}{\partial u}\left(x+2z\frac{\partial z}{\partial y}\right)+\frac{\partial f}{\partial v}\left(1+\frac{\partial z}{\partial y}\right)=0 \tag{1.46}$$

By eliminating $\frac{\partial f}{\partial u}$ and $\frac{\partial f}{\partial v}$ from equations (1.45) and (1.46), we get

$$\left(1+\frac{\partial z}{\partial y}\right)\left(y+2z\frac{\partial z}{\partial x}\right)=\left(1+\frac{\partial z}{\partial x}\right)\left(x+2z\frac{\partial z}{\partial y}\right) \tag{1.47}$$

On rearranging equation (1.47), we get

$(2z-x)\frac{\partial z}{\partial x}+(y-2z)\frac{\partial z}{\partial y}=x-y$, which is required PDE.

Example:

Form a partial differential equation by elimination of the arbitrary function from $f\left(x^2+y^2+z^2,z^2-2xy\right)=0$.

Solution:

Here $f\left(x^2+y^2+z^2,z^2-2xy\right)=0$ (1.48)

Let $u=x^2+y^2+z^2, v=z^2-2xy$ then equation (1.48) reduces to

$$f(u,v)=0 \tag{1.49}$$

Differentiating equation (1.49) partially w.r.t x and y, we get

$$\frac{\partial f}{\partial u}\left(\frac{\partial u}{\partial x}+\frac{\partial u}{\partial z}.\frac{\partial z}{\partial x}\right)+\frac{\partial f}{\partial v}\left(\frac{\partial v}{\partial x}+\frac{\partial v}{\partial z}.\frac{\partial z}{\partial x}\right)=0 \tag{1.50}$$

$$\frac{\partial f}{\partial u}\left(\frac{\partial u}{\partial y}+\frac{\partial u}{\partial z}.\frac{\partial z}{\partial y}\right)+\frac{\partial f}{\partial v}\left(\frac{\partial v}{\partial y}+\frac{\partial v}{\partial z}.\frac{\partial z}{\partial y}\right)=0 \tag{1.49}$$

respectively.

Here $\frac{\partial u}{\partial x}=2x, \frac{\partial u}{\partial y}=2y, \frac{\partial u}{\partial z}=2z, \frac{\partial v}{\partial x}=-2y, \frac{\partial u}{\partial y}=-2x, \frac{\partial v}{\partial z}=2z$.

Therefore, equations (1.50) and (1.51) are reduced to

$$\frac{\partial f}{\partial u}\left(2x+2z\frac{\partial z}{\partial x}\right)+\frac{\partial f}{\partial v}\left(2y+\frac{\partial z}{\partial x}\right)=0 \tag{1.52}$$

$$\frac{\partial f}{\partial u}\left(2y+2z\frac{\partial z}{\partial y}\right)+\frac{\partial f}{\partial v}\left(-2x+2z\frac{\partial z}{\partial y}\right)=0 \tag{1.53}$$

By eliminating $\frac{\partial f}{\partial u}$ and $\frac{\partial f}{\partial v}$ from equations (1.52) and (1.53), we get

$$\frac{2x+2z\frac{\partial z}{\partial x}}{2y+2z\frac{\partial z}{\partial y}}=\frac{-2y+2z\frac{\partial z}{\partial x}}{-2x+2z\frac{\partial z}{\partial y}} \tag{1.54}$$

On rearranging Eq. (1.47), we get

$(x+y)\left[z(p-q)+(y-x)\right]=0$, which is required PDE.

Exercise 1.1

Find the order and degree of the following partial differential equations:

1. $\left(\frac{\partial^3 f}{\partial x^3}\right)^2+xy\frac{\partial^3 f}{\partial x^2\partial y}+2\frac{\partial^3 f}{\partial x\partial y^2}+\frac{\partial^3 f}{\partial y^3}=0$
2. $\left(\frac{\partial^3 f}{\partial x^3}\right)^2+2\frac{\partial^2 f}{\partial x\partial y}+\frac{\partial f}{\partial y}=0$
3. $\left(\frac{\partial^3 f}{\partial x^3}\right)^2+2\left(\frac{\partial^2 f}{\partial x\partial y}\right)^3+\frac{\partial f}{\partial y}=0$

4. $\left(\frac{\partial^3 f}{\partial x^3}\right)^2 + 2\left(\frac{\partial^2 f}{\partial x \partial y}\right)^{\frac{3}{2}} + \frac{\partial f}{\partial y} = 0$

5. $x^2 \frac{\partial^3 f}{\partial x^3} + xy \frac{\partial^3 f}{\partial x^2 \partial y} + 2\frac{\partial^3 f}{\partial x \partial y^2} + \frac{\partial^3 f}{\partial y^3} = 0$

Obtain the partial differential equation by eliminating arbitrary constant or function from the following:

1.	$z = ax^2 + by^2$	**Ans:** $2z = xp + yq$
2.	$(x+a)^2 + (y+b)^2 + z^2 = 4$	**Ans:** $z^2\left(p^2 + q^2 + 1\right) = 4$
3.	$x^2 + y^2 + (z-a)^2 = b^2$	**Ans:** $xq - yp = 0$
4.	$ax + by + cz = 1$	**Ans:** $r = 0, s = 0, t = 0$
5.	$z = f(x)g(y)$	**Ans:** $pq = zs$
6.	$z = f(x+y).g(x-y)$	**Ans:** $(r-t)z = (p+q)(p-q)$
7.	$z = yf(x) + xg(y)$	**Ans:** $xys = px + qy - z$
8.	$z = f(x+at) + g(x-at)$	**Ans:** $t = a^2 r$
9.	$f\left(x^2 + y^2 + z^2, z^2 - 2xy\right) = 0$	**Ans:** $(x+y)\left[z(q-p) + (y-x)\right] = 0$
10.	$z = ae^{-b^2 t}\cos bx$	**Ans:** $p = t$ where $p = \frac{\partial z}{\partial t}, t = \frac{\partial^2 z}{\partial x^2}$
11.	$z = xy + y\sqrt{x^2 + a^2} + b$	**Ans:** $pq = py + qx$
12.	All plane having equal x and y intercepts i.e. $\frac{x}{a} + \frac{y}{b} + \frac{z}{c} = 1.$	**Ans:** $p = q$
13.	$z = f\left(x^2 - y^2\right)$	**Ans:** $yq + qx = 0$
14.	$z = f(\sin x + \cos y)$	**Ans:** $p\sin y + q\cos x = 0$
15.	$z = e^{ax+by} f(ax - by)$	**Ans:** $bq + aq = 2abz$

16. $z = xf\left(\frac{y}{x}\right) + yg(x)$ **Ans:** $x\frac{\partial^3 z}{\partial x \partial y^2} + \frac{\partial^2 z}{\partial y^2} + y\frac{\partial^3 z}{\partial y^3} = 0$

17. $f\left(x^2 + y^2, z - xy\right) = 0$ **Ans:** $xq - yp = x^2 - y^2$

18. $f\left(x + y + z, x^2 + y^2 - z^2\right) =$ **Ans:** $(y+z)p - (z+x)q = x - y$

1.5 Initial and Boundary Condition

Initial and boundary conditions are fundamental in PDEs as they provide essential information about the problem, ensure uniqueness and physical relevance of the solution, guide the behaviour of the system, enable the validity of numerical methods, and uphold conservation laws.

Initial condition: A condition that describes the state of the system at the beginning time or beginning moment within the PDE's domain is known as an initial condition for a partial differential equation (PDE). It offers details on the dependent variable(s) distribution at the beginning of the spatial domain or at the beginning of time.

Boundary condition: A boundary condition for a partial differential equation specifies the behaviour of the solution function or its derivatives at the boundaries of the domain in which the PDE is defined. These conditions are essential for determining a unique solution to the PDE by providing additional constraints on the solution's behaviour at the edges of the domain. The following are some typical PDE boundary condition types:

Dirichlet Condition: The Dirichlet boundary condition is a type of boundary condition named after Peter Gustav Lejeune Dirichlet (1805–1859). It is also known as first type of the boundary condition and is defined as condition specifies the value of the solution at the boundary of the domain. Mathematically, it is expressed as $u(x, y, z, ...) = g(x, y, z, ...)$ where $g(x, y, z, ...)$ is a given function representing the prescribed values.

Neumann Condition: The Neumann boundary condition is a type of boundary condition named after Carl Neumann (1832 – 1925). It is also known as second type of the boundary condition, and it specifies the derivative of the solution normal to the boundary. Mathematically, it is expressed as $\frac{\partial u}{\partial n} = h(x, y, z, \ldots\ldots)$ where is a given function representing the prescribed flux normal to the boundary i.e., When imposed on an ordinary (ODE) or a partial differential equation (PDE), it specifies the values that the derivative of a solution is going to take on the boundary of the domain.

Robin condition: The Robin boundary condition is a type of boundary condition named after Victor Gustave Robin (1855–1897). It is a linear combination of a Dirichlet condition and a Neumann condition by the solution $u(x, y)$is a function of x and y on the boundaryΛ; i.e. $a(x, y)u(x, y) + b(x, y)\frac{\partial u(x, y)}{\partial n} = c(x, y)$,where *a(x, y), b(x, y)* and *c(x, y)* are given functions.

Mixed Condition: The mixed boundary condition consists of applying different types of boundary conditions in different parts of the domain. It is important to notice that boundary conditions must be applied on the whole boundary: the "free" boundary is anyways subjected to a homogeneous Neumann condition.

The mixed boundary condition differs from the Robin condition because the latter consists of different types of boundary conditions applied to the same region of the boundary, while the mixed condition implies different types of B.C. applied to different parts of the boundary.

Cauchy Conditions: The Cauchy boundary condition is a condition on both the unknown field and its derivatives. It differs from the Robin condition because the Cauchy condition implies the imposition of two constraints (1 Dirichlet B.C. + 1 Neumann B.C.), while the Robin condition implies only one constraint on the linear combination of the unknown function and its derivatives.

1.5.1 Importance of initial and Boundary condition

Initial and boundary conditions are crucial for solving partial differential equations (PDEs) because they help define a unique solution within the given domain and time frame. Following are few points that shows that initial and boundary condition plays vital role in the physical phenomena:

Unique solution: PDEs often have multiple solutions and specifying initial and boundary conditions narrows down the set of possible solutions to find a unique one that satisfies both the differential equation and the given conditions. Without these conditions, the solution space would be too vast, leading to ambiguity in the solution.

Physical Relevance: The physical scenario that is being modelled is reflected in the initial and boundary conditions. They capture the initial conditions of the system and the boundary conditions, which describe how the system interacts with its surroundings or limits. Solutions that contradict actual occurrences may result from ignoring these requirements.

Determining Behaviour: Initial conditions determine the behaviour of the system at the starting time, while boundary conditions govern its behaviour at the spatial boundaries. Together, they provide constraints that guide the evolution of the solution over time and space. They help determine how the system responds to external influences and evolves over time.

Physical Interpretation: Initial and boundary conditions often have clear physical interpretations. For example, in heat transfer problems, the initial temperature distribution and boundary temperatures directly relate to the physical temperature of the system at different times and locations.

Conservation Laws: In many physical systems, conservation laws play a crucial role. Initial and boundary conditions ensure that these conservation laws are satisfied throughout the solution process. For example, in fluid dynamics, boundary conditions ensure the conservation of mass, momentum, and energy.

Chapter 2

First Order Partial Differential Equation and its Solution

2.1 Introduction

Partial differential equations (PDEs) are essential tools for comprehending a wide range of physical processes, including the flow of fluids, electromagnetism, and heat diffusion. Among the diverse classes of PDEs, first-order partial differential equations hold particular significance due to their fundamental role in modelling dynamic systems. Comprehending these equations is essential to understanding the complicated behaviours of systems in physics, engineering, and other fields.

First-order PDEs provide a solid foundation for examining systems with temporal or spatial fluctuations by capturing interactions between numerous variables and their corresponding partial derivatives. These equations arise in many situations, like wave propagation, transport phenomena, and conservation laws.

In this chapter we discuss the solution of linear and nonlinear partial differential equation. Section 2.2 discusses the solution of linear PDE of first and higher order. Section 2.3 deals with the solution of nonlinear PDE of first order.

2.2 Linear Partial Differential equation of First and Higher Order

The general form of first order partial differential equation is $F\left(x, y, z, \frac{\partial z}{\partial x}, \frac{\partial z}{\partial y}\right) = 0$ where x, y are two independent variables and z is dependent variable and $p = \frac{\partial z}{\partial x}, q = \frac{\partial z}{\partial y}$.

2.3 Solution of First order Partial Differential Equation

2.3.1 Direct Integration Method

If the dependent variables z occurs only in the partial derivatives of a partial differential equations, then the solution can be obtained by direct integration method.

Example:

Solve $\dfrac{\partial z}{\partial y} = x^3$ (2.1)

Solution:

Integrating equation (2.1) partially w.r.t to y, we obtain $z = x^3 y + \phi(x)$ where $\phi(x)$ the arbitrary function of x.

Example:

Solve $\dfrac{\partial^2 z}{\partial x^2} = -a^2 x$ with condition $x = 0, \dfrac{\partial z}{\partial x} = a \sin y$ and $\dfrac{\partial z}{\partial y} = 0$.

Solution:

Here $\dfrac{\partial^2 z}{\partial x^2} = -a^2 x$. (2.2)

Integrating equation (2.2) partially with respect to x, we get

$$\frac{\partial z}{\partial x} = -\frac{a^2 x^2}{2} + g(y) \qquad (2.3)$$

Again, integrating equation (2.3) with respect to x, we get

$$z=-\frac{a^2x^3}{6}+xg(y)+h(y) \tag{2.4}$$

Differentiating equation (2.4) partially w.r.t x, we get.

$$\frac{\partial z}{\partial x}=-\frac{a^2x^2}{2}+g(y) \tag{2.5}$$

Now it is given that at $x=0, \frac{\partial z}{\partial x}=a\sin y$, equation (2.4) reduces to

$a \sin y=-\frac{a^2(0)}{2}+g(y) \Rightarrow g(y)=a \sin y$.

Differentiating equation (2.4) partially w.r.t y, we get

$$\frac{\partial z}{\partial y}=xg'(y)+h'(y) \tag{2.6}$$

Using the given condition $x=0, \frac{\partial z}{\partial y}=0$, equation (2.6) reduces to

$0=0.g'(y)+h'(y)$

$\Rightarrow h'(y)=0 \Rightarrow h(y)=c$

By substituting the value of $g(y)$ and $h(y)$ in equation (2.4) ,we get

$z=-\frac{a^2x^3}{6}+xa\sin y+c$, which is the required solution where c is the arbitrary constant.

Example:

Solve $\frac{\partial^2 z}{\partial x \partial y}=x^2-y^2$.

Solution:

Here $\frac{\partial^2 z}{\partial x \partial y} = x^2 - y^2$. (2.7)

Integrating equation (2.7) with respect to x, we get

$$\frac{\partial z}{\partial y} = \frac{x^3}{3} - y^2 x + g(y). \quad (2.8)$$

Again integrating equation (2.8) with respect to y, we get $z = \frac{1}{3}x^3 y - \frac{1}{3}y^3 x + f(y) + g(x)$, which is the solution of given partial differential equation where $f(y)$ and $g(x)$ are arbitrary functions.

Example:

Solve $\frac{\partial z}{\partial x} = 6x - 3y$ and $\frac{\partial z}{\partial y} = -3x - 4y$ using the direct method.

Solution:

Here,

$$\frac{\partial z}{\partial x} = 6x - 3y, \quad (2.9)$$

$$\frac{\partial z}{\partial y} = -3x - 4y \quad (2.10)$$

Integrating equation (2.9) partially w.r.t to x, we get

$$z = 3x^2 - 3xy + f(y) \quad (2.11)$$

Again differentiating equation (2.11) partially w.r.t to y, we get

$$\frac{\partial z}{\partial y} = -3x + f'(y) \tag{2.12}$$

Comparing equation s. (2.10) and (2.12), we get

$$-3x + f'(y) = -3x - 4y \Rightarrow f'(y) = -4y \Rightarrow f(y) = -2y^2 + c.$$

Hence from equation (2.11), the general solution is $z = 3x^2 - 3xy - 2y^2 + c$ where *c* is an arbitrary constant.

2.3.2 Lagrange's Linear Partial Differential Equation

The general form of a quasi linear PDE of the first order is

$$P(x,y,z)\frac{\partial z}{\partial x} + Q(x,y,z)\frac{\partial z}{\partial y} = R(x,y,z)$$

This equation is known as Lagrange's linear equation. If *P* and *Q* are independent of *z* and *R* is linear in *z*, then this equation is linear equation. The general solution of the Lagrange's linear PDE

$$Pp + Qq = R \tag{2.13}$$

is the given by the equation

$$f(u,v) = 0 \tag{2.14}$$

Working Rule for obtaining general solution

Step I:

Rewrite the equation in the standard form $Pp + Qq = R$

Step II:

Form the Lagrange's auxiliary equation $\frac{dx}{P}=\frac{dy}{Q}=\frac{dz}{R}$ (2.15)

Step III:

Find complete solution $u(x,y,z)=c_1$ and $v(x,y,z)=c_2$ form the simultaneous equation of the form $\frac{dx}{P}=\frac{dy}{Q}=\frac{dz}{R}$

Case I: One of the variables is either absent or cancels out from the set of auxiliary equations.

Case II: If $u(x,y,z)=c_1$ is known but $v(x,y,z)=c_2$ is not possible by **Case I,** then use $u(x,y,z)=c_1$ to get $v(x,y,z)=c_2$.

Case III: Introducing Lagrange's multipliers P,Q,R which are function of x,y,z or constant, each fraction in step-III is equal to

$$\frac{P_1dx+Q_1dy+R_1dz}{P_1P+Q_1Q+R_1R} \tag{2.16}$$

If P,Q,R are chosen such that $P_1P+Q_1Q+R_1R=0$ then $P_1dx+Q_1dy+R_1dz=0$ which can be integrated.

Case IV: Multiplier's may be chosen (more than once) such that the numerator $P_1dx+Q_1dy+R_1dz=0$ is an exact differential of the denominators $P_1P+Q_1Q+R_1R$.

Example:

Solve $px+qy=2z$.

Solution:

Here,

$$px+qy=2z \quad (2.17)$$

Equation (2.17) is a linear PDE of first order of the form $Pp+Qq=R$ with $P=x,Q=y,R=2z$. Hence, the Lagrange's auxiliary equation is:

$$\frac{dx}{x}=\frac{dy}{y}=\frac{dz}{2z} \quad (2.18)$$

From equation (2.18), we have:

$$\frac{dx}{x}=\frac{dy}{y} \quad (2.19)$$

Integrating equation (2.18) both sides, we get

$$\log x=\log y+\log c_1$$

$$\log x=\log(yc_1)$$

$$x=yc_1$$

$$\frac{x}{y}=c_1 \quad (2.20)$$

From equation (2.18), we also have $\frac{dx}{x}=\frac{dz}{2z}$.

Integrating equation (2.18) both sides, we get

$$2\log x=\log z+\log c_2$$

$$\log x^2=\log(zc_2)$$

$$x^2=zc_2$$

$$\frac{x^2}{z}=c_2 \tag{2.21}$$

From equations (2.20) and (2.21), the general solution is

$$F\left(\frac{x}{y},\frac{x^2}{z}\right)=0.$$

Example:

Solve $p-q=\log(x+y)$.

Solution:

Here:

$$p-q=\log(x+y) \tag{2.22}$$

equation (2.22) is a linear PDE of first order of the form $Pp+Qq=R$ with $P=1,\ Q=-1,\ R=\log(x+y)$. Hence, the Lagrange's auxiliary equation are

$$\frac{dx}{1}=\frac{dy}{-1}=\frac{dz}{\log(x+y)} \tag{2.23}$$

From equation (2.23), we have

$$\frac{dx}{1}=\frac{dy}{-1} \tag{2.24}$$

Integrating both the side, we get

$$x=-y+c_1 \Rightarrow x+y=c_1 \tag{2.25}$$

From equation (2.23), we have

$$\frac{dx}{1} = \frac{dz}{\log(x+y)} \tag{2.26}$$

Using equation (2.25), equation (2.26) reduces to

$$\frac{dx}{1} = \frac{dz}{\log c_1} \Rightarrow \log c_1 \, dx = dz.$$

Integrating both sides, we get

$$x \log c_1 = z + c_2 \Rightarrow x \log(x+y) = z + c_2$$

$$x \log(x+y) - z = c_2 \tag{2.27}$$

From equation (2.25) and (2.27), the general solution is:

$$F\left(x+y, x\log(x+y) - z\right) = 0.$$

Example:

Solve $yzp - xzq = xy$.

Solution:

Here:

$$yzp - xzq = xy \tag{2.28}$$

Equation (2.28) is a linear PDE of first order of the form $Pp + Qq = R$ with $P = yz, Q = -xz, R = xy$. Hence, the Lagrange's auxiliary equation are

$$\frac{dx}{yz} = \frac{dy}{-xz} = \frac{dz}{xy} \tag{2.29}$$

From equation (2.29), we have

$$\frac{dx}{yz} = \frac{dy}{-xz} \Rightarrow \frac{dx}{y} = \frac{dy}{-x} \Rightarrow xdx = -ydy .$$

Integrating both the sides, we get

$$\frac{x^2}{2} = \frac{-y^2}{2} + c_1 \Rightarrow x^2 + y^2 = 2c_1 \Rightarrow x^2 + y^2 = C_1 \qquad (2.30)$$

From equation (2.29), we have:

$$\frac{dx}{yz} = \frac{dz}{xy} \Rightarrow \frac{dx}{z} = \frac{dz}{x} \Rightarrow x\,dx = zdz \qquad (2.31)$$

Integrating equation (2.31) both sides, we get

$$\frac{x^2}{2} = \frac{z^2}{2} + c_2 \Rightarrow x^2 - y^2 = 2c_2 \Rightarrow x^2 - z^2 = C_2 \qquad (2.32)$$

From equations (2.30) and (2.32), the general solution is

$$F\left(x^2 + y^2, x^2 - y^2\right) = 0 .$$

Example:

Solve $p\sqrt{x} + q\sqrt{y} = \sqrt{z}$.

Solution:

Here:

$$p\sqrt{x} + q\sqrt{y} = \sqrt{z} \qquad (2.33)$$

equation (2.33) is linear PDE of first order of the form $Pp+Qq=R$ with $P=\sqrt{x}, Q=\sqrt{y}, R=\sqrt{z}$. Hence, the Lagrange's auxiliary equation is:

$$\frac{dx}{\sqrt{x}}=\frac{dy}{\sqrt{y}}=\frac{dz}{\sqrt{z}} \tag{2.34}$$

From equation (2.35), we have $\frac{dx}{\sqrt{x}}=\frac{dy}{\sqrt{y}}$,

Integrating both the sides, we get $\frac{x^{\frac{1}{2}}}{\frac{1}{2}}=\frac{y^{\frac{1}{2}}}{\frac{1}{2}}+c_1 \Rightarrow 2x^{\frac{1}{2}}=2y^{\frac{1}{2}}+\frac{1}{2}c_1'$,

$$\Rightarrow x^{\frac{1}{2}}=y^{\frac{1}{2}}+c_1, \left(\because c_1=\frac{1}{2}c_1'\right)$$

$$\Rightarrow x^{\frac{1}{2}}-y^{\frac{1}{2}}=c_1 \tag{2.35}$$

From equation (2.34), we have $\frac{dy}{\sqrt{y}}=\frac{dz}{\sqrt{z}}$,

Integrating both the side, we get $\frac{y^{\frac{1}{2}}}{\frac{1}{2}}=\frac{z^{\frac{1}{2}}}{\frac{1}{2}}+c_2' \Rightarrow 2y^{\frac{1}{2}}=2z^{\frac{1}{2}}+\frac{1}{2}c_2'$,

$$\Rightarrow y^{\frac{1}{2}}=z^{\frac{1}{2}}+c_2, \left(\because c_2=\frac{1}{2}c_2'\right)$$

$$\Rightarrow y^{\frac{1}{2}}-z^{\frac{1}{2}}=c_2 \tag{2.36}$$

From equations (2.35) and (2.36), the general solution is

$$F\left(x^{\frac{1}{2}}-y^{\frac{1}{2}}, y^{\frac{1}{2}}-z^{\frac{1}{2}}\right)=0.$$

Example:

Solve $z\left(z^2+xy\right)\left(px-qy\right)=x^4$.

Solution:

Here:

$$z\left(z^2+xy\right)\left(px-qy\right)=x^4 \qquad (2.37)$$

Rearranging equation (2.37), it reduces to $xz\left(z^2+xy\right)p-yz\left(x^2+xy\right)q=x^4$, which is a linear PDE of first order of the form $Pp+Qq=R$ with $P=xz\left(z^2+xy\right), Q=-yz\left(z^2+xy\right), R=x^4$. Hence, the Lagrange's auxiliary equation is:

$$\frac{dx}{xz\left(z^2+xy\right)}=\frac{dy}{-yz\left(x^2+xy\right)}=\frac{dz}{x^4} \qquad (2.38)$$

From equation (2.38), we have $\dfrac{dx}{xz\left(z^2+xy\right)}=\dfrac{dy}{-yz\left(x^2+xy\right)}$

$$\Rightarrow \frac{dx}{x}=\frac{dy}{-y}.$$

Integrating both the sides, we get

$$\log x=-\log y+\log c_1 \Rightarrow \log(xy)=\log c_1,$$

$$\Rightarrow xy=c_1 \qquad (2.39)$$

From equation (2.38), we have:

$$\frac{dx}{xz\left(z^2+xy\right)}=\frac{dz}{x^4}\Rightarrow x^3dx=z\left(z^2+xy\right)dz\,,$$

$$\Rightarrow x^3dx=z\left(z^2+c_1\right)dz\,.$$ (Using equation (2.39))

Integrating both the sides, we get $\Rightarrow x^3dx=\left(z^3+c_1z\right)dz$,

$$\Rightarrow\frac{x^4}{4}=\left(\frac{z^4}{4}+c_1\frac{z^2}{2}\right)+c_2\,,$$

$$\Rightarrow\frac{x^4}{4}=\left(\frac{z^4}{4}+(xy)\frac{z^2}{2}\right)+c_2\,.$$ (From equation (2.39)) (2.40)

From equations (2.39) and (2.40), the general solution is

$$F\left(xy,\frac{x^4}{4}-\frac{z^4}{4}-\frac{xyz^2}{2}\right)=0\,.$$

Example:

Solve $(z-y)p+(x-z)q=y-x$.

Solution:

Here:

$$(z-y)p+(x-z)q=y-x \tag{2.41}$$

Equation (2.41) is a linear PDE of first order of the form $Pp+Qq=R$ with $P=z-y$, $Q=x-z$, $R=y-x$. Hence, the Lagrange's auxiliary equation is:

$$\frac{dx}{z-y}=\frac{dy}{x-z}=\frac{dz}{y-x} \qquad (2.42)$$

From equation (2.42), we have $\frac{dx+dy+dz}{z-y+x-z+y-x}=\frac{dx+dy+dz}{0}$

(Choosing Multipliers as x, y, z)

$\Rightarrow dx+dy+dz=0$

Integrating the above equation on both sides, we get

$$x+y+z=c_1 \qquad (2.43)$$

From equation (2,42), we have:

$$\frac{xdx+ydy+zdz}{x(z-y)+y(x-z)+z(y-x)}=\frac{xdx+ydy+zdz}{0},$$

(Choosing multipliers as x, y, z)

$\Rightarrow xdx+ydy+zdz=0.$

Integrating both sides, we get $\frac{x^2}{2}+\frac{y^2}{2}+\frac{z^2}{2}=\frac{c_2}{2}$,

$$\Rightarrow x^2+y^2+z^2=c_2 \qquad (2.44)$$

From equation s. (2.43) and (2.44), the general solution is

$F\left(x+y+z, x^2+y^2+z^2\right)=0.$

Example:

Solve $(y+z)p+(z+x)q=x+y.$

Solution:

Here:

$$(y+z)p+(z+x)q=x+y \tag{2.45}$$

Equation (2.45) is a linear PDE of first order of the form $Pp+Qq=R$ with $P=y+z$, $Q=z+x$, $R=x+y$. Hence, the Lagrange's auxiliary equation is:

$$\frac{dx}{y+z}=\frac{dy}{z+x}=\frac{dz}{x+y} \tag{2.46}$$

From equation (2.46), we have:

$$\frac{dx-dy}{(y+z)-(z+x)}=\frac{dx-dz}{(y+z)-(x+y)}$$

$$\Rightarrow \frac{dx-dy}{y-x}=\frac{dx-dz}{z-x},$$

$$\Rightarrow \frac{dx-dy}{x-y}=\frac{dx-dz}{x-z}.$$

Integrating both sides, we get:

$$\log(x-y)=\log(x-z)+\log c_1,$$

$$\Rightarrow \log\left(\frac{x-y}{x-z}\right)=\log c_1,$$

$$\Rightarrow \frac{x-y}{x-z}=c_1 \tag{2.47}$$

From equation (2.46), we have

$$\frac{dx-dy}{(y+z)-(z+x)}=\frac{dx+dy+dz}{(y+z)+(z+x)+(x+y)},$$

$$\Rightarrow \frac{dx-dy}{y-x}=\frac{dx+dy+dz}{2(x+y+z)},$$

$$\Rightarrow \frac{dx+dy+dz}{2(x+y+z)}+\frac{dx-dy}{x-y}=0.$$

Integrating both the sides, we get

$$\frac{\log(x+y+z)}{2}+\log(x-y)=\log c_2,$$

$$\Rightarrow \log(x+y+z)^{\frac{1}{2}}(x-y)=\log c_2$$

$$\Rightarrow (x+y+z)^{\frac{1}{2}}(x-y)=c_2 \tag{2.48}$$

From equation (2.47) and equation (2.48), the general solution is

$$F\left(\frac{x-y}{x-z},(x+y+z)^{\frac{1}{2}}(x-y)\right)=0.$$

Example:

Solve $\left(z^2-2yz-y^2\right)p+(xy+zx)q=xy-zx$.

Solution:

Here:

$$\left(z^2-2yz-y^2\right)p+(xy+zx)q=xy-zx \tag{2.49}$$

Equation (2.49) is a linear PDE of first order of the form $Pp+Qq=R$ with $P=z^2-2yz-y^2,\ Q=xy+zx, R=xy-zx$. Hence, the Lagrange's auxiliary equation is:

$$\frac{dx}{z^2-2yz-y^2}=\frac{dy}{xy+zx}=\frac{dz}{xy-zx} \tag{2.50}$$

From equation (2.50), we have $\frac{dy}{xy+zx}=\frac{dz}{xy-zx}$,

$$\Rightarrow \frac{dy}{y+z}=\frac{dz}{y-z},$$

$$\Rightarrow \frac{dy}{dz}=\frac{y+z}{y-z} \tag{2.51}$$

The given equation is a homogeneous differential equation. Take

$$y=vz \Rightarrow \frac{dy}{dz}=v+z\frac{du}{dz} \tag{2.52}$$

Substituting equation (2.52) in equation (2.51), we get

$$v+z\frac{du}{dz}=\frac{vz+z}{vz-z},$$

$$\Rightarrow z\frac{du}{dz}=\frac{v+1}{v-1}-v,$$

$$\Rightarrow z\frac{du}{dz}=\frac{v+1-v^2+v}{v-1},$$

$$\Rightarrow z\frac{du}{dz}=\frac{-\left(v^2-2v-1\right)}{v-1},$$

$$\Rightarrow \frac{v-1}{v^2-2v-1}dv = \frac{-dz}{z},$$

$$\Rightarrow \frac{2(v-1)}{2(v^2-2v-1)}dv = \frac{-dz}{z}.$$

Integrating both sides, we get

$$\frac{1}{2}\log(v^2-2v-1) = -\log z + \log c_1,$$

$$\Rightarrow \log(v^2-2v-1)^{\frac{1}{2}} = \log\left(\frac{c_1}{z}\right),$$

$$\Rightarrow z(v^2-2v-1)^{\frac{1}{2}} = c_1 \tag{2.53}$$

We know that

$$y = vz \Rightarrow v = \frac{y}{z} \tag{2.54}$$

Substituting value from equation (2.54) in equation (2.53), we get

$$z\left(\left(\frac{y}{z}\right)^2 - 2\frac{y}{z} - 1\right)^{\frac{1}{2}} = c_1 \tag{2.55}$$

From equation (2.50), we have

$$xdx + ydy + zdz = x(z^2-2yz-y^2) + y(xy+zx) + z(xy-zx),$$

(Choosing multiplier as x, y, z)

$$xdx + ydy + zdz = 0.$$

Integrating both the sides, we get

$$\frac{x^2}{2}+\frac{y^2}{2}+\frac{z^2}{2}=\frac{c_2}{2},$$

$$\Rightarrow x^2+y^2+z^2=c_2 \tag{2.56}$$

From equations (2.55) and (2.56) the general solution is

$$F\left(z\left(\left(\frac{y}{z}\right)^2-2\frac{y}{z}-1\right)^{\frac{1}{2}},x^2+y^2+z^2\right)=0.$$

Example:

Solve $\left(\frac{y^2z}{x}\right)p+zxq=y^2.$

Solution: Here:

$$\left(\frac{y^2z}{x}\right)p+zxq=y^2 \tag{2.57}$$

This is a linear PDE of first order of the form $Pp+Qq=R$ with $P=\frac{y^2z}{x},Q=zx,R=y^2$. Hence, the Lagrange's auxiliary equation is

$$\frac{dx}{\frac{y^2z}{x}}=\frac{dy}{zx}=\frac{dz}{y^2} \tag{2.58}$$

From equation (2.58), we have:

$$\frac{dx}{\frac{y^2 z}{x}} = \frac{dy}{zx},$$

$$\Rightarrow x^2 dx = y^2 dy.$$

Integrating both the sides, we get

$$\frac{x^3}{3} = \frac{y^3}{3} + c_1',$$

$$\Rightarrow x^3 = y^3 + 3c_1',$$

$$\Rightarrow x^3 - y^3 = c_1 \; (c_1 = 3c_1'). \tag{2.59}$$

From equation (2.58), we have $\frac{dx}{\frac{y^2 z}{x}} = \frac{dz}{y^2}$,

$$\Rightarrow xdx = zdz.$$

Integrating both sides, we get:

$$\frac{x^2}{2} = \frac{z^2}{2} + c_2',$$

$$\Rightarrow x^2 = z^2 + 2c_2',$$

$$\Rightarrow x^2 - z^2 = c_2 \left(c_2 = 2c_2'\right). \tag{2.60}$$

From equations (2.59) and (2.60), the general solution of is $F\left(x^3 - y^3, x^2 - z^2\right) = 0$.

Example:

Solve $\left(x^2-yz\right)p+\left(y^2-zx\right)q=z^2-xy$.

Solution:

Here:

$$\left(x^2-yz\right)p+\left(y^2-zx\right)q=z^2-xy \tag{2.61}$$

Equation (2.61) is a linear PDE of first order of the form $Pp+Qq=R$ with $P=x^2-yz,\ \ Q=y^2-zx,\ \ R=z^2-xy$. Hence, the Lagrange's auxiliary equation is

$$\frac{dx}{x^2-yz}=\frac{dy}{y^2-zx}=\frac{dz}{z^2-xy} \tag{2.62}$$

From equation (2.62), we have:

$$\frac{dx-dy}{\left(x^2-yz\right)-\left(y^2-zx\right)}=\frac{dy-dz}{\left(y^2-zx\right)-\left(z^2-xy\right)},$$

$$\frac{dx-dy}{x^2-y^2+z(x-y)}=\frac{dy-dz}{y^2-z^2+x(y-z)},$$

$$\frac{dx-dy}{(x-y)(x+y)+z(x-y)}=\frac{dy-dz}{(y-z)(y+z)+x(y-z)},$$

$$\frac{dx-dy}{(x-y)(x+y+z)}=\frac{dy-dz}{(y-z)(x+y+z)},$$

$$\frac{dx-dy}{x-y}=\frac{dy-dz}{y-z}.$$

Integrating both sides of the above equation, we get

$$\log(x-y)=\log(y-z)+\log c_1,$$

$$\log\left(\frac{x-y}{y-z}\right)=\log c_1,$$

$$\Rightarrow \frac{x-y}{y-z}=c_1 \tag{2.63}$$

From equation (2.62), we have

$$\frac{dx-dz}{\left(x^2-yz\right)-\left(z^2-xy\right)}=\frac{dy-dz}{\left(y^2-zx\right)-\left(z^2-xy\right)},$$

$$\frac{dx-dz}{x^2-z^2+y(x-z)}=\frac{dy-dz}{y^2-z^2+x(y-z)},$$

$$\frac{dx-dz}{(x-z)(x+z)+y(x-z)}=\frac{dy-dz}{(y-z)(y+z)+x(y-z)},$$

$$\frac{dx-dz}{(x-z)(x+y+z)}=\frac{dy-dz}{(y-z)(x+y+z)},$$

$$\frac{dx-dy}{x-z}=\frac{dy-dz}{y-z}.$$

Integrating both sides, we get:

$$\log(x-z)=\log(y-z)+\log c_2,$$

$$\log\left(\frac{x-z}{y-z}\right)=\log c_2,$$

$$\Rightarrow \frac{x-z}{y-z} = c_2 \tag{2.64}$$

From equation (2.63) and (2.64), the general solution is

$$F\left(\frac{x-y}{y-z}, \frac{x-z}{y-z}\right) = 0.$$

Example:

Solve $px(x+y) = qy(x+y) - (2x+2y+z)(x-y)$.

Solution:

Here, $px(x+y) = qy(x+y) - (2x+2y+z)(x-y)$

This is a linear PDE of first order of the form $Pp + Qq = R$ with
$P = x(x+y),\ Q = -y(x+y),\ R = -(2x+2y+z)(x-y)$.
Hence, the Lagrange's auxiliary equation is:

$$\frac{dx}{x(x+y)} = \frac{dy}{-y(x+y)} = \frac{dz}{-(2x+2y+z)(x-y)} \tag{2.65}$$

From equation (2.65), we have:

$$\frac{dx}{x(x+y)} = \frac{dy}{-y(x+y)},$$

$$\frac{dx}{x} = \frac{dy}{-y}.$$

Integrating both sides, we get:

$$\log x = -\log y + \log c_1 ,$$

$$\log(xy) = \log c_1,$$

$$\Rightarrow xy = c_1 \tag{2.66}$$

From equation (2.65), we have:

$$\frac{dx+dy}{x(x+y)-y(x+y)} = \frac{dy-dz}{-(x-y)(2x+2y+z)},$$

$$\frac{dx+dy}{(x-y)(x+y)} = \frac{dz}{-(x-y)(2x+2y+z)},$$

$$\frac{dx+dy}{(x+y)} = \frac{dz}{-(2x+2y+z)},$$

$$(2x+2y+z)(dx+dy) = -(x+y)dz$$

$$(2x+2y+z)(dx+dy)+(x+y)dz = 0,$$
$$(x+y+z+x+y)(dx+dy)+(x+y)dz = 0,$$

$$(x+y+z)(dx+dy)+(x+y)(dx+dy)+(x+y)dz = 0,$$

$$(x+y+z)d(x+y)+(x+y+z)(dx+dy+dz) = 0,$$

$$d\left((x+y+z)(x+y)\right) = 0.$$

Integrating both the sides, we get

$$(x+y+z)(x+y) = c_2 \tag{2.68}$$

From equations (2.66) and (2.68), the general solution is

$F\left(xy,(x+y)(x+y+z)\right)=0$.

Example:

Solve $\frac{y-z}{yz}+\frac{z-x}{zx}q=\frac{x-y}{xy}$.

Solution:

Here, $\frac{y-z}{yz}+\frac{z-x}{zx}q=\frac{x-y}{xy}$.

This is a linear PDE of the first order of the form $Pp+Qq=R$ with $P=\frac{y-z}{yz}, Q=\frac{z-x}{zx}, R=\frac{x-y}{xy}$. Hence, the Lagrange's auxiliary equation is $\frac{dx}{\frac{y-z}{yz}}=\frac{dy}{\frac{z-x}{zx}}=\frac{dz}{\frac{x-y}{xy}}$.

$$\frac{yz\,dx}{y-z}=\frac{zxdy}{z-x}=\frac{xydz}{x-y} \tag{2.69}$$

Choosing multipliers as x, y and z, we get $x(y-z)+y(z-x)+z(x-y)=0$, therefore from equation (2.69), we get:

$x(yz\,dx)+y(zxdy)+z(xydz)=0$,

$xyz\,dx+yzxdy+zxydz=0$,

$dx+dy+dz=0$.

Integrating both the sides, we get:

$$x+y+z=c_1 \tag{2.70}$$

Choosing multipliers as x, y and z, we get $(y-z)+(z-x)+(x-y)=0$, therefore from equation (2.70), we have:

$$yz\,dx + zxdy + xydz = 0,$$

$$yz\,dx + zxdy + xydz = 0 \Rightarrow d(xyz) = 0.$$

Integrating both the sides of the above equation, we get:

$$xyz = c_2 \tag{2.71}$$

From equations (2.70) and (2.71), the general solution is $F(x+y+z, xyz)=0$.

Example:

Solve $p+3q = 5z + \tan(y-3x)$.

Solution:

Here $p+3q = 5z + \tan(y-3x)$.

This is a linear PDE of first order of the form $Pp+Qq=R$ with $P=1, Q=3,\ R=3z+\tan(y-3x)$. Hence, the Lagrange's auxiliary equation is

$$\frac{dx}{1} = \frac{dy}{3} = \frac{dz}{3z+\tan(y-3x)} \tag{2.72}$$

From equation (2.72), we have $\frac{dx}{1} = \frac{dy}{3}$.

Integrating both sides, we get:

$$x = \frac{1}{3} y + c_1^{'}.$$

$$y - 3x = c_1 \left(-3c_1^{'} = c_1\right). \tag{2.73}$$

From equation (2.72), we have:

$$\frac{dx}{1} = \frac{dz}{3z + \tan\left(y - 3x\right)}.$$

Substituting $y - 3x = c_1$ from equation (2.73), the above equation reduces to $\frac{dx}{1} = \frac{dz}{3z + c_1}$. Integrating both the sides, we get:

$$x = \int \frac{1}{5z + \tan c_1} dz,$$

$$x = \frac{1}{5} \log\left(5z + \tan c_1\right) + c_2,$$

$$x - \frac{1}{5} \log\left(5z + \tan\left(y - 3x\right)\right) = c_2 \tag{2.74}$$

From equations (2.73) and (2.74), the general solution is

$$F\left(y - 3x, x - \frac{1}{5} \log\left(5z + \tan\left(y - 3x\right)\right)\right) = 0.$$

Exercise

Solve the following partial differential equation:

$$y^2 p - xyq = x\left(z - 2y\right)$$

(**Ans** $F\left(x^2 + y^2, yz - y^2\right) = 0$

$px+qy=z$

(**Ans** $F\left(\frac{x}{y},\frac{x}{z}\right)=0$)

$yzp+zxq=xy$

(**Ans** $F\left(x^2-y^2,z^2-y^2\right)=0$)

$\tan x\, p+\tan y\, q=\tan z$

(**Ans** $F\left(\frac{\sin x}{\sin y},\frac{\sin y}{\sin z}\right)=0$)

$2p+3q=1$

(**Ans** $F\left(3x-2y,y-3z\right)=0$)

$\left(x^2+2y^2\right)p-xyq=xz$

(**Ans** $F\left(yz,x^2y^2+y^4\right)=0$)

$p+q=1$

(**Ans** $F\left(x-y,y-z\right)=0$)

$px-qy=xy$

(**Ans** $F\left(yz,x^2y^2+y^4\right)=0$)

$p-2q=3z+\tan\left(y+2x\right)$

(**Ans** $F\left(2x+y,3x-\ln\left(3z+\tan\left(y+2x\right)\right)\right)=0$)

$px\left(z-2y^2\right)=\left(z-qy\right)\left(z-y^2-2x^3\right)$

(**Ans** $F\left(\frac{z}{y},\frac{1}{x}\left(y^2-z-x^3\right)\right)=0$)

$p+3q=3z+\cot\left(y-3x\right)$

(**Ans** $F\left(y-3x, x-\log\left(z+\cot\left(y-3x\right)\right)\right)=0$)

$$z^2-pz+qz+\left(x+y\right)^2=0$$

(**Ans** $F\left(x+y, e^{2y}\left(z^2+\left(x+y\right)^2\right)\right)=0$)

$$px+qy=xyz^2\left(x^2-y^2\right)$$

(**Ans** $F\left(x^2-y^2, y^2\left(x^2-y^2\right)+\frac{2}{z}\right)=0$)

$$\left(mz-ny\right)p+\left(nx-lz\right)q=ly-mx$$

(**Ans** $F\left(x^2+y^2+z^2, lx+my+nz\right)=0$)

$$z\left(x+y\right)p+z\left(x-y\right)p=x^2+y^2$$

(**Ans** $F\left(x^2-y^2-z^2, 2xy-z^2\right)=0$)

$$\left(\frac{b-c}{a}\right)yzp+z\left(\frac{c-a}{b}\right)zxq=\left(\frac{a-b}{c}\right)xy$$

(**Ans** $F\left(ax^2+by^2+cz^2, a^2x^2+b^2y^2+c^2z^2\right)=0$)

$$p\cos\left(x+y\right)+q\sin\left(x+y\right)=z$$

(**Ans** $F\left(\cot\left(\frac{x+y}{2}+\frac{\pi}{8}\right), e^{y-x}\left(\sin\left(x+y\right)+\cos\left(x+y\right)\right)\right)=0$)

$$p+q=x+y+z$$

(**Ans** $F\left(x-y, e^{-x}\left(2+x+y+z\right)\right)=0$)

$$px+qy=z-\alpha\sqrt{x^2+y^2+z^2}$$

(**Ans** $F\left(\frac{x}{y}, \frac{x^{1-\alpha}}{z+\sqrt{x^2+y^2+z^2}}\right)=0$)

$x^2p+y^2q=nxy$

(**Ans** $F\left(\frac{y-x}{xy}, z-\frac{nxy}{y-x}\log\left(\frac{y}{x}\right)\right)=0$)

2.4 Non- Linear Partial Differential Equation of First Order

In this section we discuss the solution of the first order nonlinear partial differential equation.

2.4.1 PDE of the form $F(p,q)=0$

For the equation of the form $F(p,q)=0$, the variables *x, y* and *z* do not occur explicitly.

If $z=ax+by+c$ is a trial solution of $F(p,q)=0$. Then $\frac{\partial z}{\partial x}=a, \frac{\partial z}{\partial y}=b$.Hence, $F(a,b)=0$ and $z=ax+by+c$ will be a solution if $F(a,b)=0$. Now solving for *b* from $F(a,b)=0$, we obtain $b=f(a)$. Therefore, $z=ax+f(a)y+c$ is the complete integral. To determine the general solution, we substitute $c=g(a)$ or $z=ax+f(a)y+g(a)$.

For the above equations, we are finding for the general solution. To find the singular integral, we eliminate a and c from equation $z=ax+f(a)y+c$ and the equations $\frac{\partial z}{\partial a}=0$ and $\frac{\partial z}{\partial c}=0$. Since $\frac{\partial z}{\partial c}=0$, there is no singular integral.

Example:

Solve $p^3 - q^3 = 0$.

Solution:

Here $p^3 - q^3 = 0$.

Let the complete solution be $z = ax + by + c$, where a and b are arbitrary constants. Differentiating above equation partially w.r.t to x and y, we get $p = \frac{\partial z}{\partial x} = a$ and $q = \frac{\partial z}{\partial y} = b$. By substituting the value of p and q in given equation, we get $a^3 - b^3 = 0 \Rightarrow a = b$.

Substituting value from above equation we get $z = ax + ay + c$ is a solution of given PDE.

Example:

Solve $p^2 - q^2 = 1$.

Solution:

Here $p^2 - q^2 = 1$.

Let the complete solution be $z = ax + by + c$, where a and b are arbitrary constants. If we take partial derivative of above equation w.r.t to *x* and *y*, we get $p = \frac{\partial z}{\partial x} = a$ and $q = \frac{\partial z}{\partial y} = b$. Now substituting the value of $p = a$ and $q = b$ in given equation, we get

$a^2 - b^2 = 1 \Rightarrow b^2 = a^2 - 1 \Rightarrow b = \sqrt{a^2 - 1}$.

Substituting value $b = \sqrt{a^2 - 1}$ in $z = ax + by + c$, we get

$z = ax + \sqrt{a^2 - 1}y + c$,

which is a solution of PDE.

Example:

Solve $p^2 + q^2 = npq$.

Solution:

Here $p^2 + q^2 = npq$.

Let the complete solution be $z = ax + by + c$, where *a* and *b* are arbitrary constant. Differentiating partially above equation w.r.t to *x* and *y*, we get $p = \frac{\partial z}{\partial x} = a$ and $q = \frac{\partial z}{\partial y} = b$. Now substituting the value of $p = a$ and $q = b$, we get

$$a^2 + b^2 = nab \Rightarrow b = \frac{na \pm \sqrt{n^2a^2 - 4a^2}}{2} \Rightarrow b = \frac{a}{2}\left[n \pm \sqrt{n^2 - 4}\right].$$

Substituting value $b = \frac{a}{2}\left[n \pm \sqrt{n^2 - 4}\right]$ in $z = ax + by + c$, we get $z = ax + \frac{a}{2}\left[n \pm \sqrt{n^2 - 4}\right]y + c$ is a complete solution of PDE.

Example:

Solve $(x + y)(p + q)^2 + (x - y)(p - q)^2 = 1$.

Solution:

Here $(x + y)(p + q)^2 + (x - y)(p - q)^2 = 1$.

To reduce the above equation in $F(p,q)=0$, put $x+y=X^2$ and $x-y=Y^2$. Thus $X=\sqrt{x+y}$ and $Y=\sqrt{x-y}$. Now

$$p=\frac{\partial z}{\partial x}=\frac{\partial z}{\partial X}.\frac{\partial X}{\partial x}+\frac{\partial z}{\partial Y}.\frac{\partial Y}{\partial x},$$

$$=\frac{\partial z}{\partial X}.\frac{1}{2\sqrt{x+y}}+\frac{\partial z}{\partial Y}.\frac{1}{2\sqrt{x-y}}.$$

Substituting $X=\sqrt{x+y}$ and $Y=\sqrt{x-y}$ in above equation, we get

$$p=\frac{1}{2X}\frac{\partial z}{\partial X}+\frac{1}{2Y}\frac{\partial z}{\partial Y} \tag{2.75}$$

Similarly, $q=\frac{\partial z}{\partial y}=\frac{\partial z}{\partial X}.\frac{\partial X}{\partial y}+\frac{\partial z}{\partial Y}.\frac{\partial Y}{\partial y},$

$$=\frac{\partial z}{\partial X}.\frac{1}{2\sqrt{x+y}}+\frac{\partial z}{\partial Y}.\frac{1}{2\sqrt{x-y}}.$$

Substituting $X=\sqrt{x+y}$ and $Y=\sqrt{x-y}$ in above equation, we get

$$q=\frac{1}{2X}\frac{\partial z}{\partial X}-\frac{1}{2Y}\frac{\partial z}{\partial Y} \tag{2.76}$$

By adding equations (2.75) and (2.76), we get

$$p+q=\frac{1}{X}\frac{\partial z}{\partial X} \tag{2.77}$$

And by subtracting equations (2.75) and (2.76), we get

$$p-q=\frac{1}{Y}\frac{\partial z}{\partial Y} \tag{2.78}$$

Thus, the given equation $(x+y)(p+q)^2+(x-y)(p-q)^2=1$ reduces to

$$X^2\left(\frac{1}{X}\frac{\partial z}{\partial X}\right)^2+Y^2\left(\frac{1}{Y}\frac{\partial z}{\partial Y}\right)^2=1,$$

$$\left(\frac{\partial z}{\partial X}\right)^2+\left(\frac{\partial z}{\partial Y}\right)^2=1 \tag{2.79}$$

Let the complete solution be $z=aX+bY+c$ of equation (1.131) where a and b are arbitrary constant. Hence $a^2+b^2=1$ $\Rightarrow b=\sqrt{a^2-1}$. Substituting $b=\sqrt{a^2-1}$ in $z=aX+bY+c$, we get $z=aX+\sqrt{1-a^2}Y+c$. Now substituting the value of $X=\sqrt{x+y}$ and $Y=\sqrt{x-y}$ in $z=aX+\sqrt{1-a^2}Y+c$, we get

$$z=a\sqrt{x+y}+\sqrt{1-a^2}\sqrt{x-y}+c,$$

which, is a complete solution of PDE.

Example:

Solve $(x-y)(px-qy)=(p-q)^2$.

Solution:

Here

$$(x-y)(px-qy)=(p-q)^2 \tag{2.80}$$

is given PDE. To reduce equation (2.80) to $F(p,q)=0$, put $x+y=u$ and $xy=v$. $\therefore p=\frac{\partial z}{\partial x}=\frac{\partial z}{\partial u}.\frac{\partial u}{\partial x}+\frac{\partial z}{\partial v}.\frac{\partial v}{\partial x}$,

$$p=\frac{\partial z}{\partial u}+\frac{\partial z}{\partial v}.y \tag{2.81}$$

Similarly, $q=\frac{\partial z}{\partial y}=\frac{\partial z}{\partial u}.\frac{\partial u}{\partial y}+\frac{\partial z}{\partial v}.\frac{\partial v}{\partial y}$,

$$q=\frac{\partial z}{\partial u}+\frac{\partial z}{\partial v}.x \tag{2.82}$$

Subtracting equation (2.81) and equation (2.82), we get

$$p-q=(y-x)\frac{\partial z}{\partial v} \tag{2.83}$$

and

$$xp-yq=(x-y)\frac{\partial z}{\partial u} \tag{2.84}$$

Substituting values from equations (2.83) and (2.84) in equation (2.80), we get

$$(x-y)^2\frac{\partial z}{\partial u}=(y-x)^2\left(\frac{\partial z}{\partial u}\right)^2,$$

$$\Rightarrow\frac{\partial z}{\partial u}=\left(\frac{\partial z}{\partial u}\right)^2 \tag{2.85}$$

Let the complete solution of equation (2.85) be $z=au+bv+c$ where a and b are arbitrary constants. Now $\frac{\partial z}{\partial u}=P=a$ and $\frac{\partial z}{\partial v}=Q=b$. Substituting the value of a and b in equation (2.85), we get $a=b^2$.

$\therefore z=au+bv+c\Rightarrow z=b^2u+bv+c$ is complete solution.

Hence $z=b^2(x+y)+b(xy)+c$ is a complete solution of PDE.

2.4.2 Partial differential equation is of the form $F(z,p,q)=0$

In $F(z,p,q)=0$ form does not involve the independent variables x and y. Assume $q=ap$ and substitute in given PDE, we get $F(z,p,ap)=0$, if we solve this equation for p, we get $p=\phi(z)$.

Now, $dz=\frac{\partial z}{\partial x}dx+\frac{\partial z}{\partial y}dy=pdx+qdy=pdx+ap\,dy$,

$$dz=p(dx+ady)=\phi(z)(dx+ady),$$

$$dx+ady=\frac{dz}{\phi(z)}.$$

By integrating on both the sides, we get

$$x+ay=\int\frac{dz}{\phi(z)}+b.$$

where a and b are arbitrary constants. This will give you a solution of the given PDE.

Example:

Solve $z=p^2+q^2$.

Solution:
Here $z=p^2+q^2$.

Let $q=ap$ then the given equation reduce to

$$z=p^2+a^2p^2,$$

$$\Rightarrow z = p^2\left(1+a^2\right),$$

$$p^2 = \frac{z}{\left(1+a^2\right)},$$

$$p = \sqrt{\frac{z}{\left(1+a^2\right)}} \tag{2.86}$$

Again, we know that

$$dz = pdx + qdy,$$

$$dz = pdx + apdy,$$

$$dz = p\left(dx + ady\right),$$

$$dz = \sqrt{\frac{z}{\left(1+a^2\right)}}\left(dx + ady\right) \text{ (from equation (2.86))}$$

$$\sqrt{\left(1+a^2\right)}\frac{dz}{\sqrt{z}} = \left(dx + ady\right),$$

Integrating above equation on both the sides, we get

$x + ay = \sqrt{1+a^2}\,2\sqrt{z} + c$, where c is any arbitrary constant.

$$\left(x + ay - c\right) = \sqrt{1+a^2}\,2\sqrt{z},$$

$\left(x + ay - c\right)^2 = 4\left(1+a^2\right)z$ which is the complete solution.

Example:

Solve $zpq = p + q$.

Solution:

Here:

$$zpq = p + q \tag{2.87}$$

Let $q = ap$ then the equation (2.87) reduces to

$$zp(ap) = p + ap,$$

$$\Rightarrow zap = 1 + a,$$

$$p = \frac{1+a}{az}.$$

Again, we know that $dz = pdx + qdy$,

$$= pdx + apdy \ ,$$

$$dz = p(dx + ady),$$

$$dz = \frac{1+a}{az}(dx + ady),$$

$$az\,dz = (1+a)(dx + a\,dy),$$

Integrating both sides, we get $a\frac{z^2}{2} = (1+a)(x+ay)+c$,
which is the complete solution, where c is any arbitrary constant.

Example:

Solve $z^2\left(p^2z^2+q^2\right)=1$.

Solution:

Here

$$z^2\left(p^2z^2+q^2\right)=1 \tag{2.88}$$

Let $q=ap$ then the given equation reduce to

$$z^2\left(p^2z^2+a^2p^2\right)=1,$$

$$\Rightarrow p^2=\frac{1}{z^2\left(z^2+a^2\right)},$$

$$\Rightarrow p=\frac{1}{z\sqrt{\left(z^2+a^2\right)}} \tag{2.89}$$

Again, we know that

$$dz=pdx+qdy \ =pdx+apdy \ ,$$

$$dz=p\left(dx+ady\right),$$

$$dz=\frac{1}{z\sqrt{\left(z^2+a^2\right)}}\left(dx+ady\right),$$

$$z\sqrt{z^2+a^2}\,dz=dx+a\,dy,$$

Integrating both the sides, we get $\int z\sqrt{z^2+a^2}\,dz=\int dx+a\int dy$,

$$\frac{1}{3}\left(z^2+a^2\right)^{\frac{3}{2}}=x+ay+c,$$

$\left(z^2+a^2\right)^3=9\left(x+ay+c\right)^2$, which is the solution of PDE equation.

Example:

Solve $p^2z^2+q^2=p^2q$.

Solution:

Here

$$p^2z^2+q^2=p^2q \tag{2.90}$$

Let $q=ap$ then equation (2.90) reduces to

$$p^2z^2+(ap)^2=p^2ap,$$

$$\Rightarrow p=\frac{z^2+a^2}{a} \tag{2.91}$$

Again, we know that

$$dz=pdx+qdy\;=pdx+apdy\,,$$

$$dz=p\left(dx+ady\right),$$

$dz=\dfrac{z^2+a^2}{a}\left(dx+ady\right)$ (From equation (2.91))

$$\frac{a}{z^2+a^2}dz = dx + ady,$$

Integrating both the sides, we get

$$\int \frac{a}{z^2+a^2}dz = \int dx + a\int dy + c,$$

$$\tan^{-1}\left(\frac{z}{a}\right) = x + ay + c,$$

$$\frac{z}{a} = \tan\left(x + ay + c\right),$$

$z = a\tan\left(x+ay+c\right)$, which is the solution of PDE, where c is an arbitrary constant.

Example:

Solve $p^2x^2 = z\left(z - qy\right)$.

Solution:

Here:

$$p^2x^2 = z\left(z-qy\right) \tag{2.92}$$

To reduce above equation in the form $F\left(z,p,q\right)=0$, put $X = \ln x$, $Y = \ln y$ so that

$$dX = \frac{dx}{x}, dY = \frac{dy}{y}.$$

Now $\dfrac{\partial z}{\partial x} = \dfrac{\partial z}{\partial X}.\dfrac{\partial X}{\partial x} = \dfrac{1}{x}\dfrac{\partial z}{\partial X}$ (2.93)

Similarly, $\frac{\partial z}{\partial y}=\frac{1}{y}\frac{\partial z}{\partial Y}$ (2.94)

Rewriting equation (2.92) using equations (2.93) and (2.94) as follows:

$$\left(z\frac{\partial z}{\partial x}\right)^2=z\left(z-y\frac{\partial z}{\partial y}\right)\Rightarrow\left(\frac{\partial z}{\partial X}\right)^2=z\left(z-\frac{\partial z}{\partial Y}\right). \tag{2.95}$$

Let $P=\frac{\partial z}{\partial X},Q=\frac{\partial z}{\partial Y}$ then equation (2.95) reduces to

$$P^2=z(z-Q) \tag{2.96}$$

which is of the form $F(z,P,Q)=0$. Let $Q=aP$, thus equation (2.96) is reduced to

$$P^2=z(z-aP),$$

$$\Rightarrow P^2+azP-z^2=0,$$

$$\Rightarrow P=\frac{-az\pm\sqrt{a^2z^2+4z^2}}{2},$$

$$\Rightarrow P=z\left(\frac{-a\pm\sqrt{a^2+4}}{2}\right),$$

$$\Rightarrow P=zk \text{ , where } k=\left(\frac{-a\pm\sqrt{a^2+4}}{2}\right),$$

Again, we know that

$$dz=PdX+QdY \;=PdX+aPdY \;,$$

$$dz = zk\left(dX + adY\right), \quad \left(\because\ P = zk\right)$$

$$\frac{dz}{zk} = dX + adY,$$

Integrating on both the sides, we get

$$\frac{1}{k}\log z = X + aY,$$

$$\log z^{\frac{1}{k}} = \ln x + a\ln y + \ln c,$$

$$\log z^{\frac{1}{k}} = \ln\left(xy^{a}c\right),$$

$z^{\frac{1}{k}} = xy^{a}c$, which is the solution of PDE.

Example:

Solve $9\left(p^2z^2 + q^2\right) = 4$.

Solution:
Here

$$9\left(p^2z^2 + q^2\right) = 4 \tag{2.97}$$

Let $q = ap$, then equation (2.97) reduces to

$$9\left(p^2z^2 + (ap)^2\right) = 4,$$

$$\Rightarrow p^2 = \frac{4}{9\left(z^2 + a^2\right)},$$

$$\Rightarrow p=\pm\frac{2}{3\sqrt{z^2+a^2}} \tag{2.98}$$

Again, we know that

$$dz = pdx + qdy \;= pdx + apdy\ ,$$

$$dz = p(dx + ady),$$

$$dz = \pm\frac{2}{3\sqrt{z^2+a^2}}(dx+ady) \text{(From (1.149))}$$

$$\sqrt{z^2+a^2}\,dz = \frac{2}{3}dx + \frac{2}{3}ady.$$

Integrating on both sides, we get

$$\int\sqrt{z^2+a^2}\,dz = \int\frac{2}{3}dx + \int\frac{2}{3}ady,$$

$\left(z^2+a^2\right)^{\frac{3}{2}} = x + ay + c$, which is the solution of PDE.

2.4.3 Partial differential equation of the form $f(x,p)=g(y,q)$

In PDE of the form $f(x,p)=g(y,q)$, there is no z term in the equation. Assume $f(x,p)=g(y,q)=a=\text{constant}$. Solving each equation for p and q, we get

$$p = f_1(x,a) \text{ and } q = g_1(y,a) \tag{2.99}$$

Now

$$dz = p\,dx + q\,dy,$$

$dz = f_1(x,a)\,dx + g_1(y,a)\,dy$. (From equation (2.99))

Now integrating on both sides, we get

$z = \int f_1(x,a)\,dx + \int g_1(y,a)\,dy + c$, which gives the required solution.

Example:

Solve $p + q = \sin x + \sin y$.

Solution:

Here:

$$p + q = \sin x + \sin y \qquad (2.100)$$

Equation (2.100) can be written as $p - \sin x = q - \sin y = a$.

Hence:

$$p = a + \sin x \text{ and } q = \sin y - a. \qquad (2.101)$$

Again, we know that $dz = p\,dx + q\,dy$.

Now substituting values of p and q from equation (2.101) in the above equation, we get

$$dz = (\sin x + a)\,dx + (\sin y - a)\,dy.$$

Integrating on both the side, we get

$z = \int (\sin x + a)\,dx + \int (\sin y - a)\,dy + c_1$ where c_1 is an arbitrary constant.

$$z = -\cos x + ax - \cos y - ay + c_1,$$

$z = -(\cos x + \cos y) + a(x - y) + c_1$, which is required solution of given PDE.

Example:

Solve $yp = 2xy + \log q$.

Solution:

Here

$$yp = 2xy + \log q \tag{2.102}$$

Equation (2.102) can be written as

$$p = 2x + \frac{\log q}{y} \Rightarrow p - 2x = \frac{\log q}{y} = a \tag{2.103}$$

From equation (2.103), we get

$$p = a + 2x \tag{2.104}$$

and

$$\frac{\log q}{y} = a \Rightarrow \log q = ay \Rightarrow q = e^{ay} \tag{2.105}$$

Again, we know that $dz = p\,dx + q\,dy$.

Now substituting values of p and q from equations (2.104) and (2.105) in above equation, we get

$$dz = (a + 2x)dx + e^{ay}\,dy.$$

Integrating on both the side, we get

$$z=\int(a+2x)\,dx+\int e^{ay}\,dy,$$

$z=ax+x^2+\dfrac{e^{ay}}{a}+c_1$, which required solution to equation (2.102).

Example:

Solve $z\left(p^2-q^2\right)=(x-y)$.

Solution: Here

$$z\left(p^2-q^2\right)=(x-y) \tag{2.106}$$

Equation (2.106) can be written as

$$zp^2-zq^2=x-y\Rightarrow\left(\sqrt{z}p\right)^2-\left(\sqrt{z}q\right)^2=x-y \tag{2.107}$$

Take $Z=\dfrac{2}{3}z^{\frac{3}{2}}\Rightarrow dZ=\sqrt{z}dz$.

Differentiating partially w.r.t x and y, we get

$$\frac{\partial Z}{\partial x}=\frac{\partial z}{\partial x}\sqrt{z},\frac{\partial Z}{\partial y}=\frac{\partial z}{\partial y}\sqrt{z}.$$

Equation (2.107) reduces to $\left(\dfrac{\partial Z}{\partial x}\right)^2-\left(\dfrac{\partial Z}{\partial y}\right)^2=x-y$.

Taking $P=\dfrac{\partial Z}{\partial x}$ and $Q=\dfrac{\partial Z}{\partial y}$, above equation reduces to

$P^2-Q^2=x-y\Rightarrow P^2-x=Q^2-y=a$ (say) and we get

$P=\sqrt{a+x}$ and $Q=\sqrt{a+y}$.

Substituting values of P and Q in $dZ=Pdx+Q\,dy$, we get

$$dZ=\sqrt{a+x}\,dx+\sqrt{a+y}\,dy.$$

Integrating on both the sides, we get

$$Z=\frac{2(a+x)^{\frac{3}{2}}}{3}+\frac{2(a+y)^{\frac{3}{2}}}{3}+c,$$

$$\frac{2}{3}z^{\frac{3}{2}}=\frac{2(a+x)^{\frac{3}{2}}}{3}+\frac{2(a+y)^{\frac{3}{2}}}{3}+c. \quad (\because Z=\frac{2}{3}z^{\frac{3}{2}}) \tag{2.108}$$

where a and c are arbitrary constants.

Example:

Solve $zpy^2=x\left(y^2+z^2q^2\right)$.

Solution:

Here:

$$zpy^2=x\left(y^2+z^2q^2\right) \tag{2.109}$$

To reduce equation (2.109) in standard form put

$Z=\frac{z^2}{2}\Rightarrow\frac{\partial Z}{\partial x}=z\frac{\partial z}{\partial x}$ and $\frac{\partial Z}{\partial y}=z\frac{\partial z}{\partial y}\Rightarrow dZ=z\,dz$

Again $zp=z\frac{\partial z}{\partial x}=P$ and $zq=z\frac{\partial z}{\partial y}=Q$

Substituting the value of zp and zq in equation (2.109), we get

$$Py^2 = x\left(y^2 + Q^2\right),$$

$$\Rightarrow \frac{P}{x} = \frac{\left(y^2 + Q^2\right)}{y^2} = a \text{ (say)},$$

$$\frac{P}{x} = a \quad \text{and} \quad \frac{\left(y^2 + Q^2\right)}{y^2} = a,$$

$$P = ax \quad \text{and} \quad 1 + \frac{Q^2}{y^2} = a,$$

$$P = ax \quad \text{and} \quad Q^2 = y^2\left(a - 1\right),$$

$$P = ax \text{ and } Q = y\sqrt{\left(a - 1\right)}.$$

We know that $dZ = P\,dx + Q\,dy$. Substituting the value of P and Q in $dZ = P\,dx + Q\,dy$, we get $dZ = ax\,dx + y\sqrt{a-1}\,dy$,
$zdz = ax\,dx + y\sqrt{a-1}\,dy$. $\left(\because dZ = z\,dz\right)$

Now integrating on both the sides, we get

$$\frac{z^2}{2} = \frac{ax^2}{2} + \sqrt{a-1}y^2 + c,$$

$z^2 = ax^2 + 2\sqrt{a-1}y^2 + 2c_1$, which is required solution and c_1 is an arbitrary constant.

2.4.4 Clairaut's equation

The equation of the form $z = px + qy + f(p,q)$ is known as Clairaut equation and its complete solution is given by $z = ax + by + f(a,b)$ which is obtained by replacing p by a and q by b in the Clairaut's equation.

Example:

Solve $z = px + qy + 3p^{\frac{1}{3}}q^{\frac{1}{3}}$.

Solution:

Here $z = px + qy + 3p^{\frac{1}{3}}q^{\frac{1}{3}}$.

This PDE is of the form $z = px + qy + f(p,q)$. Hence its complete solution is given by $z = ax + by + 3a^{\frac{1}{3}}b^{\frac{1}{3}}$.

Example:

Solve $z = px + qy + \ln pq$.

Solution:

Here $z = px + qy + \ln(pq)$

This PDE is of the form $z = px + qy + f(p,q)$.Hence its complete solution is given by $z = ax + by + \ln(ab)$.

Example:

Find the complete and singular solution of $z = px + qy + \sqrt{1 + p^2 + q^2}$.

Solution: Here $z = px + qy + \sqrt{1+p^2+q^2}$

This PDE is of the form $z = px + qy + f(p,q)$.Hence its complete solution is given by $z = ax + by + \sqrt{1+a^2+b^2}$.

Now differentiating w.r.t to a and b, we obtain

$$0 = x + \frac{a}{\sqrt{1+a^2+b^2}} \Rightarrow x = -\frac{a}{\sqrt{1+a^2+b^2}} \tag{2.110}$$

$$0 = y + \frac{b}{\sqrt{1+a^2+b^2}} \Rightarrow y = -\frac{b}{\sqrt{1+a^2+b^2}} \tag{2.111}$$

From above equations, we get $x^2 + y^2 = \dfrac{a^2+b^2}{1+a^2+b^2}$.

Hence, $1-(x^2+y^2) = \dfrac{1}{1+a^2+b^2} \Rightarrow \sqrt{1+a^2+b^2} = \dfrac{1}{\sqrt{1-x^2-y^2}}$

$\therefore a = \dfrac{-x}{\sqrt{1-x^2-y^2}}$ (From equation (2.110))

and $b = \dfrac{-y}{\sqrt{1-x^2-y^2}}$. (From equation (2.111))

Substituting the value of a and b, we obtain

$$z = -\frac{x^2}{\sqrt{1-x^2+y^2}} - \frac{y^2}{\sqrt{1-x^2-y^2}} + \frac{1}{\sqrt{1-x^2-y^2}},$$

$$z = \frac{-x^2-y^2+1}{\sqrt{1-x^2-y^2}} = \sqrt{1-x^2-y^2},$$

$z^2 = 1-x^2-y^2$, which required a singular solution.

Exercise

Solve the following partial differential equation

$$\frac{\partial^2 z}{\partial x \partial y} = e^{-y}.\cos x$$

(**Ans** $z = -e^{-y}\sin y + \int g(y)\,dy + h(x)$)

$$\frac{\partial^2 u}{\partial x \partial y} = 4x\sin(3xy)$$

(**Ans** $u = 24x\cos(3xy) - 36x^2\sin(3xy) + f(x) + g(x)$)

$\frac{\partial^2 z}{\partial y^2} + z = 0$ at $y = 0, z = e^x$ and $\frac{\partial z}{\partial y} = 1$

(**Ans** $u = \sin y + e^x \cos y$)

$\frac{\partial^2 z}{\partial x^2} = ax^2$, given that when $x = 0, \frac{\partial z}{\partial x} = a\sin y$ and $\frac{\partial z}{\partial y} = 0$

(**Ans** $z = \frac{a^3x^3}{6} + ax\sin y + c$)

Solve the following PDE which are in the form $f(p,q) = 0$

$p + \sin q = 0$

(**Ans** $z = -x\sin b + by + c$)

$pq = k$

(**Ans** $z = ax + k\left(\frac{y}{a}\right) + c$)

$(p+q)(p-q) = \lambda$

(**Ans** $z = ax + \left(\sqrt{a^2 - \lambda}\right)y + c$)

$$p^2+q^2=m^2$$

(**Ans** $z=ax+\left(\sqrt{m^2-a^2}\right)y+c$)

$$p^{\frac{1}{2}}+q^{\frac{1}{2}}=1$$

(**Ans** $z=ax+\left(1-\sqrt{a}\right)y+c$)

$$p=e^q$$

(**Ans** $z=ax+y\ln a+c$)

$$p=2q^2+1$$

(**Ans** $z=\left(2b^2+1\right)x+by+c$)

Solve the following PDE which are in the form $f(z,p,q)=0$

$$z=p^2+q^2$$

(**Ans** $4\left(1+a^2\right)z=\left(x+ay+b\right)^2$)

$$p^2z^2+q^2=1$$

(**Ans** $az\left(1+a^2z\right)^{\frac{1}{2}}-\ln\left(az+\left(1+a^2z^2\right)^{\frac{1}{2}}\right)=2a\left(ax+y+b\right)$)

$$p\left(1+q\right)=qz$$

(**Ans** $\ln\left(az-1\right)=\left(x+ay+b\right)$)

$$4\left(1+z^3\right)=qz^4pq$$

(**Ans** $a\left(1+z^3\right)=\left(x+ay+b\right)^2$)

$$p\left(z+p\right)+q=0$$

(**Ans** $a\left(1+z^3\right)=\left(x+ay+b\right)^2$)

$$q^2 = z^2 p^2 \left(1 - p^2\right)$$

(**Ans** $a^2 z^2 = \left(y + ax + c\right)^2 + 1$))

$$z^2 = 1 + p^2 + q^2$$

(**Ans** $z = \cosh\left(x + ay + \frac{c}{\sqrt{1+a^2}}\right)$))

$$z^2\left(1 + p^2 + q^2\right) = 1$$

(**Ans** $\left(a^2+1\right)\left(1-z^2\right) = \left(x + ay + b\right)^2$))

$$z^2\left(p^2 x^2 + q^2\right) = 1$$

(**Ans** $z^2\sqrt{a^2+1} = \pm 2\left(\ln x + ay\right) + b$))

Find the complete integral of the following PDE form

$$f\left(x, p\right) = g\left(y, q\right)$$

$$\sqrt{p} + \sqrt{q} = x + y$$

(**Ans** $z = \frac{\left(a+x\right)^3}{3} + \frac{\left(y-a\right)^3}{3} + b$)

$$\sqrt{p} - \sqrt{q} + 3x = 0$$

(**Ans** $z = -\frac{\left(9-3x\right)^3}{9} + a^2 y + b$)

$$pq = xy$$

(**Ans** $z = \frac{a}{2}x^2 + \frac{1}{2a}y^2 + b$)

$$p^2 \pm q^2 = x \pm y$$

(**Ans** $z=\frac{2(a+x)^{\frac{3}{2}}}{3}+\frac{2(\mp a+y)^{\frac{3}{2}}}{3}+b$)

$p^2y(1+x^2)=qx^2$

(**Ans** $z=a\sqrt{1+x^2}+\frac{1}{2}a^2y^2+b$)

$pe^y=qe^x$

(**Ans** $z=a(e^x+e^y)+b$)

$\frac{x}{p}+\frac{y}{q}=1$

(**Ans** $z=\frac{x^2}{2(a+1)}+\frac{y^2}{2a}+b$)

$q=xyp^2$

(**Ans** $(2z-ay^2-2b)^2=16ax$)

$p-q=x^2+y^2$

(**Ans** $z=\frac{1}{3}(x^3-y^3)+a(x+y)+b$)

$z(p^2-q^2)=x-y$

(**Ans** $z^{\frac{3}{2}}=(a+x)^{\frac{3}{2}}+(a+y)^{\frac{3}{2}}+c$))

Find the complete integral of the following PDE which is of the form $z=px+qy+f(p,q)$**:**

$z=px+qy\pm pq$

(**Ans** $z=ax+by\pm ab$)

$$(px+qy-z)^2 = q(1+p^2+q^2)$$

(Ans $z = ax+by \pm \sqrt{1+a^2+b^2}$ **)**

$$(p+q)(z-px-qy) = 1$$

(Ans $z = ax+by+\dfrac{1}{a+b}$**)**

$$pqz = p^2(xq+p^2)+q^2(yq+p^2)$$

(Ans $z = ax+by+\left(\dfrac{a^3}{b}+\dfrac{b^3}{a}\right)$**)**

$$2q(z-px-qy) = 1+q^2$$

(Ans $z = ax+by+\dfrac{b^2+1}{2b}$**)**

2.5 Charpit's Method: General Method to Solve the PDE of First Order

Charpit's method is a general method to find the complete solution of the first order non-linear PDE. Let the general PDE be

$$f(x,y,z,p,q) = 0 \tag{2.112}$$

where $p = \dfrac{\partial z}{\partial x}$ and $q = \dfrac{\partial z}{\partial y}$.Since z depends on x and y, we have

$$dz = \frac{\partial z}{\partial x}dx + \frac{\partial z}{\partial y}dy \Rightarrow dz = pdx + qdy \tag{2.113}$$

The main idea in Charpit's method is to find another relation between variables x, y, z , p and q. Let the relation be

$$F(x,y,z,p,q) = 0 \tag{2.114}$$

Solving equations (2.112) and (2.114), we get the value of p and q. These values of p and q are substituted in equation (2.113), it become integrable. To find $F(x,y,z,p,q)=0$, differentiating $f(x,y,\mathrm{z},\mathrm{p},\mathrm{q})=0$ and $F(x,y,z,p,q)=0$ partially w.r.t. x , we get

$$\frac{\partial f}{\partial x}+\frac{\partial f}{\partial z}p+\frac{\partial f}{\partial p}\frac{\partial p}{\partial x}+\frac{\partial f}{\partial q}\frac{\partial q}{\partial x}=0 \tag{2.115a}$$

and

$$\frac{\partial F}{\partial x}+\frac{\partial F}{\partial z}p+\frac{\partial F}{\partial p}\frac{\partial p}{\partial x}+\frac{\partial F}{\partial q}\frac{\partial q}{\partial x}=0 \tag{2.115b}$$

and differentiating $f(x,y,\mathrm{z},\mathrm{p},\mathrm{q})=0$ and $F(x,y,z,p,q)=0$ partially w.r.t. y ,we get

$$\frac{\partial f}{\partial y}+\frac{\partial f}{\partial z}q+\frac{\partial f}{\partial p}\frac{\partial p}{\partial y}+\frac{\partial f}{\partial q}\frac{\partial q}{\partial y}=0 \tag{2.116a}$$

and

$$\frac{\partial F}{\partial y}+\frac{\partial F}{\partial z}q+\frac{\partial F}{\partial p}\frac{\partial p}{\partial y}+\frac{\partial F}{\partial q}\frac{\partial q}{\partial y}=0 \tag{2.116b}$$

Eliminating $\frac{\partial p}{\partial x}$ from equation .(2.115 a) and (2.115b), by multiplying equation (2.115a) with $\frac{\partial F}{\partial p}$ and equation .(2.115b) with $\frac{\partial f}{\partial p}$ and subtracting, we get

$$\left(\frac{\partial f}{\partial x}.\frac{\partial F}{\partial p}-\frac{\partial F}{\partial x}.\frac{\partial f}{\partial p}\right)+\left(\frac{\partial f}{\partial z}.\frac{\partial F}{\partial p}-\frac{\partial F}{\partial z}.\frac{\partial f}{\partial p}\right)p+\left(\frac{\partial f}{\partial q}.\frac{\partial F}{\partial p}-\frac{\partial F}{\partial q}.\frac{\partial f}{\partial p}\right)\frac{\partial q}{\partial x}=0 \tag{2.117}$$

Similarly, eliminating $\frac{\partial q}{\partial y}$ from equations (2.116a) and (2.116b), we get

$$\left(\frac{\partial f}{\partial y}\cdot\frac{\partial F}{\partial q}-\frac{\partial F}{\partial y}\cdot\frac{\partial f}{\partial q}\right)+\left(\frac{\partial f}{\partial z}\cdot\frac{\partial F}{\partial q}-\frac{\partial F}{\partial z}\cdot\frac{\partial f}{\partial q}\right)q+\left(\frac{\partial f}{\partial p}\cdot\frac{\partial F}{\partial q}-\frac{\partial F}{\partial p}\cdot\frac{\partial f}{\partial q}\right)\frac{\partial q}{\partial y}=0 \qquad (2.118)$$

Adding equation (2.117) and equation (2.118) and using

$$\frac{\partial q}{\partial x}=\frac{\partial}{\partial x}\left(\frac{\partial z}{\partial y}\right)=\frac{\partial^2 z}{\partial x\partial y}=\frac{\partial^2 z}{\partial y\partial x}=\frac{\partial}{\partial y}\left(\frac{\partial z}{\partial x}\right)=\frac{\partial p}{\partial y},$$

$$\left(\frac{\partial f}{\partial x}+p\frac{\partial f}{\partial z}\right)\frac{\partial F}{\partial p}+\left(\frac{\partial f}{\partial y}+q\frac{\partial f}{\partial z}\right)\frac{\partial F}{\partial q}+\left(-p\frac{\partial f}{\partial p}-q\frac{\partial f}{\partial q}\right)\frac{\partial F}{\partial z}+\left(-\frac{\partial f}{\partial p}\right)\frac{\partial F}{\partial x}+\left(-\frac{\partial f}{\partial q}\right)\frac{\partial F}{\partial y}=0$$

Rearranging terms in the above equation, we get

$$\frac{\partial f}{\partial p}\frac{\partial F}{\partial x}+\frac{\partial f}{\partial q}\frac{\partial F}{\partial y}+\left(p\frac{\partial f}{\partial p}+q\frac{\partial f}{\partial q}\right)\frac{\partial F}{\partial z}-\left(\frac{\partial f}{\partial x}+p\frac{\partial f}{\partial z}\right)\frac{\partial F}{\partial p}-\left(\frac{\partial f}{\partial y}+q\frac{\partial f}{\partial z}\right)\frac{\partial F}{\partial q}=0$$

$$f_p\frac{\partial F}{\partial x}+f_q\frac{\partial F}{\partial y}+\left(pf_p+qf_q\right)\frac{\partial F}{\partial z}-\left(f_x+pf_z\right)\frac{\partial F}{\partial p}-\left(f_y+qf_z\right)\frac{\partial F}{\partial q}=0 \qquad (2.119)$$

where $f_p=\frac{\partial f}{\partial p}$, $f_q=\frac{\partial f}{\partial q}$, $f_x=\frac{\partial f}{\partial x}$, $f_y=\frac{\partial f}{\partial y}$ and $f_z=\frac{\partial f}{\partial z}$.

Equation (2.119) is Lagrange's linear partial differential equation in F . Its subsidiary equations are

$$\frac{dx}{f_p}=\frac{dy}{f_q}=\frac{dz}{pf_p+qf_q}=\frac{dp}{-\left(f_x+pf_z\right)}=\frac{dq}{-\left(f_y+qf_z\right)}=\frac{dF}{0}.$$

These equations are known as Charpit's equations. Solving these equations, *p* and *q* are obtained. The simplest of the relations should be taken to obtain *p* and *q* easily.

Working rules while using Charpit's method:

Step I. Transfer all the terms of the given equation to L.H.S. and denote the entire expression by f .

Step II. Write down the subsidiary equations or Charpit's auxiliary equations i.e.,

$$\frac{dp}{\frac{\partial f}{\partial x}+p\frac{\partial f}{\partial z}}=\frac{dp}{\frac{\partial f}{\partial y}+q\frac{\partial f}{\partial z}}=\frac{dz}{-p\left(\frac{\partial f}{\partial p}\right)-q\left(\frac{\partial f}{\partial q}\right)}=\frac{dx}{\frac{\partial f}{\partial p}}=\frac{dy}{\frac{\partial f}{\partial q}}=\frac{dF}{0}$$

Step III. Using the value of f in step I write down the values of $\frac{\partial f}{\partial x},\frac{\partial f}{\partial y},...$ etc. occurring in step II and put these in Charpit's auxiliary equations.

Step IV. After simplifying the step III, select two proper fractions so that the resulting integral may come out to be the simplest relation involving at least one of p and q .

Step V. The simplest relation of step IV is solved along with the given equation to determine p and q . But these values of p and q in $dz = pdx + qdy$ which on integration gives the complete integral of the equation.

Example:

Solve $p^2x+q^2y=u$.

Solution:

To find a general solution, we proceed as follows:

Step I. Computing f_x, f_y, f_u, f_p, f_q

Set $f = p^2x+q^2y-u=0$.

Then $f_x = p^2$, $f_y = q^2$, $f_u = -1$, $f_p = 2px$, $f_q = 2qy$.

Hence, $pf_p + qf_q = 2p^2x + 2q^2y$, $-(f_x + pf_u) = -p^2 + p$, $-(f_y + qf_u) = -q^2 + q$.

Step II. Writing subsidiary equations or Charpit's equations and finding a solution $g(x, y, u, p, q, a)$.

The Charpit's equations (or auxiliary) equations are:

$$\frac{dx}{f_p} = \frac{dy}{f_q} = \frac{du}{pf_p + qf_q} = \frac{dp}{-(f_x + pf_u)} = \frac{dq}{-(f_y + qf_u)},$$

$$\frac{dx}{2px} = \frac{dy}{2qy} = \frac{du}{2(p^2x + q^2y)} = \frac{dp}{-p^2 + p} = \frac{dq}{-q^2 + q}.$$

From which it follows that

$$\frac{p^2dx + 2pxdp}{2p^3x + 2p^2x - 2p^3x} = \frac{q^2dy + 2qydq}{2q^3y + 2q^2y - 2q^3y},$$

$$\Rightarrow \quad \frac{p^2dx + 2pxdp}{p^2x} = \frac{q^2dy + 2qydq}{q^2y},$$

On integrating, we obtain

$$\log(p^2x) = \log(q^2y) + \log a,$$

$\Rightarrow$ $p^2x = aq^2y$, where a is an arbitrary constant.

Step III. Solving for p and q

$$p^2x + q^2y = u, \quad p^2x = aq^2y$$

$$\Rightarrow \left(aq^2y\right)+q^2y=u \;\Rightarrow\; q^2y\left(1+a\right)=u$$

$$\Rightarrow \qquad q^2=\frac{u}{y\left(1+a\right)} \;\Rightarrow\; \qquad q=\left[\frac{u}{y\left(1+a\right)}\right]^{1/2}$$

And

$$p^2=aq^2\frac{y}{x}=a\frac{u}{y\left(1+a\right)}\frac{y}{x}=\frac{au}{\left(1+a\right)x}$$

$$\Rightarrow \qquad p=\left[\frac{au}{\left(1+a\right)x}\right]^{1/2}$$

Step IV. Writing $du=p\left(x,y,u,a\right)dx+q\left(x,y,u,a\right)dy$ and finding its solution

Writing $du=\left[\frac{au}{\left(1+a\right)x}\right]^{1/2}dx+\left[\frac{u}{\left(1+a\right)y}\right]^{1/2}dy$

$$\Rightarrow \left(\frac{1+a}{u}\right)^{1/2}du=\left(\frac{a}{x}\right)^{1/2}dx+\left(\frac{a}{y}\right)^{1/2}dy$$

Integrate to have $\left[\left(1+a\right)u\right]^{1/2}=\left(ax\right)^{1/2}+\left(y\right)^{1/2}+b$

Example:

Solve $z=px+qy+p^2+q^2$. (2.120)

Solution:

Here $z=px+qy+p^2+q^2$.

The subsidiary equations of (2.120) are

$$\frac{dx}{f_p}=\frac{dy}{f_q}=\frac{dz}{pf_p+qf_q}=\frac{dp}{-(f_x+pf_z)}=\frac{dq}{-(f_y+qf_z)} \qquad (2.121)$$

Here $z=px+qy+p^2+q^2 \Rightarrow f: z-px-qy-p^2-q^2=0$.
Hence, $f_x=-p, f_y=-q, f_z=1, f_p=-x-2p, f_q=-y-2q$.

By using above values equation (2.121) reduces to

$$\frac{dx}{-x-2p}=\frac{dy}{-y-2q}=\frac{dz}{-p(x+2p)-q(y+2q)}=\frac{dp}{0}=\frac{dq}{0}.$$

From above equation, we get $dp=0 \Rightarrow p=a$ and $dq=0 \Rightarrow q=b$

Substituting the value of p and q in given equation, we get $z=ax+by+a^2+b^2$, which is the required solution.

Example:

Solve $p^2x+q^2y=u$.

Solution: To find a general solution, we proceed as follows:

Step I. Computing f_x, f_y, f_u, f_p, f_q

Set $f=p^2x+q^2y-u=0$.

Then $f_x=p^2$, $f_y=q^2$, $f_u=-1$, $f_p=2px$, $f_q=2qy$

And hence, $pf_p+qf_q=2p^2x+2q^2y$, $-(f_x+pf_u)=-p^2+p$, $-(f_y+qf_u)=-q^2+q$

Step II. (Writing Charpit's equations and finding a solution $g(x,y,u,p,q,a)$).

The Charpit's equations (or auxiliary) equations are:

$$\frac{dx}{f_p}=\frac{dy}{f_q}=\frac{du}{pf_p+qf_q}=\frac{dp}{-(f_x+pf_u)}=\frac{dq}{-(f_y+qf_u)}$$

$$\Rightarrow \quad \frac{dx}{2px}=\frac{dy}{2qy}=\frac{du}{2(p^2x+q^2y)}=\frac{dp}{-p^2+p}=\frac{dq}{-q^2+q} \tag{2.122}$$

From which it follows that

$$\frac{p^2dx+2pxdp}{2p^3x+2p^2x-2p^3x}=\frac{q^2dy+2qydq}{2q^3y+2q^2y-2q^3y}$$

$$\Rightarrow \quad \frac{p^2dx+2pxdp}{p^2x}=\frac{q^2dy+2qydq}{q^2y}$$

On integrating, we obtain

$$\log(p^2x)=\log(q^2y)+\log a$$

$\Rightarrow p^2x=aq^2y$ Where a is an arbitrary constant.

Step III. Solving for p and q

$$p^2x+q^2y=u\ ,\ p^2x=aq^2y \tag{2.123}$$

$$\Rightarrow \quad (aq^2y)+q^2y=u \Rightarrow q^2y(1+a)=u$$

$$\Rightarrow \quad q^2=\frac{u}{y(1+a)} \Rightarrow \quad q=\left[\frac{u}{y(1+a)}\right]^{1/2}$$

And

$$p^2 = aq^2 \frac{y}{x} = a \frac{u}{y(1+a)} \frac{y}{x} = \frac{au}{(1+a)x}$$

$$\Rightarrow \qquad p = \left[\frac{au}{(1+a)x} \right]^{1/2}$$

Step IV. Writing $du = p(x, y, u, a)dx + q(x, y, u, a)dy$ and finding its solution

Writing $du = \left[\frac{au}{(1+a)x} \right]^{1/2} dx + \left[\frac{u}{(1+a)y} \right]^{1/2} dy$

$$\Rightarrow \qquad \left(\frac{1+a}{u} \right)^{1/2} du = \left(\frac{a}{x} \right)^{1/2} dx + \left(\frac{a}{y} \right)^{1/2} dy$$

Integrating both sides, we have

$$\left[(1+a)u \right]^{1/2} = (ax)^{1/2} + (y)^{1/2} + b$$

which is the required solution.

Example:

Solve $z^2 = pqxy$.

Solution:

Here:

$$z^2 = pqxy \tag{2.124}$$

The subsidiary equations of (2.124) are

$$\frac{dx}{f_p}=\frac{dy}{f_q}=\frac{dz}{pf_p+qf_q}=\frac{dp}{-(f_x+pf_z)}=\frac{dq}{-(f_y+qf_z)} \tag{2.125}$$

Here $z^2 = pqxy \Rightarrow f : z^2 - pqxy = 0$.

Hence, $f_x=-pqy, f_y=-pqx, f_z=2z, f_p=-qxy, f_q=-pxy$.

By using above results, equation (2.120) reduces to

$$\frac{dx}{-qxy}=\frac{dy}{-pxy}=\frac{dz}{-2pqxy}=\frac{dp}{-(-pqy+p2z)}=\frac{dq}{-(-pqx+2qz)}.$$

From above equation, we get

$$\frac{pdx+xdp}{-pqxy+xpqy-2pxz}=\frac{qdy+ydq}{-qpxy+ypdx-2ydz},$$

$$\frac{pdx+xdp}{-2pxz}=\frac{qdy+ydq}{-2yzq},$$

$$\frac{d(xp)}{xp}=\frac{d(yq)}{yq}.$$

Integrating on both sides, we get

$$\log(xp)=\log(yq)+\log a,$$

$$\Rightarrow \frac{xp}{yq}=a,$$

$$\Rightarrow \frac{xp}{ya}=q \tag{2.126}$$

Using the value of q in equation (2.126) , we get

$$z^2 = p\left(\frac{xp}{yq}\right)xy = \frac{p^2x^2}{q} \Rightarrow p = \sqrt{a}\,\frac{z}{x} \tag{2.127}$$

From equations (2.126) and (2.127) , we get

$$q = \frac{xp}{ya} = \frac{x}{ya}\left(\frac{\sqrt{a}z}{x}\right) = \frac{z}{\sqrt{a}y} \tag{2.128}$$

Substituting the value of p and q from equation s. (2.127) and (2.128) in $dz = pdx + qdy$, we get

$$dz = \sqrt{a}\,\frac{z}{x}\,dx + \frac{1}{\sqrt{a}}\frac{z}{y}\,dy,$$

$$\Rightarrow \frac{dz}{z} = \frac{\sqrt{a}}{x}\,dx + \frac{1}{\sqrt{a}}\frac{1}{y}\,dy.$$

Now integrating on both the sides, we get

$$\log z = \sqrt{a}\log x + \frac{1}{\sqrt{a}}\log y + \log c$$

$$\Rightarrow \frac{z}{x^{\sqrt{a}}\,y^{\frac{1}{\sqrt{a}}}} = c$$

$\Rightarrow z = c\,x^{\sqrt{a}}\,y^{\frac{1}{\sqrt{a}}}$, which is the required solution.

Example:

Solve $q = 3p^2$.

Solution:

Here:

$$q = 3p^2 \tag{2.129}$$

The subsidiary equations of (2.129) are

$$\frac{dx}{f_p} = \frac{dy}{f_q} = \frac{dz}{pf_p + qf_q} = \frac{dp}{-(f_x + pf_z)} = \frac{dq}{-(f_y + qf_z)}.$$

Here $q = 3p^2 \Rightarrow f : 3p^2 - q = 0$.

Hence, $f_x = 0, f_y = 0, f_z = 0, f_p = 6p, f_q = -1$.

By using the above results, we get

$$\frac{dx}{6p} = \frac{dy}{-1} = \frac{dz}{6p^2 - q} = \frac{dp}{0} = \frac{dq}{0}.$$

From the above equation, we get $dp = 0 \Rightarrow p = a$ and $dq = 0 \Rightarrow q = b$. Substituting the value of p and q in $dz = pdx + qdy$, we get $dz = adx + 3a^2 dy$.

Integrating both the sides, we get $z = ax + 3a^2 y + c$, which is the required solution.

Example:

Solve $2xz - px^2 - 2qxy + pq = 0$.

Solution:

Here

$$2xz - px^2 - 2qxy + pq = 0 \tag{2.130}$$

The subsidiary equations of (2.130) are

$$\frac{dx}{f_p}=\frac{dy}{f_q}=\frac{dz}{pf_p+qf_q}=\frac{dp}{-(f_x+pf_z)}=\frac{dq}{-(f_y+qf_z)}.$$

Here $2xz-px^2-2qxy+pq=0$ so $f_x=2z-2px-2qy$, $f_y=-2qx,\ f_z=2x,\ f_p=-x^2+q,\ f_q=-2xy+p$.

By using above results, we get

$$\frac{dx}{-x^2+q}=\frac{dy}{-2xy+p}=\frac{dz}{px^2+2pq-2xyq}=\frac{dp}{2z-2qy}=\frac{dq}{0}.$$

From above equation, we get $dq=0\Rightarrow q=a$.

Substituting the value of $q=a$ in equation (2.130), we get

$$2xz-px^2-2axy+pa=0,$$

$$\Rightarrow p=\frac{2xz-2axy}{x^2-a}.$$

We know that $dz=pdx+qdy$. Substituting the value of p and q in $dz=pdx+qdy$, we get

$$dz=\left(\frac{2xz-2axy}{x^2-a}\right)dx+ady,$$

$$dz-ady=\frac{2x(z-ay)}{x^2-a}dx,$$

$$\frac{dz-ady}{z-ay}=\frac{2x}{x^2-a}dx.$$

Integrating on both sides, we get

$$\log(z-ay)=\log(x^2-a)+\log c,$$

$$\log\left(\frac{z-ay}{x^2-a}\right)=\log c,$$

$$\frac{z-ay}{x^2-a}=c,$$

$z=ay+c(x^2-a)$, which is required complete solution.

Example:

Solve $2(z+xp+yq)=yp^2$.

Solution:

Here

$$2(z+xp+yq)=yp^2 \tag{2.131}$$

The subsidiary equations of (2.131) are

$$\frac{dx}{f_p}=\frac{dy}{f_q}=\frac{dz}{pf_p+qf_q}=\frac{dp}{-(f_x+pf_z)}=\frac{dq}{-(f_y+qf_z)}$$

Here $f:2(z+xp+yq)=yp^2 \Rightarrow 2(z+xp+yq)-yp^2=0$,

so $\quad f_x=2p,\ f_y=2q-p^2,\ f_z=2,\quad f_p=2x-2yp,\quad f_q=2y.$

By using above results, we get

$$\frac{dx}{2x-2yp}=\frac{dy}{2y}=\frac{dz}{2xp-2yp^2+2qy}=\frac{dp}{-(2p+2p)}=\frac{dq}{-(2q-p^2+2q)}$$

$$\Rightarrow \frac{dx}{x-yp}=\frac{dy}{y}=\frac{dz}{xp-yp^2+qy}=\frac{dp}{-2p}=\frac{dq}{-\left(2q-\frac{p^2}{2}\right)}.$$

From above equation, we get

$$\frac{dy}{y}=\frac{dp}{-2p}\Rightarrow 2\frac{dy}{y}+\frac{dp}{p}=0\Rightarrow 2\log y+\log p=\log a$$

$$\Rightarrow \log y^2+\log p=\log a\Rightarrow py^2=a\Rightarrow p=\frac{a}{y^2}$$

Substituting the value of p in $2(z+xp+yq)-yp^2=0$, we get

$$2(z+xp+yq)=yp^2,$$

$$\Rightarrow 2yq=yp^2-2z-2xp,$$

$$\Rightarrow 2yq=y\left(\frac{a}{y^2}\right)^2-2z-2x\left(\frac{a}{y^2}\right),$$

$$\Rightarrow q=\frac{a^2}{2y^4}-\frac{z}{y}-\frac{ax}{y^3}.$$

We know that $dz=pdx+qdy$. Substituting the value of p and q in $dz=pdx+qdy$, we get

$$dz=\frac{a}{y^2}dx+\left(\frac{a^2}{2y^4}-\frac{z}{y}-\frac{ax}{y^3}\right)dy,$$

$$2y^4dz=2ay^2dx+a^2dy-2y^3zdy-2axydy,$$

$$\frac{ydz+zdy}{y}=\frac{aydx-axdy}{y^3}+\frac{a^2}{2y^4}dy,$$

$$\left(ydz+zdy\right)=\frac{\left(aydx-axdy\right)}{y^2}+\frac{a^2}{2y^3}dy,$$

$$d\left(yz\right)=ad\left(\frac{x}{y}\right)+\frac{a^2}{2}\frac{dy}{y^3}.$$

Integrating on both sides, we get

$yz=a\left(\frac{x}{y}\right)+\frac{a^2}{2}\left(-\frac{1}{2y^2}\right)+b$, which is the required complete solution.

Example:

Solve $p^2x+q^2y=z$.

Solution:

Here:

$$p^2x+q^2y=z \tag{2.132}$$

The subsidiary equations of (2.132) are

$$\frac{dx}{f_p}=\frac{dy}{f_q}=\frac{dz}{pf_p+qf_q}=\frac{dp}{-\left(f_x+pf_z\right)}=\frac{dq}{-\left(f_y+qf_z\right)}.$$

Here $p^2x+q^2y=z \Rightarrow f:p^2x+q^2y-z=0$ so
$f_x=p^2, f_y=q^2, f_z=-1, f_p=2px, f_q=2qy$.

By using the above results, we get

$$\frac{dx}{2px}=\frac{dy}{2qy}=\frac{dz}{2\left(p^2x+q^2y\right)}=\frac{dp}{-p^2+p}=\frac{dq}{-q^2+q}.$$

From the above equation, we get

$$\frac{2pxdx+p^2dx}{2p^2x}=\frac{2qydy+q^2dy}{2q^2y}\Rightarrow\frac{d\left(p^2x\right)}{p^2x}=\frac{d\left(q^2y\right)}{q^2y}$$

Integrating on both sides, we get

$$\log\left(p^2x\right)=\log\left(q^2y\right)+\log a\Rightarrow p^2x=a.q^2y \tag{2.133}$$

Using above result in equation (2.132), we get

$$aq^2y+q^2y=z\Rightarrow q^2=\frac{z}{(1+a)y}\Rightarrow q=\pm\sqrt{\frac{z}{(1+a)y}}$$

Substituting value of equation (2.133), we get $p^2x=ay\frac{z}{(1+a)y}$,

$$\Rightarrow p^2=a\frac{z}{x(1+a)}\Rightarrow p=\pm\sqrt{\frac{az}{(1+a)x}}.$$

Substituting the value of p and q in $dz=pdx+qdy$, we get

$$dz=\pm\sqrt{\frac{az}{(1+a)x}}dx\pm\sqrt{\frac{z}{(1+a)y}}dy,$$

Separating the variables, we get

$$\pm\frac{\sqrt{1+a}dz}{\sqrt{z}}=\pm\sqrt{a}\frac{dx}{\sqrt{x}}\pm\frac{dy}{\sqrt{y}}.$$

Integrating above equation on both the side, we get

$$\pm\sqrt{1+a}\sqrt{z}=\sqrt{ax}+\sqrt{y}+b,$$

$$\Rightarrow(1+a)z=\left(\sqrt{ax}+\sqrt{y}+b\right)^2.$$

Hence, the complete solution is $(1+a)z=\left(\sqrt{ax}+\sqrt{y}+b\right)^2$.

Example:

Solve $\left(p^2+q^2\right)y=qz$.

Solution:

Here:

$$\left(p^2+q^2\right)y=qz \tag{2.134}$$

The subsidiary equations of (2.133) are

$$\frac{dx}{f_p}=\frac{dy}{f_q}=\frac{dz}{pf_p+qf_q}=\frac{dp}{-(f_x+pf_z)}=\frac{dq}{-\left(f_y+qf_z\right)}.$$

Here $\left(p^2+q^2\right)y=qz\Rightarrow f:\left(p^2+q^2\right)y-qz=0$ so $f_x=0$, $f_y=p^2+q^2$, $f_z=-q$, $f_p=2py$, $f_q=2qy-z$.

By using above results, we get

$$\frac{dx}{2py}=\frac{dy}{2qy-z}=\frac{dz}{2y\left(p^2+q^2\right)-qz}=\frac{dp}{pq}=\frac{dq}{-p^2}.$$

From above equation, we get $\dfrac{dp}{pq}=\dfrac{dq}{-p^2}\Rightarrow pdp+qdq=0$.

Integrating on both sides, we get

$$\frac{p^2}{2}+\frac{q^2}{2}=c_1^{'} \Rightarrow p^2+q^2=2c_1^{'} \Rightarrow p^2+q^2=a^2.(2c_1^{'}=a^2) \quad (2.135)$$

Using equations (2.134) and (2.135), we get, $a^2y=qz \Rightarrow q=\frac{a^2y}{z}$.

Substituting $q=\frac{a^2y}{z}$ in $p^2+q^2=a^2$, we get

$$p^2+\left(\frac{a^2y}{z}\right)^2=a^2 \Rightarrow p^2=a^2-\frac{a^4y^2}{z^2}=\frac{a^2z^2-a^4y^2}{z^2},$$

$$\Rightarrow p=\pm\frac{a}{z}\sqrt{z^2-a^2y^2}\ .$$

Substituting the value of p and q in $dz=pdx+qdy$, we get

$$dz=\pm\frac{a}{z}\sqrt{z^2-a^2y^2}\,dx\pm\frac{a^2y}{z}dy,$$

$$\Rightarrow \frac{zdz-a^2ydy}{\pm\sqrt{z^2-a^2y^2}}=adx.$$

Integrating on both sides, we get $\pm\sqrt{z^2-a^2y^2}=ax+b$,

$$\Rightarrow z^2-a^2y^2=(ax+b)^2,$$

$$\Rightarrow z^2=(ax+b)^2+a^2y^2.$$

Hence, the complete solution is $z^2=(ax+b)^2+a^2y^2$.

Example:

Solve $px^5 - 4q^2x^2 + 6x^2z - 2 = 0$.

Solution:

Here:

$$px^5 - 4q^2x^2 + 6x^2z - 2 = 0 \tag{2.136}$$

The subsidiary equations of (2.136) are

$$\frac{dx}{f_p} = \frac{dy}{f_q} = \frac{dz}{pf_p + qf_q} = \frac{dp}{-(f_x + pf_z)} = \frac{dq}{-(f_y + qf_z)}.$$

Here $f : px^5 - 4q^2x^2 + 6x^2z - 2 = 0$ so $f_x = 5px^4 - 8q^2x + 12xz$, $f_y = 0$, $f_z = 6x^2$, $f_p = x^5$, $f_q = -8qx^2$

By using above results, we get

$$\frac{dx}{x^5} = \frac{dy}{-8qx^2} = \frac{dz}{px^5 - 8q^2x^2} = \frac{dp}{-(5px^4 - 8q^2x + 12xz + 6px^2)} = \frac{dq}{-6qx^2}$$

From above equation, we get $\frac{dq}{-6qx^2} = \frac{dy}{-8qx^2} \Rightarrow 4dq = 3dy$.

Integrating on both sides, we get

$$4q = 3y + 3a \Rightarrow q = \frac{3}{4}(y + a) \tag{2.137}$$

Using the equations (2.136) and (2.137), we get

$$p=\frac{\frac{9x^2}{4}(y+a)^2-6x^2z+2}{x^5}.$$

Substituting the value of p and q in $dz=pdx+qdy$, we get

$$dz=\frac{9}{4x^3}(y+a)^2\,dx-\frac{6z}{x^3}dx+\frac{2}{x^5}dx+\frac{3}{4}(y+a)\,dy,$$

$$\frac{6z}{x^3}dx+dz=\frac{9}{4x^3}(y+a)^2\,dx+\frac{3}{4}(y+a)\,dy+\frac{2}{x^5}dx \qquad (2.138)$$

From LHS of above equation is of the form $M\,dx+Ndy$ where $M=\frac{6z}{x^3}, N=1,$

we have $\frac{1}{N}\left(\frac{\partial M}{\partial z}-\frac{\partial N}{\partial x}\right)=\frac{6}{x^3}$.

Hence, I.F $e^{\int Pdx}=e^{\int\frac{6}{x^3}dx}=e^{-\frac{3}{x^2}}$. Now multiplying equation (2.138) by $e^{-\frac{3}{x^2}}$, we get

$$\begin{aligned} e^{-\frac{3}{x^2}}\frac{6z}{x^3}dx+e^{-\frac{3}{x^2}}\,dz &= e^{-\frac{3}{x^2}}\frac{9}{4x^3}(y+a)^2dx+e^{-\frac{3}{x^2}}\frac{3}{4}(y+a)dy \\ &\quad + e^{-\frac{3}{x^2}}\frac{2}{x^5}dx \end{aligned}$$

$$d\left(ze^{-\frac{3}{x^2}}\right)=\frac{3}{8}d\left((y+a)^2e^{-\frac{3}{x^2}}\right)+\frac{2}{x^5}e^{-\frac{3}{x^2}}dx.$$

Integrating on both the side, we get

$$ze^{-\frac{3}{x^2}}=\frac{3}{8}(y+a)^2e^{-\frac{3}{x^2}}+\int\frac{1}{x^2}e^{-\frac{3}{x^2}}\frac{1}{x^3}dx,$$

$$ze^{-\frac{3}{x^2}} = \frac{3}{8}(y+a)^2 e^{-\frac{3}{x^2}} - \frac{1}{9}\left(\left(-\frac{3}{x^2}\right)e^{-\frac{3}{x^2}} - e^{-\frac{3}{x^2}}\right) + b,$$

$z = \dfrac{3(y+a)^2}{8} + \dfrac{1}{3x^2} + \dfrac{1}{9} + be^{\frac{3}{x^2}}$ which is the complete integral.

Example:

Solve $xp + 3yq = 2\left(z - x^2q^2\right)$.

Solution:

Here

$$xp + 3yq = 2\left(z - x^2q^2\right) \tag{2.139}$$

The subsidiary equations of (2.139) are

$$\frac{dx}{f_p} = \frac{dy}{f_q} = \frac{dz}{pf_p + qf_q} = \frac{dp}{-(f_x + pf_z)} = \frac{dq}{-\left(f_y + qf_z\right)}.$$

Here $xp + 3yq = 2\left(z - x^2q^2\right) \Rightarrow f : xp + 3yq - 2\left(z - x^2q^2\right) = 0$ so
$f_x = p + 4q^2x,\ f_y = 3q,\ f_z = -2,\ f_p = x,\ f_q = 3y + 4x^2q$.

By using above results, we get

$$\frac{dx}{x} = \frac{dy}{3y + 4x^2q} = \frac{dz}{px + q\left(3y + 4x^2q\right)} = \frac{dp}{p + 4xq^2 - 4xq^2} = \frac{dq}{-q}.$$

From above equation, we get $\dfrac{dq}{-q} = \dfrac{dx}{x}$.

Integrating on both sides, we get

$$\log q = -\log x + \log a \Rightarrow \frac{q}{x} = a \Rightarrow q = ax.$$

Substituting $q = ax$ in equation (2.139), we get

$$xp + 3y\left(\frac{a}{x}\right) - 2z + 2x^2\left(\frac{a^2}{x^2}\right) = 0.$$

$$\Rightarrow p = \frac{2\left(z - a^2\right)}{x} - \frac{3ay}{x^2}.$$

Substituting the value of p and q in $dz = pdx + qdy$, we get

$$dz = \left\{\frac{2\left(z - a^2\right)}{x} - \frac{3ay}{x^2}\right\}dx + \frac{a}{x}dy,$$

$$x^2 dz = 2x\left(z - a^2\right)dx - 3aydx + axdy,$$

$$\frac{x^2 dz - 2x\left(z - a^2\right)dx}{x^4} = -\frac{3aydx}{x^4} + \frac{a}{x^3}dy,$$

$$d\left(\frac{z - a^2}{x^2}\right) = d\left(\frac{ay}{x^3}\right).$$

Integrating on both sides, we get

$\frac{z - a^2}{x^2} = \frac{ay}{x^3} + b$, which is the required solution.

Exercise

Solve the following PDE

$px + qy = pq$

(Ans $az = \frac{1}{2}(ax + y)^2 = b$**)**

$z = pq$

(Ans $z = (x + a)(y + b)$**)**

$zpq = p + q$

(Ans $z^2 = 2(1 + a)\left(x + \frac{y}{a}\right) + b$**)**

$2xz - px^2 - 2qxy + pq = 0$

(Ans $z - ay = b(x^2 - a)$**)**

$qz - p^2y - q^2y = 0$

(Ans $z^2 = a\left(y^2 + (x + b)^2\right)$**)**

$(x^2 - y^2)pq - xy(p^2 - q^2) = 1$

(Ans $z = \frac{a}{2}\log(x^2 + y^2) + \frac{1}{a}\tan^{-1}\left(\frac{y}{x}\right) + b$**)**

$z^2(p^2z^2 + q^2) = 1$

(Ans $a^4(ax + y)^2 = \left\{(a^2z^2 + 1)\sqrt{a^2z^2 + 1} + 3a^2b\right\}^2$**)**

$q + zp = p^2$

(Ans $z = \frac{x^2}{4} \pm \frac{1}{2}\left[\frac{x\sqrt{x^2 + 4a}}{2} + \frac{4a}{2}\log\left(x + \sqrt{x^2 + 4a}\right) + ay + b\right]$**)**

$xpq + yq^2 = 1$

(Ans $z = \frac{x}{a} + \frac{ay}{x} + b$**)**

$(p+q)+(px+qy)=1$

(Ans $\left(z\sqrt{1+a}-b\right)^2 = 4(ax+y)$**)**

$16p^2z^2 + 9q^2z^2 + 4z^2 - 4 = 0$

(Ans $\left(\sqrt{16a^2+9}\sqrt{1-z^2}\right) - b = \pm 2(ax+b)$**)**

$yz - p(xy+q) - qy = 0$

(Ans $(z-ax)(y+a)^a = be^y$**)**

$z - q^2y - p^2x = 0$

(Ans $\sqrt{1+a}\sqrt{z} = \sqrt{a}\sqrt{x} + \sqrt{y} + b$**)**

Chapter 3

Higher Order Partial Differential Equation

3.1 Introduction

Partial Differential Equations (PDEs) arise naturally when modeling physical or biological systems, particularly under the assumptions of continuity and differentiability of the quantities involved. They are mathematical expressions involving functions of several variables and their partial derivatives, where the goal is to determine one or more of the functions, subject to specific initial or boundary conditions.

In problems involving PDEs, the primary concern is the existence of solutions. In certain cases, this involves finding explicit formulas for the solutions. However, in most cases, such explicit constructions are not feasible. Consequently, one often relies on analytical proofs of existence or nonexistence, as well as qualitative analyses to understand the properties of the solutions.

The field of partial differential equations is both vast and complex, with its complexity arising from the wide variety of PDE types and the numerous techniques used to analyze them. PDEs can be linear or nonlinear, involve a single equation or a system of equations, and range from first-order to higher-order formulations. Furthermore, PDEs are commonly classified based on the behavior of their solutions into three major types: elliptic, parabolic, and hyperbolic equations. Representative examples include Laplace's and Poisson's equations (elliptic), the heat or diffusion equation (parabolic), and the wave equation (hyperbolic), all of which are foundational in mathematical physics.

In this chapter, we have focused on the study and solution methods of higher-order differential equations, tailored specifically for engineering and university-level students.

3.2 Linear Partial Differential Equations of second order and higher order with constant coefficients

A partial differential equation in which the dependent variable and its derivatives appear only in the first degree and are not multiplied together, their coefficients being constants or functions of x and y, is known as a linear partial differential equation. The general form of such an equation is:

$$\left(A_0\frac{\partial^n Z}{\partial x^n}+A_1\frac{\partial^{n-1}Z}{\partial y\partial x^{n-1}}+A_2\frac{\partial^{n-2}Z}{\partial y^2\partial x^{n-2}}+\ldots\ldots+A_n\frac{\partial^n Z}{\partial y^n}\right) \tag{3.1}$$

$$+\left(B_0\frac{\partial^{n-1}Z}{\partial x^{n-1}}+B_1\frac{\partial^{n-2}Z}{\partial y\partial x^{n-2}}+B_2\frac{\partial^{n-3}Z}{\partial y^2\partial x^{n-3}}+\ldots\ldots+B_n\frac{\partial^{n-1}Z}{\partial y^{n-1}}\right)$$

$$+\ldots\ldots+\left[M_0\frac{\partial z}{\partial x}+M_1\frac{\partial z}{\partial y}\right]+N_0\,z=f(x,y)$$

where the coefficients $A_0, A_1, \ldots\ldots, A_n, B_0, B_1, \ldots\ldots, B_n, M_0, M_1, N_0$ are all constants, then equation (3.1) is called a linear partial differential equation with constant coefficients. For convenience $\frac{\partial}{\partial x}$ and $\frac{\partial}{\partial y}$ will be denoted by D and D' respectively. We shall restrict ourselves to consider only the linear partial differential equations of higher order with constant coefficients. These can be classified into the following two categories: (a) homogeneous (b) non-homogeneous. An equation in which the derivatives involved are all the same order, is called homogeneous $\frac{\partial^2 z}{\partial x^2}+2\frac{\partial^2 z}{\partial x\partial y}+\frac{\partial^2 z}{\partial y^2}=0$ is homogeneous of order two, whereas $\frac{\partial^2 z}{\partial x\partial y}+\frac{\partial^2 z}{\partial y^2}=f(x,y)$ is certainly non-homogeneous.

3.3 Complementary Function of PDE

We know that the general form of a linear partial differential equation is given by

$$A_0 \frac{\partial^n z}{\partial x^n} + A_1 \frac{\partial^{n-1} z}{\partial y \partial x^{n-1}} + A_2 \frac{\partial^{n-2} z}{\partial y^2 \partial x^{n-2}} + \ldots\ldots + A_n \frac{\partial^n z}{\partial y^n} = f(x,y)$$

$$A_0 \frac{\partial^n z}{\partial x^n} + A_1 \frac{\partial^{n-1} z}{\partial y \partial x^{n-1}} + A_2 \frac{\partial^{n-2} z}{\partial y^2 \partial x^{n-2}} + \ldots\ldots + A_n \frac{\partial^n z}{\partial y^n} = f(x,y) \tag{3.2}$$

where the coefficients $A_n, A_{n-1}, A_{n-2}, \ldots\ldots, A_1, A_0$ are constants or functions of x and y. If $A_n, A_{n-1}, A_{n-2}, \ldots\ldots, A_1, A_0$ are all constants, then (3.2) is called a linear partial differential equation with constant coefficients. For convenience $\frac{\partial}{\partial x}$ and $\frac{\partial}{\partial y}$ will be denoted by D and $D^{'}$ respectively. Therefore equation (3.2) can be written as

$$A_0 D^n z + A_1 D^{n-1} D^{'} z + A_2 D^{n-2} D^{2'} z + \ldots\ldots. + A_n D^{'n} z = f(x,y)$$

$$\left(A_0 D^n + A_1 D^{n-1} D^{'} + A_2 D^{n-2} D^{2'} + \ldots\ldots. + A_n D^{'n}\right) z = f(x,y) \tag{3.3}$$

The complementary function of equation (3.3) is general solution given by

$$\left(A_0 D^n + A_1 D^{n-1} D^{'} + A_2 D^{n-2} D^{2'} + \ldots\ldots. + A_n D^{'n}\right) z = 0 \tag{3.4}$$

Or

$$\left[\left(D - m_1 D^{'}\right)\left(D - m_2 D^{'}\right)\left(D - m_3 D^{'}\right)\ldots\ldots\ldots.\left(D - m_n D^{'}\right)\right] z = 0,$$

where the coefficients $m_n, m_{n-1}, m_{n-2}, \ldots\ldots, m_1, m_0$ are constants. Clearly, the solution of any of equations $\left(D-m_1D'\right)=0, \left(D-m_2D'\right)=0, \left(D-m_3D'\right)=0, \ldots\ldots\ldots\left(D-m_nD'\right)=0$ are also solution of equation (3.4). So, we now prove that the general solution of $\left(D-mD'\right)z=0$ is $z=F(y+mx)$ where F is an arbitrary function.

Now, $\left(D-mD'\right)z=0$ or $\frac{\partial z}{\partial x}-m\frac{\partial z}{\partial y}=0$ or $p-mq=0$,

which is Lagrange's linear partial differential equation of the form $Pp+Qq=R$. The Lagrange's auxiliary equation is

$$\frac{dx}{1}=\frac{dy}{-m}=\frac{dz}{0}$$
$$(a) \quad (b) \quad (c)$$

Taking the fraction (a) and (b) of above equation, we get $dy+mdx=0$.

And integrating it, we get $y+mx=c_1$ (3.5)

From (c), we get $dz=0$ so that $z=c_2$. (3.6)

From equations (3.5) and (3.6), the general solution of

$\left(D-mD'\right)z=0$ or $\frac{\partial z}{\partial x}-m\frac{\partial z}{\partial y}=0$ or $p-mq=0$ is $z=\phi(y+mx)$, where *F* is an arbitrary function. Assume that a solution of equation (3.4) is of the form $z=\phi(y+mx)$. So, we have

$$Dz=\frac{\partial z}{\partial x}=m\phi'(y+mx),$$

$$D^2 z = \frac{\partial^2 z}{\partial x^2} = m^2 \phi''(y+mx),$$

....

$$D^n z = \frac{\partial^n z}{\partial x^n} = m^n \phi^n (y+mx),$$

and $$D' z = \frac{\partial z}{\partial y} = \phi'(y+mx),$$

$$D'^2 z = \frac{\partial^2 z}{\partial y^2} = \phi''(y+mx),$$

....

$$D'^n z = \frac{\partial^n z}{\partial y^n} = \phi^n (y+mx),$$

Substitute these values in equation (3.4) and simplifying, we get

$$\left(A_0 m^n + A_1 m^{n-1} + A_2 m^{n-2} + \ldots\ldots + A_n\right)\phi^n\left(y+mx\right) = 0$$

which is true if m is a root of the equation

$$A_0 m^n + A_1 m^{n-1} + A_2 m^{n-2} + \ldots\ldots + A_n = 0. \tag{3.7}$$

Equation (3.7) is known as the auxiliary equation and is obtained by putting $D = m$ and $D' = 1$in $F\left(D, D'\right) = 0$.

Let $m_1, m_2, m_3, \ldots\ldots, m_n$ be the *n* roots of the equation (3.7). Two cases arise.

Case I - When all roots are distinct

Let $m_1, m_2, m_3, \ldots\ldots, m_n$ be the *n* distinct roots of the equation (3.7).Then the complementary function corresponding to $m = m_r$ is $z = \phi(y + m_i x)$ for $i = 1, 2, 3, 4, \ldots\ldots\ldots\ldots, n$. Since equation (3.4) is linear, the sum of the solution is also a solution.

C.F $=\phi_1(y+m_1x)+\phi_2(y+m_2x)+\phi_3(y+m_3x)+.......+\phi_n(y+m_nx)$ (3.8)

where $\phi_1,\phi_2,........,\phi_n$ are arbitrary function.

Case II - Repeated Roots

Let $m_1=m_2=m$ and $m_3,m_4,.....,m_n$ be distinct roots of the equation (3.7). Now, we consider

$$\left(D-mD^{'}\right)\left(D-mD^{'}\right)z=0. \tag{3.9}$$

Let $\left(D-mD^{'}\right)z=u$, equation (3.9) changes to

$$\left(D-mD^{'}\right)u=0. \tag{3.10}$$

Then $\left(D-mD^{'}\right)u=0$ or $\frac{\partial u}{\partial x}-m\frac{\partial u}{\partial y}=0$ or $p-mq=0$,

which is a Lagrange's linear partial differential equation with auxiliary equation

$$\underset{(a)}{\frac{dx}{1}}=\underset{(b)}{\frac{dy}{-m}}=\underset{(c)}{\frac{dz}{0}}$$

Taking the fraction (a) and (b) of above equation , we get $dy+mdx=0$. And integrating it , we get

$$y+mx=c_1. \tag{3.11}$$

From (c), we get

$$du=0 \text{ so that } u=c_2. \tag{3.12}$$

From equations (3.11) and (3.12), the general solution of

$$\left(D-mD^{'}\right)u=0 \text{ or } \frac{\partial u}{\partial x}-m\frac{\partial u}{\partial y}=0 \text{ or } p-mq=0$$

is $u=\phi(y+mx)$, where ϕ is an arbitrary function. Putting the value of $u=\phi(y+mx)$ in $\left(D-mD^{'}\right)z=u$, we get

$$\left(D-mD^{'}\right)z=\phi\left(y+mx\right),$$

$$\frac{\partial z}{\partial x}-m\frac{\partial z}{\partial y}=\phi\left(y+mx\right),$$

which is Lagrange's linear partial differential equation and auxiliary equations being

$$\frac{dx}{1}=\frac{dy}{-m}=\frac{dz}{\phi(y+mx)}$$

From the first two fractions of the above equation gives

$$y+m\,x=a\,, \tag{3.13}$$

where *a* is an arbitrary constant.

Now, from the first and third fraction we get, $\frac{dx}{1}=\frac{dz}{\phi(y+mx)}$ Putting $y+m\,x=a$ in $\frac{dx}{1}=\frac{dz}{\phi(y+mx)}$,

we get $\frac{dx}{1}=\frac{dz}{\phi(a)}$,

$$dz-\phi\left(a\right)dx=0,$$

$$z-x\phi\left(a\right)=b, \tag{3.14}$$

where b is an arbitrary constant. From equation (3.13), equation (3.14) changes to

$$z - x\phi(y+mx) = b. \tag{3.15}$$

From equations (3.14) and (3.15), we get

$$z = x\phi(y+mx) + \psi(y+mx).$$

Relabeling

$\phi(y+mx) = \phi_2(y+mx)$ and $\psi(y+mx) = \phi_1(y+mx)$

for convenience, we get

$$z = \phi_1(y+mx) + x\phi_2(y+mx),$$

where ϕ_1 and ϕ_2 are any arbitrary functions.

Hence, complementary solution with $m_1 = m_2 = m$ and $m_3, m_4, \ldots\ldots, m_n$ be distinct roots of the equation (3.7) is given by

$$\text{C.F} = \phi_1(y+mx) + x\phi_2(y+mx) + \phi_3(y+m_3x) + \ldots\ldots + \phi_n(y+m_nx)$$

where $\phi_1, \phi_2, \phi_3, \ldots\ldots, \phi_n$ are arbitrary functions.

Similarly, $m_1 = m_2 = m_3$ and $m_4, m_5, \ldots\ldots, m_n$ be distinct roots of the equation (3.7) are given by

$m_1 = m_2 = m_3$ then,

$$\text{C.F} = \phi_1(y+m_1x) + x\phi_2(y+m_1x) + x^2\phi_3(y+m_1x) + \phi_4(y+m_4x) + \ldots\ldots + \phi_n(y+m_nx),$$

where $\phi_1, \phi_2, \phi_3, \ldots\ldots, \phi_n$ are arbitrary functions.

In general, if root *m* of equation (3.7) is repeated *'n'* times, the corresponding part of the complementary function is
C.F $= \phi_1(y+m_1x) + x\phi_2(y+m_1x) + x^2\phi_3(y+m_1x) + x^3\phi_4(y+m_4x) + ... + x^{n-1}\phi_n(y+m_nx)$,

Example:

Solve $2\frac{\partial^2 z}{\partial x^2} + 5\frac{\partial^2 z}{\partial x \partial y} + 2\frac{\partial^2 z}{\partial y^2} = 0$.

Solution:

For convenience $\frac{\partial}{\partial x}$ and $\frac{\partial}{\partial y}$ will be denoted by D and D' respectively.

Therefore, $2\frac{\partial^2 z}{\partial x^2} + 5\frac{\partial^2 z}{\partial x \partial y} + 2\frac{\partial^2 z}{\partial y^2} = 0$ reduces to
$\left(2D^2 + 5DD' + 2D'^2\right)z = 0$.

Now, the auxiliary equation for $\left(2D^2 + 5DD' + 2D'^2\right)z = 0$ is

$$(2m^2 + 5m + 2) = 0 \Rightarrow (2m+1)(m+2) = 0 \Rightarrow m = -2, -\frac{1}{2}$$

Hence the C.F function is $z = \phi_1(y - \frac{1}{2}x) + \phi_2(y - 2x)$
which is a required solution, where ϕ_1 and ϕ_2 are arbitrary functions.

Example:

Solve $\frac{\partial^3 z}{\partial x^3} - 3\frac{\partial^3 z}{\partial x^2 \partial y} + 2\frac{\partial^3 z}{\partial x \partial y^2} = 0$.

Solution:

For convenience $\frac{\partial}{\partial x}$ and $\frac{\partial}{\partial y}$ will be denoted by D and D' respectively.

Therefore, $\frac{\partial^3 z}{\partial x^3} - 3\frac{\partial^3 z}{\partial x^2 \partial y} + 2\frac{\partial^3 z}{\partial x \partial y^2} = 0$ reduces to
$\left(D^3 - 3D^2D' + 2DD'^2\right)z = 0.$

The auxiliary equations for $\left(D^3 - 3D^2D' + 2DD'^2\right)z = 0$ is

$m^3 - 3m^2 + 2 = 0,$

$\Rightarrow m(m-1)(m-2) = 0,$

$\Rightarrow m = 0, 1, 2.$

Hence, the C.F function is

$z = \phi_1(y) + \phi_1(y+x) + \phi_3(y+2x)$

which is a required solution, where ϕ_1, ϕ_2 and ϕ_3 are arbitrary functions.

Example:

Solve $\frac{\partial^4 z}{\partial x^4} - \frac{\partial^4 z}{\partial y^4} = 0.$

Solution:

For convenience $\frac{\partial}{\partial x}$ and $\frac{\partial}{\partial y}$ will be denoted by D and D' respectively.

Therefore, $\frac{\partial^4 z}{\partial x^4}-\frac{\partial^4 z}{\partial y^4}=0$ reduces to $\left(D^4-D'^4\right)z=0$. Now, the auxiliary equations for $\left(D^4-D'^4\right)z=0$ is

$$(m^4-1)=0 \Rightarrow (m^2-1)(m^2+1)=0 \Rightarrow (m^2-1)(m+i)(m-i)=0 \Rightarrow m=1,\,-1,\,i,-i$$

Hence the C.F function is

$$z=\phi_1(y+x)+\phi_2(y-x)+\phi_3(y+ix)++\phi_4(y-ix)$$

which is a required solution, where ϕ_1, ϕ_2, ϕ_3 and ϕ_4 are arbitrary functions.

Example:

Solve $\frac{\partial^3 z}{\partial x^3}-4\frac{\partial^3 z}{\partial x^2\partial y}+4\frac{\partial^2 z}{\partial x\partial y}=0$.

Solution:

For convenience $\frac{\partial}{\partial x}$ and $\frac{\partial}{\partial y}$ will be denoted by D and D' respectively.

Therefore, $\frac{\partial^3 z}{\partial x^3}-4\frac{\partial^3 z}{\partial x^2\partial y}+4\frac{\partial^2 z}{\partial x\partial y}=0$ reduces to $(D^3-4D^2D'+4DD'^2)z=0$.

The auxiliary equations for $\left(D^3-4D^2D'+4DD'^2\right)z=0$ is

$$(m^3-4m^2+4m)=0 \Rightarrow m(m^2-4m+4)=0 \Rightarrow m(m-2)^2=0 \Rightarrow m=0,2,2$$

Hence the C.F function is

$$z=\phi_1(y)+\phi_2(y+2x)+x\phi_3(y+2x)$$

which is a required solution, where ϕ_1, ϕ_2 and ϕ_3 are arbitrary functions.

Example:

Solve $\frac{\partial^4 z}{\partial x^4} - 2\frac{\partial^4 z}{\partial x^3 \partial y} + 2\frac{\partial^4 z}{\partial x^2 \partial y^2} - \frac{\partial^4 z}{\partial y^4} = 0$.

Solution:

For convenience $\frac{\partial}{\partial x}$ and $\frac{\partial}{\partial y}$ will be denoted by D and D' respectively.

Therefore, $\frac{\partial^4 z}{\partial x^4} - 2\frac{\partial^4 z}{\partial x^3 \partial y} + 2\frac{\partial^4 z}{\partial x^2 \partial y^2} - \frac{\partial^4 z}{\partial y^4} = 0$ reduces to

$$\left(D^4 - 2D^3 D^{2'} + 2D^2 D' - D'^4\right) z = 0.$$

The auxiliary equations for

$\left(D^4 - 2D^3 D^{2'} + 2D^2 D' - D'^4\right) z = 0$ is

$$(m^4 - 2m^3 + 2m - 1) = 0,$$

$$\Rightarrow m^3(m+1) - m^2(m+1) - m(m+1) + (m+1) = 0$$

$$\Rightarrow (m+1)(m-1)^3 = 0 \Rightarrow m = 1, 1, 1, -1$$

Hence the C.F function is

$$z = \phi_1(y+x) + x\phi_2(y+x) + x^2\phi_3(y+x) + \phi_4(y-x)$$

which is a required solution, where ϕ_1, ϕ_2, ϕ_3 and ϕ_4 are arbitrary functions.

Example:

Solve $\frac{\partial^4 z}{\partial x^4}+\frac{\partial^4 z}{\partial y^4}=0$.

Solution:

For convenience $\frac{\partial}{\partial x}$ and $\frac{\partial}{\partial y}$ will be denoted by D and $D^{'}$ respectively.

Therefore, $\frac{\partial^4 z}{\partial x^4}+\frac{\partial^4 z}{\partial y^4}=0$ reduces to $\left(D^4+D^{'4}\right)z=0$.The auxiliary equations for $\left(D^4+D^{'4}\right)z=0$ is

$$
\begin{aligned}
(m^4+1)=0 \;\; &\Rightarrow (m^4+1)+2m^2=2m^2\\
&\Rightarrow (m^2+1)^2=2m^2\\
&\Rightarrow (m^2-\sqrt{2}m+1)(m^2+\sqrt{2}m+1)=0\\
&\Rightarrow (m^2-\sqrt{2}m+1)=0,\;\; (m^2+\sqrt{2}m+1)=0\\
&\Rightarrow m=\frac{-1+i}{\sqrt{2}},\frac{1+i}{\sqrt{2}},\frac{-1-i}{\sqrt{2}},\frac{1-i}{\sqrt{2}}
\end{aligned}
$$

$$\Rightarrow (m^2-\sqrt{2}m+1)=0,\;\; (m^2+\sqrt{2}m+1)=0$$

$$\Rightarrow m=\frac{-1+i}{\sqrt{2}},\frac{1+i}{\sqrt{2}},\frac{-1-i}{\sqrt{2}},\frac{1-i}{\sqrt{2}}$$

Hence the C.F function is

$$z=\phi_1\left(y+\left(\frac{-1+i}{\sqrt{2}}\right)x\right)+\phi_2\left(y+\left(\frac{1+i}{\sqrt{2}}\right)x\right)+\phi_3\left(y+\left(\frac{-1-i}{\sqrt{2}}\right)x\right)+\phi_4\left(y+\left(\frac{1-i}{\sqrt{2}}\right)x\right)$$

which is a required solution, where ϕ_1, ϕ_2, ϕ_3 and ϕ_4 are arbitrary functions.

Example:

Solve $\frac{\partial^5 z}{\partial x^3 \partial y^2}+\frac{\partial^5 z}{\partial x^2 \partial y^3}=0$.

Solution:

For convenience $\frac{\partial}{\partial x}$ and $\frac{\partial}{\partial y}$ will be denoted by D and $D^{'}$ respectively.

Therefore, $\frac{\partial^5 z}{\partial x^3 \partial y^2}+\frac{\partial^5 z}{\partial x^2 \partial y^3}=0$ reduces to

$\left(D^3 D^{'2}+D^2 D^{'3}\right)z=0$.

The auxiliary equations for $\left(D^3 D^{'2}+D^2 D^{'3}\right)z=0$ is

$(m^3+m^2)=0 \;\Rightarrow m^2(m+1)=0 \Rightarrow m=0,\ 0,-1,$

Hence the part of C.F function is $\phi_1(y)+x\phi_2(y)+\phi_3(y-x)$

Again $D^{'2}=0$ will give us $\phi_4(x)+y\phi_5(x)$

Hence the CF of this equation is
$z=\phi_1(y)+x\phi_2(y)+\phi_3(y-x)+\phi_4(x)+y\phi_5(x)$

which is a required solution, where ϕ_1, ϕ_2, ϕ_3, ϕ_4 and ϕ_5 are arbitrary functions.

Example:

Solve $\frac{\partial^3 z}{\partial x^3}-4\frac{\partial^3 z}{\partial x^2 \partial y}+4\frac{\partial^3 z}{\partial x \partial y^2}=0$.

Solution:

For convenience $\frac{\partial}{\partial x}$ and $\frac{\partial}{\partial y}$ will be denoted by D and D' respectively.

Therefore, $\frac{\partial^3 z}{\partial x^3}-4\frac{\partial^3 z}{\partial x^2\partial y}+4\frac{\partial^3 z}{\partial x\partial y^2}=0$ reduces to

$\left(D^3D'^2-4D^2D'+4DD'^2\right)z=0.$

For complementary function, the auxiliary equation is

$(m^3-4m^2+4m)=0 \Rightarrow m(m^2-4m+4)=0 \Rightarrow m(m-2)^2=0 \Rightarrow m=0,\ 2,2$

Hence the C.F function is $CF=\phi_1(y)+\phi_2(y+2x)+x\phi_3(y+2x)$

which is a required solution, where ϕ_1, ϕ_2 and ϕ_3 are arbitrary functions.

EXERCISE

	Solve:	Ans:
1.	$\frac{\partial^2 z}{\partial x^2}+4\frac{\partial^2 z}{\partial x\partial y}-5\frac{\partial^2 z}{\partial y^2}=0$	$z=\phi_1(y+x)+\phi_2(y-5x)$
2.	Solve: $\left(D^3-(D')^3\right)z=0$	Ans: $z=\phi_1(y+x)+\phi_2\left(y+\left(\frac{-1+i\sqrt{3}}{2}\right)x\right)+\phi_3\left(y+\left(\frac{-1-i\sqrt{3}}{2}\right)x\right)$
3.	Solve: $\left(D^2-2DD'+(D')^2\right)z=0$	Ans: $z=\phi_1(y+x)+x\phi_2(y+x)$

4. Solve: $\frac{\partial^3 z}{\partial x^3}-2\frac{\partial^3 z}{\partial x^2\partial y}-\frac{\partial^3 z}{\partial x\partial y^2}+2\frac{\partial^3 z}{\partial y^3}=0$ Ans: $z=\phi_1(y+x)+\phi_2(y-x)+\phi_2(y+2x)$

5. Solve: $(D^3-6D^2D^{'}+12D(D')^2$ Ans: $z=\phi_1(y+2x)+x\phi_2(y+2x)+x^2\phi_3(y+2x)$

6. Solve: $\frac{\partial^4 z}{\partial x^4}-\frac{\partial^4 z}{\partial y^4}=0$ Ans: $z=\phi_1(y+x)+\phi_2(y-x)+\phi_3(y+ix)+\phi_4(y-ix)$

7. Solve: $\frac{\partial^4 z}{\partial x^4}+\frac{\partial^4 z}{\partial y^4}=2\frac{\partial^4 z}{\partial x^2\partial y^2}$ Ans: $z=\phi_1(y+x)+\phi_2(y-x)+x\phi_3(y+x)+x\phi_4(y-2x)$

3.4 Particular Integral of homogenous linear partial differential equation

Consider the homogenous linear PDE $F\left(D,D^{'}\right)z=f\left(x,y\right)$. The inverse operator $\frac{1}{F(D,D^{'})}f(x,y)$ of the operator $F\left(D,D^{'}\right)$ is given by $f(D,D^{'})\left(\frac{1}{f(D,D^{'})}f(x,y)\right)=f(x,y)$ Therefore the particular integral of $F\left(D,D^{'}\right)z=f\left(x,y\right)$ is given by $\frac{1}{F(D,D^{'})}f(x,y)$.

3.5 Method of finding Particular Integral

In this section the shortcut method is discussed to evaluate the particular integral of partial differential equation.

3.5.1 Shortcut Method When $f\left(x,y\right)$ is of the form $f\left(ax+by\right)$

The method under consideration is based on the following theorem.
Theorem: If $F\left(D,D^{'}\right)$ be homogeneous function of D and $D^{'}$ of degree *n*, then

$$\frac{1}{F(D,D')} f^{(n)}(ax+by) = \frac{1}{F(a,b)} f(ax+by),\ F(a,b) \neq 0,$$

$f^{(n)}$ being the nth order derivative w.r.t. $ax+by$ as a whole.

Proof:

By direct differentiation, we have

$D^r f(ax+by) = a^r f^{(r)}(ax+by), b^n f^n(ax+by),$

$D'^s f(ax+by) = b^s f^{(s)}(ax+by)$
And $D^r D'^s f(ax+by) = a^r b^s f^{(r+s)}(ax+by)$

Since $F(D, D')$ is homogeneous function of degree *n*, so we have

$F(D,D') f(ax+by) = F(a,b) f^{(n)}(ax+by)$

Operating both sides of above equation by $\frac{1}{F(D,D')}$,

we have, $f(ax+by) = F(a,b) \frac{1}{F(D,D')} f^{(n)}(ax+by)$

Since $F(a,b) \neq 0$,dividing both sides of above equation by $F(a,b)$, we get

$$\frac{1}{F(D,D')} f^{(n)}(ax+by) = \frac{1}{F(a,b)} f(ax+by)$$

An important deduction from the result:

Putting $ax+by=t$, gives $\frac{1}{F(D,D')} f^{(n)}(t) = \frac{1}{F(a,b)} f(t)$

Integrating both sides n times w.r.t. 't', we have

$$\frac{1}{F(D,D')}f(t)=\frac{1}{F(a,b)}\int\int....\int f(t)\,dt\,dt....dt$$

where $ax+by=t$.

When $F(a,b)=0$, then the above theorem does not hold good. In such a case the new method is based on the following theorem. Note that $F(a,b)=0$ if and only if $(bD-aD')$ is a factor $F(D,D')$.

Theorem: $\frac{1}{(bD-aD')^n}f(ax+by)=\frac{x^n}{b^n n!}f(ax+by)$

Proof: Consider the equation $(bD-aD')z=x^r f(ax+by)$

$$bp-aq=x^r f(ax+by)$$

Lagrange's subsidiary equations for above equation are

$$\frac{dx}{b}=\frac{dy}{-a}=\frac{dz}{x^r f(ax+by)}$$

Taking the first two fractions, we have $adx+bdy=0$ so that

$$ax+by=c_1$$

Taking the first and third members

$$\frac{dx}{b}=\frac{dz}{x^r f(c_1)}\left(\because ax+by=c_1\right),$$

$$dz=\frac{x^r f(c_1)}{b}dx,$$

Integrating we have

$$z=\frac{x^{r+1}f(c_1)}{b(r+1)}=\frac{x^{r+1}f(ax+by)}{b(r+1)}$$

$$z=\frac{x^{r+1}f(ax+by)}{b(r+1)}$$

is a solution of

$$(bD-aD')z=x^{r}f(ax+by)$$

$$\therefore (bD-aD')\frac{x^{r+1}f(ax+by)}{b(r+1)}=x^{r}f(ax+by)$$

$$\therefore \frac{1}{(bD-aD')}x^{r}f(ax+by)=\frac{x^{r+1}f(ax+by)}{b(r+1)}$$

Hence, if $z=\frac{1}{(bD-aD')^{n}}f(ax+by)$, then we have

$$z=\frac{1}{(bD-aD')^{n-1}}\left[\frac{1}{(bD-aD')}x^{0}f(ax+by)\right]$$

$$z=\frac{1}{(bD-aD')^{n-1}}\frac{x}{b}f(ax+by) \text{ for } r=0$$

$$z=\frac{1}{b}\frac{1}{(bD-aD')^{n-2}}\left[\frac{1}{(bD-aD')}xf(ax+by)\right]$$

$$z=\frac{1}{b}\frac{1}{(bD-aD')^{n-2}}\left[\frac{x^{2}}{2b}f(ax+by)\right] \text{ for } r=1$$

$$z=\frac{1}{2!b^{2}}\frac{1}{(bD-aD')^{n-2}}x^{2}f(ax+by)$$

$$z = \frac{1}{n!b^n}\frac{1}{(bD - aD')^{n-n}}x^n f(ax+by) = \frac{1}{n!b^n}x^n f(ax+by)$$

[after repeated $n-2$ times more]

Working rule for finding the particular integral where

$$f(x,y) = f(ax+by)$$

The following rules will be used depending upon the situation in hand.

Formula (i). When $F(a,b) \neq 0$ and $F(D,D')$ is a homogeneous function of degree n, then

$$\text{P.I.} = \frac{1}{F(D,D')}f(ax+by) = \frac{1}{F(a,b)}\int\int....\int f(t)\,dt\,dt....dt$$

where $ax+by=t$

Note that R.H.S. contains a multiple integral of nth order.

Formula (ii). When $F(a,b)=0$, we have

$$\frac{1}{(bD-aD')^n}f(ax+by) = \frac{1}{n!b^n}x^n f(ax+by)$$

Example: Solve $\frac{\partial^2 z}{\partial x^2} + 3\frac{\partial^2 z}{\partial x \partial y} + 2\frac{\partial^2 z}{\partial y^2} = x+y$.

Solution: Here, $\frac{\partial^2 z}{\partial x^2} + 3\frac{\partial^2 z}{\partial x \partial y} + 2\frac{\partial^2 z}{\partial y^2} = x+y$,

$$\therefore (D^2 + 3DD' + 2D'^2)z = x+y,$$

For complementary function, the auxiliary equation is

$$(m^2+3m+2)=0 \Rightarrow m=-1,-2$$

Hence the C.F function is $CF=\phi_1(y-x)+\phi_2(y-2x)$

Again, for the particular integral

$$P.I=\frac{1}{(D^2+3DD'+2D'^2)}(x+y),$$

$$=\frac{1}{(1^2+3(1)(1)+2(1)^2)}\iint t\,dt\,dt \text{ where } t=x+y,$$

$$=\frac{1}{6}\int\frac{t^2}{2}dt=\frac{1}{12}\frac{t^3}{3}=\frac{t^3}{36},$$

$$=\frac{(x+y)^3}{36}(\because t=x+y).$$

Hence, the general solution is

$$z=\phi_1(y-x)+\phi_2(y-2x)+\frac{(x+y)^3}{36},$$

which is a required solution, where ϕ_1 and ϕ_2 are arbitrary functions.

Example:

Solve $\frac{\partial^2 z}{\partial x^2}-2\frac{\partial^2 z}{\partial x\partial y}+\frac{\partial^2 z}{\partial y^2}=\tan(x+y)$.

Solution:

Here, $\frac{\partial^2 z}{\partial x^2}-2\frac{\partial^2 z}{\partial x\partial y}+\frac{\partial^2 z}{\partial y^2}=\tan(x+y)$.

$\therefore (D^2-2DD^{'}+D^{'2})z=\tan(x+y)$.

For complementary function, the auxiliary equation is

$(m^2-2m+1)=0 \Rightarrow m=1,1$.

Hence the C.F function is $CF=\phi_1(y+x)+x\phi_2(y+x)$.

Again, for the particular integral

$$P.I=\frac{1}{(D^2-2DD^{'}+D^{'2})}\tan\left(x+y\right)=\frac{1}{(D-D^{'})^2}\tan\left(x+y\right),$$

$$=\frac{x^2}{1^2\times 2!}\tan\left(x+y\right)=\frac{x^2}{2}\tan\left(x+y\right).$$

Hence, the general solution is

$$z=\phi_1(y+x)+x\phi_2(y+x)+\frac{x^2}{2}\tan\left(x+y\right),$$

which is a required solution, where ϕ_1 and ϕ_2 are arbitrary functions.

Example:

Solve $\frac{\partial^3 z}{\partial x^3}-4\frac{\partial^3 z}{\partial x^2\partial y}+4\frac{\partial^3 z}{\partial x\partial y^2}=\cos(2x+y)$.

Solution:

Here, $\frac{\partial^3 z}{\partial x^3} - 4\frac{\partial^3 z}{\partial x^2 \partial y} + 4\frac{\partial^3 z}{\partial x \partial y^2} = \cos(2x+y)$.

$\therefore (D^3 - 4D^2D' + 4DD'^2)z = \cos(2x+y)$.

For complementary function, the auxiliary equation is

$(m^3 - 4m^2 + 4m) = 0 \Rightarrow m(m-2)^2 \Rightarrow m = 0, 2, 2.$

Hence the C.F function is

$CF = \phi_1(y) + \phi_2(y+2x) + x\phi_3(y+2x)$.

Again, for Particular integral

$$P.I = \frac{1}{(D^3 - 4D^2D' + 4DD'^2)}\cos(2x+y),$$

$$= \frac{1}{D(D-2D')^2}\cos(2x+y),$$

$$= \frac{1}{2}\frac{1}{(D-2D')^2}\sin(2x+y),$$

$$= \frac{1}{2}\frac{1}{(D-2D')^2}\sin(2x+y),$$

$$= \frac{1}{2}\frac{x^2}{2!}\sin(2x+y),$$

$$\left(\because \frac{1}{\left(bD - aD'\right)^r} f(ax+by) = \frac{x^r}{b^r r!} f(ax+by)\right)$$

$$= \frac{x^2}{4}\sin(2x+y).$$

Hence the general solution is

$$z = \phi_1(y) + \phi_2(y+2x) + x\phi_3(y+2x) + \frac{x^2}{4}\sin(2x+y),$$

which is a required solution, where ϕ_1, ϕ_2 and ϕ_3 are arbitrary functions.

Example:

Solve $\frac{\partial^3 z}{\partial x^3} - 4\frac{\partial^3 z}{\partial x^2 \partial y} + 4\frac{\partial^3 z}{\partial x \partial y^2} = 2\sin(3x+2y)$.

Solution:

Here, $\frac{\partial^3 z}{\partial x^3} - 4\frac{\partial^3 z}{\partial x^2 \partial y} + 4\frac{\partial^3 z}{\partial x \partial y^2} = 2\sin(3x+2y)$.

$$\therefore (D^3 - 4D^2D' + 4DD'^2)z = 2\sin(3x+2y).$$

For complementary function, the auxiliary equation is

$$(m^3 - 4m^2 + 4m) = 0 \Rightarrow m(m-2)^2 \Rightarrow m = 0, 2, 2.$$

Hence the C.F function is

$$CF = \phi_1(y) + \phi_2(y+2x) + x\phi_2(y+2x).$$

Again, for the particular integral,

$$P.I = \frac{1}{(D^3 - 4D^2D' + 4DD'^2)} 2\sin(3x+2y),$$

$$= 2\frac{1}{D(D-2D')^2}\sin(3x+2y),$$

$$=\frac{2}{3}\frac{1}{(3-2(2))^2}\cos(3x+2y),$$

$$=\frac{2}{3}\cos(3x+2y).$$

Hence the general solution is

$$z=\phi_1(y)+\phi_2(y+2x)+x\phi_2(y+2x)+\frac{2}{3}\cos(3x+2y),$$

which is a required solution, where ϕ_1, ϕ_2 and ϕ_3 are arbitrary functions.

3.5.2 Shortcut method when $f(x,y)$ is of the form e^{ax+by}

We have, $D^r F(x,y)=a^r F(x,y)$

$$D'^r F(x,y)=b^r F(x,y)$$

$$\therefore f(D,D')e^{ax+by}=f(a,b)e^{ax+by}$$

$$\therefore \frac{1}{f(D,D')}f(D,D')e^{ax+by}=\frac{1}{f(D,D')}f(a,b)e^{ax+by}$$

$$\therefore e^{ax+by}=f(a,b)\frac{1}{f(D,D')}e^{ax+by}$$

$$\therefore \frac{1}{f(a,b)}e^{ax+by}=\frac{1}{f(D,D')}e^{ax+by}$$

$$\Rightarrow \frac{1}{f(D,D')}e^{ax+by}=\frac{1}{f(a,b)}e^{ax+by},\quad f(a,b)\neq 0.$$

If $f(a,b)=0$ that means $D=a,\ D'=b$ satisfies

$f(D,D')=0$ that means $\left(D-\frac{a}{b}D'\right)^k$ is a factor of

$f(D,D')$ for some $k\in N$ such that

$f(D,D')=\left(D-\frac{a}{b}D'\right)^k \phi(D,D')$, where $\phi(a,b)\neq 0$.

Hence,

$$P.I=\frac{1}{f(D,D')}e^{ax+by}=\frac{1}{\left(D-\frac{a}{b}D'\right)^k \phi(D,D')}e^{ax+by}$$

$$=\frac{1}{\left(D-\frac{a}{b}D'\right)^{k-1}\left(D-\frac{a}{b}D'\right)\phi(D,D')}e^{ax+by}$$

$$=\frac{1}{\phi(a,b)\left(D-\frac{a}{b}D'\right)^{k-1}\left(D-\frac{a}{b}D'\right)}e^{ax+by}$$

$$=\frac{1}{\phi(a,b)\left(D-\frac{a}{b}D'\right)^{k-1}}\int e^{ax+b\left(c-\frac{a}{b}\right)x}dx$$

Where $c=y+\frac{a}{b}x$

$$=\frac{1}{\phi(a,b)\left(D-\frac{a}{b}D'\right)^{k-1}}xe^{bc}=\frac{1}{\phi(a,b)\left(D-\frac{a}{b}D'\right)^{k-1}}xe^{ax+by},$$

where $c = y + \frac{a}{b}x$

proceeding in this way, we get

$$\frac{1}{f(D,D^{'})}e^{ax+by} = \frac{1}{\phi(a,b)\left(D - \frac{a}{b}D^{'}\right)^{k-2}}\frac{x^{2}}{2!}e^{ax+by}$$

In general, we get $\frac{1}{f(D,D^{'})}e^{ax+by} = \frac{1}{\phi(a,b)}\frac{x^{k}}{k!}e^{ax+by}$.

But $f(D,D^{'}) = \left(D - \frac{a}{b}D^{'}\right)^{k}\phi(D,D^{'})$

$$\therefore \left(\frac{d}{dD^{k}}f(D,D^{'})\right)_{D=a,D^{'}=b} = k!\phi(a,b)$$

$$\frac{1}{f(D,D^{'})}e^{ax+by} = \frac{x^{k}e^{ax+by}}{\left(\frac{d}{dD^{k}}f(D,D^{'})\right)_{D=a,D^{'}=b}}.$$

Example:

Solve $\frac{\partial^{2}z}{\partial x^{2}} - 2\frac{\partial^{2}z}{\partial x\partial y} + \frac{\partial^{2}z}{\partial y^{2}} = e^{x+2y}$.

Solution:

Here, $\frac{\partial^{2}z}{\partial x^{2}} - 2\frac{\partial^{2}z}{\partial x\partial y} + \frac{\partial^{2}z}{\partial y^{2}} = e^{x+2y}$.

$\therefore (D^{2} - 2DD^{'} + D^{'2})z = e^{x+2y}$.

For complementary function, the auxiliary equation is

$(m^2-2m+1)=0 \Rightarrow (m-1)^2=0 \Rightarrow m=1,1.$

Hence the C.F function is $CF=\phi_1(y+x)+x\phi_2(y+x)$.

Again, for the particular integral,

$$P.I=\frac{1}{(D^2-2DD'+D'^2)}e^{x+2y},$$

$$=\frac{1}{(1^2-2(1)(2)+(2)^2)}e^{x+2y},$$

$$=\frac{1}{1}e^{x+2y},$$

$$=e^{x+2y}.$$

Hence, the general solution is

$$z=\phi_1(y+x)+x\phi_2(y+x)+e^{x+2y},$$

which is a required solution, where ϕ_1 and ϕ_2 are arbitrary functions.

Example:

Solve $\frac{\partial^3 z}{\partial x^3}-3\frac{\partial^2 z}{\partial x^2\partial y}+4\frac{\partial^3 z}{\partial y^3}=e^{x+2y}$.

Solution:

Here, $\frac{\partial^3 z}{\partial x^3}-3\frac{\partial^2 z}{\partial x^2\partial y}+4\frac{\partial^3 z}{\partial y^3}=e^{x+2y}$.

$\therefore (D^3-3D^2D'+4D'^3)z=e^{x+2y}$.

For complementary function, the auxiliary equation is

$$(m^3 - 3m^2 + 4) = 0 \Rightarrow m = -1, 2, 2.$$

Hence the C.F function is

$$CF = \phi_1(y - x) + \phi_2(y + 2x) + x\phi_3(y + 2x)$$

Again, for the particular integral,

$$P.I = \frac{1}{(D^3 - 3D^2D' + 4D'^3)} e^{x+2y},$$

$$= \frac{1}{(1^3 - 3(1)^2(2) + 4(2)^3)} e^{x+2y},$$

$$= \frac{1}{(27)} e^{x+2y}.$$

Hence, the general solution is

$$z = \phi_1(y - x) + \phi_2(y + 2x) + x\phi_3(y + 2x) + \frac{1}{(27)} e^{x+2y},$$

which is a required solution, where ϕ_1, ϕ_2 and ϕ_3 are arbitrary functions.

Example:

Solve $\dfrac{\partial^3 z}{\partial x^3} - 2\dfrac{\partial^3 z}{\partial x^2 \partial y} - \dfrac{\partial^3 z}{\partial x \partial y^2} + 2\dfrac{\partial^3 z}{\partial y^3} = e^{x+y}$.

Solution:

Here, $\dfrac{\partial^3 z}{\partial x^3} - 2\dfrac{\partial^3 z}{\partial x^2 \partial y} - \dfrac{\partial^3 z}{\partial x \partial y^2} + 2\dfrac{\partial^3 z}{\partial y^3} = e^{x+y}$.

$$\therefore (D^3 - 2D^2D' - DD'^2 + 2D'^3)z = e^{x+y}.$$

For complementary function, the auxiliary equation is

$$(m^3 - 2m^2 - m + 2) = 0$$

$$\Rightarrow m^2(m-2) - (m-2) = 0$$

$$\Rightarrow (m-2)(m^2 - 1) = 0$$

$$\Rightarrow m = 2, 1, -1$$

Hence the C.F function is

$$CF = \phi_1(y-2x) + \phi_2(y+x) + \phi_3(y-x).$$

Again, for the particular integral,

$$P.I = \frac{1}{(D^3 - 2D^2D' - DD'^2 + 2D'^3)} e^{x+y},$$

$$= \frac{1}{(D^2 - D'^2)(D - 2D')} e^{x+y},$$

$$= \frac{1}{(D-D')(D+D')(D-2D')} e^{x+y},$$

$$= \frac{1}{(D-D')(1+1)(1-2(1))} e^{x+y},$$

$$= \frac{1}{(D-D')(2)(-1)} e^{x+y},$$

$$= \frac{1}{(-2)(D-D')} e^{x+y}.$$

Let

$$v=\frac{1}{(D-D^{'})}e^{x+y}\Rightarrow(D-D^{'})v=e^{x+y}\Rightarrow Dv-D'V=e^{x+y},$$

which is Lagrange's equation. The Lagrange's auxiliary equations are

$$\frac{dx}{1}=\frac{dy}{-1}=\frac{dv}{e^{x+y}}.$$

From first two ratio, we have $\frac{dx}{1}=\frac{dy}{-1}$.

Hence, the solution is $x+y=c,$ where c arbitrary constants.

Again, compare first and third ratio, we have $\frac{dx}{1}=\frac{dv}{e^{x+y}}$

$$\therefore\frac{dx}{1}=\frac{dv}{e^{c}}\Rightarrow dv=e^{c}dx\Rightarrow v=e^{c}x\Rightarrow v=e^{x+y}x$$

$$\therefore P.I=\frac{1}{(-2)}e^{x+y}x$$

Hence, the general solution is

$$z=\phi_1(y-2x)+\phi_2(y+x)+\phi_3(y-x)-\frac{1}{(2)}e^{x+y}x\ ,$$

which is a required solution, where ϕ_1, ϕ_2 and ϕ_3 are arbitrary functions.

Example:

Solve $\frac{\partial^2 z}{\partial x^2} - 4\frac{\partial^2 z}{\partial x \partial y} + 4\frac{\partial^2 z}{\partial y^2} = e^{2x+y}$.

Solution:

Here, $\frac{\partial^2 z}{\partial x^2} - 4\frac{\partial^2 z}{\partial x \partial y} + 4\frac{\partial^2 z}{\partial y^2} = e^{2x+y}$,

$\therefore (D^2 - 4DD' + 4D'^2)z = e^{2x+y}$.

For complementary function, the auxiliary equation is

$(m^2 - 4m + 4) = 0 \;\Rightarrow m^2(m-2) - (m-2) = 0 \Rightarrow (m-2)^2 = 0 \Rightarrow m = 2, 2$

Hence the C.F function is $CF = \phi_1(y+2x) + x\phi_2(y+2x)$

Again, for Particular integral

$$P.I = \frac{1}{(D^2 - 4DD' + 4D'^2)} e^{2x+y} = \frac{x}{(2D - 4D')} e^{2x+y},$$

$$\left(\because \frac{1}{f(D,D')} e^{ax+by} = \frac{x^k e^{ax+by}}{\left(\frac{d}{dD^k} f(D,D') \right)_{D=a, D'=b}} \right)$$

$$P.I = \frac{x^2}{(2)} e^{2x+y} = \frac{x^2}{2} e^{2x+y}.$$

Hence the general solution is

$$z = \phi_1(y+2x) + x\phi_2(y+2x) + \frac{x^2}{2} e^{2x+y},$$

which is a required solution, where ϕ_1 and ϕ_2 are arbitrary functions.

Example:

Solve $4\frac{\partial^2 z}{\partial x^2}+12\frac{\partial^2 z}{\partial x \partial y}+9\frac{\partial^2 z}{\partial y^2}=e^{3x-2y}$.

Solution:

Here, $4\frac{\partial^2 z}{\partial x^2}+12\frac{\partial^2 z}{\partial x \partial y}+9\frac{\partial^2 z}{\partial y^2}=e^{3x-2y}$.

$\therefore (4D^2+12DD'+9D'^2)z=e^{3x-2y}$.

For complementary function, the auxiliary equation is

$$(4m^2+12m+9)=0 \Rightarrow (2m+3)^2=0 \Rightarrow m=-\frac{2}{3},-\frac{2}{3}.$$

Hence the C.F function is $CF=\phi_1(2y-3x)+x\phi_2(2y-3x)$
Again, for particular integral

$$P.I=\frac{1}{(4D^2+12DD'+9D'^2)}e^{3x-2y},$$

$$=\frac{1}{(2D+3D')^2}e^{3x-2y}$$

$$(\because \frac{1}{\left(bD-aD'\right)^r}f(ax+by)=\frac{x^r}{b^r r!}f(ax+by),$$

$$=\frac{x^2}{2^2 2!}e^{3x-2y},$$

$$=\frac{x^2}{8}e^{3x-2y}.$$

Hence the general solution is

$$z=\phi_1(2y-3x)+x\phi_2(2y-3x)+\frac{x^2}{8}e^{3x-2y},$$

which is a required solution, where ϕ_1 and ϕ_2 are arbitrary functions.

3.5.3 Shortcut method when $f(x,y)$ is of the form $\sin(ax+by)$ or $\cos(ax+by)$

We know that,

$$D^2\sin(ax+by+c)=-a^2\sin(ax+by+c)$$

$$DD'\sin(ax+by+c)=-ab\sin(ax+by+c)$$

$$D'^2\sin(ax+by+c)=-b^2\sin(ax+by+c)$$

$$\therefore\frac{1}{F(D^2,DD',D'^2)}F(D^2,DD',D'^2)\sin(ax+by+c)$$
$$=\frac{1}{F(D^2,DD',D'^2)}F(-a^2,-ab,-b^2)\sin(ax+by+c)$$

$$\therefore \frac{1}{F(D^2, DD', D'^2)} \sin(ax+by+c) \, or \cos(ax+by+c)$$

$$= \frac{1}{F(-a^2, -ab, -b^2)} \sin(ax+by+c) \, or \cos(ax+by+c)$$

When, $F(-a^2, -ab, -b^2) = 0$ then

$$\frac{1}{f(D, D')} \sin(ax+by+c) \;\; or \;\; \cos(ax+by+c)$$

$$= \frac{x^k}{\left(\dfrac{d}{dD^k} f(D, D')\right)_{D^2=-a^2, D'^2=-b^2, DD'=-ab}} \sin(ax+by+c) \; or \; \cos(ax+by+c)$$

Example:

Solve $\dfrac{\partial^2 z}{\partial x^2} + 2\dfrac{\partial^2 z}{\partial x \partial y} = \sin x \cos 2y$.

Solution:

Here, $\dfrac{\partial^2 z}{\partial x^2} + 2\dfrac{\partial^2 z}{\partial x \partial y} = \sin x \cos 2y$.

$\therefore (D^2 + 2DD')z = \sin x \cos 2y$.

For complementary function, the auxiliary equation is

$(m^2 + 2m) = 0 \;\; \Rightarrow m(m+2) = 0 \Rightarrow m = 0, -2$.

Hence the C.F function is $CF = \phi_1(y) + \phi_2(y - 2x)$,

Again, for particular integral

$$P.I = \frac{1}{(D^2 + 2DD')} \sin x \cos 2y = \frac{1}{2} \frac{1}{(D^2 + 2DD')} 2\cos 2y \sin x$$

$$=\frac{1}{2}\frac{1}{(D^2+2DD')}(\sin(x+2y)+\sin(x-2y))$$

$$=\frac{1}{2}\left(\frac{1}{(D^2+2DD')}(\sin(x+2y)+\frac{1}{(D^2+2DD')}\sin(x-2y))\right)$$

$$=\frac{1}{2}\left(\frac{1}{((1)^2+2(1)(2)}(\sin(x+2y)+\frac{1}{((1)^2+2(1)(-2))}\sin(x-2y))\right)$$

$$=\frac{1}{2}\left(\frac{1}{5}(\sin(x+2y)-\frac{1}{3}\sin(x-2y))\right)$$

$$=\left(\frac{1}{10}(\sin(x+2y)-\frac{1}{6}\sin(x-2y))\right)$$

Hence the general solution is

$$z=\phi_1(y)+\phi_2(y-2x)+\left(\frac{1}{10}(\sin(x+2y)-\frac{1}{6}\sin(x-2y))\right)$$

which is a required solution, where ϕ_1 and ϕ_2 are arbitrary functions.

Example:

Solve $\frac{\partial^2 z}{\partial x^2}-2\frac{\partial^2 z}{\partial x\partial y}+\frac{\partial^2 z}{\partial y^2}=\sin(2x+3y)$

Solution:

Here, $\frac{\partial^2 z}{\partial x^2}-2\frac{\partial^2 z}{\partial x\partial y}+\frac{\partial^2 z}{\partial y^2}=\sin(2x+3y)$

$$\therefore (D^2 - 2DD' + D'^2)z = \sin(2x+3y)$$

For complementary function, the auxiliary equation is

$$(m^2 - 2m + 1) = 0 \;\Rightarrow (m-1)^2 = 0 \Rightarrow m = 1,\, 1$$

Hence the C.F function is $CF = \phi_1(y+x) + x\phi_2(y+x)$

Again, for the particular integral,

$$P.I = \frac{1}{(D^2 - 2DD' + D'^2)}\sin(2x+3y)$$

$$= \frac{1}{(-(2)^2 - 2(-(2)(3)) - (3)^2)}\sin(2x+3y)$$

$$= \frac{1}{(-4+12-9)}\sin(2x+3y)$$

$$= \frac{1}{-1}\sin(2x+3y)$$

$$= -\sin(2x+3y)$$

Hence the general solution is
$z = \phi_1(y+x) + x\phi_2(y+x) - \sin(2x+3y)$,

which is a required solution, where ϕ_1 and ϕ_2 are arbitrary functions.

Example:

Solve $\dfrac{\partial^2 z}{\partial x^2} - 2\dfrac{\partial^2 z}{\partial x \partial y} + \dfrac{\partial^2 z}{\partial y^2} = \sin x$.

Solution:

Here, $\frac{\partial^2 z}{\partial x^2} - 2\frac{\partial^2 z}{\partial x \partial y} + \frac{\partial^2 z}{\partial y^2} = \sin x$.

$\therefore (D^2 - 2DD' + D'^2)z = \sin x$

For complementary function, the auxiliary equation is

$(m^2 - 2m + 1) = 0 \Rightarrow (m-1)^2 = 0 \Rightarrow m = 1, 1$.

Hence the C.F function is $CF = \phi_1(y+x) + x\phi_2(y+x)$.

Again, for the particular integral,

$$P.I = \frac{1}{(D^2 - 2DD' + D'^2)} \sin x$$

$$= \frac{1}{(-(1)^2 - 0 + 0)} \sin x$$

$$= -\sin x$$

Hence the general solution is

$$z = \phi_1(y+x) + x\phi_2(y+x) - \sin x$$

which is a required solution, where ϕ_1 and ϕ_2 are arbitrary functions.

Example:

Solve $\frac{\partial^2 z}{\partial x^2} + 2\frac{\partial^2 z}{\partial x \partial y} + \frac{\partial^2 z}{\partial y^2} = \sin(2x+3y)$.

Solution:

Here, $\frac{\partial^2 z}{\partial x^2}+2\frac{\partial^2 z}{\partial x \partial y}+\frac{\partial^2 z}{\partial y^2}=\sin(2x+3y)$

$\therefore (D^2+2DD'+D'^2)z=\sin(2x+3y)$.

For complementary function, the auxiliary equation is

$(m^2+2m+1)=0 \ \Rightarrow (m+1)^2=0 \Rightarrow m=-1,\ -1$.

Hence the C.F function is $CF=\phi_1(y-x)+x\phi_2(y-x)$, where ϕ_1 and ϕ_2 are arbitrary functions. Again, for particular integral

$$P.I=\frac{1}{(D^2+2DD'+D'^2)}\sin(2x+3y)$$

$$=\frac{1}{(-(2)^2+2(-(2)(3))-(3)^2)}\sin(2x+3y)$$

$$=\frac{1}{(-4-12-9)}\sin(2x+3y)$$

$$=\frac{1}{-25}\sin(2x+3y)$$

Hence the general solution is

$$z=\phi_1(y-x)+x\phi_2(y-x)-\frac{1}{25}\sin(2x+3y),$$

which is a required solution, where ϕ_1 and ϕ_2 are arbitrary functions.

Example:

Find the particular Integral of $\left(\frac{\partial z}{\partial x}+1\right)\left(\frac{\partial z}{\partial x}+\frac{\partial z}{\partial y}-1\right)=\sin(x+2y)$.

Solution:

Here, $\left(\frac{\partial z}{\partial x}+1\right)\left(\frac{\partial z}{\partial x}+\frac{\partial z}{\partial y}-1\right)=\sin(x+2y)$.

$\therefore ((D+1)(D+D^{'}-1))z=\sin(x+2y)$

For particular integral

$$P.I=\frac{1}{(D+1)(D+D^{'}-1)}\sin(x+2y)$$

$$=\frac{1}{(D^2+DD^{'}+D^{'}-1)}\sin(x+2y)$$

$$=\frac{1}{(-1-(1)(2)+D^{'}-1)}\sin(x+2y)$$

$$=\frac{1}{(D^{'}-4)}\sin(x+2y)$$

$$=\frac{(D^{'}+4)}{(D^{'2}-16)}\sin(x+2y)$$

$$=\frac{(D^{'}+4)}{(-1^2-16)}\sin(x+2y)$$

$$=-\frac{1}{17}(2\cos(x+2y)+4\sin(x+2y).$$

which is the required particular integral.

Example:

Solve $\frac{\partial^2 z}{\partial x^2}+\frac{\partial^2 z}{\partial y^2}=\cos mx \cos ny$.

Solution:

Here, $\frac{\partial^2 z}{\partial x^2}+\frac{\partial^2 z}{\partial y^2}=\cos mx \cos ny$.

$\therefore (D^2+D'^2)z=\cos mx \cos ny$.

For complementary function, the auxiliary equation is

$(m^2+1)=0 \;\Rightarrow (m^2+1)=0 \Rightarrow m=i,-i$

Hence the C.F function is $CF=\phi_1(y+ix)+\phi_2(y-ix)$

Again, for particular integral

$$P.I=\frac{1}{(D^2+D'^2)}\cos mx \cos ny$$

$$=\frac{1}{2}\frac{1}{(D^2+D'^2)}2\cos mx \cos ny$$

$$=\frac{1}{2}\frac{1}{(D^2+D'^2)}(\cos(mx+ny)+\cos(mx-ny))$$

$$=\frac{1}{2}\left(\frac{1}{(D^2+D'^2)}\cos(mx+ny)+\frac{1}{(D^2+D'^2)}\cos(mx-ny)\right)$$

$$=\frac{1}{2}\left(\frac{1}{(-m^2-n^2)}\cos(mx+ny)+\frac{1}{(-m^2-n^2)}\cos(mx-ny)\right)$$

$$=\left(\frac{1}{-2(m^2+n^2)}(\cos(mx+ny)+\cos(mx-ny)\right)$$

Hence the general solution is

$$z=\phi_1(y+ix)+\phi_2(y-ix)-\frac{1}{2(m^2+n^2)}(\cos(mx+ny)+\cos(mx-ny)),$$

which is a required solution, where ϕ_1 and ϕ_2 are arbitrary functions.

Example:

Solve $\frac{\partial^2 z}{\partial x^2}-\frac{\partial^2 z}{\partial x\partial y}=\sin x\cos 2y$.

Solution:

Here, $\frac{\partial^2 z}{\partial x^2}-\frac{\partial^2 z}{\partial x\partial y}=\sin x\cos 2y$.

$$\therefore (D^2-DD')z=\sin x\cos 2y.$$

For complementary function, the auxiliary equation is

$$(m^2-m)=0 \Rightarrow m(m-1)=0 \Rightarrow m=0,1.$$

Hence the C.F function is $CF=\phi_1(y)+\phi_2(y+x)$

Again, for particular integral

$$P.I=\frac{1}{(D^2-DD')}\sin x\cos 2y$$

$$=\frac{1}{2}\frac{1}{(D^2-DD')}(2\sin x\cos 2y)$$

$$=\frac{1}{2}\frac{1}{(D^2-DD')}(\sin(x+2y)+\sin(x-2y))$$

$$=\frac{1}{2}(\frac{1}{(D^2-DD')}\sin(x+2y)+\frac{1}{(D^2-DD')}\sin(x-2y))$$

$$=\frac{1}{2}(\frac{1}{(-1^2-(-(1)(2)))}\sin(x+2y)+\frac{1}{(-1^2-(-(1)(-2))}\sin(x-2y))$$

$$=\frac{1}{2}(\frac{1}{1}\sin(x+2y)+\frac{1}{-3}\sin(x-2y))$$

$$=\left(\frac{1}{2}\sin(x+2y)-\frac{1}{6}\sin(x-2y)\right)$$

Hence, the general solution is

$$z=\phi_1(y)+\phi_2(y+x)-\left(\frac{1}{2}\sin(x+2y)-\frac{1}{6}\sin(x-2y)\right),$$

which is a required solution, where ϕ_1 and ϕ_2 are arbitrary functions.

Example:

Solve $\frac{\partial^2 z}{\partial x^2}-5\frac{\partial^2 z}{\partial x\partial y}+\frac{\partial^2 z}{\partial y^2}=\sin(4x+y)$.

Solution:

Here, $\frac{\partial^2 z}{\partial x^2}-5\frac{\partial^2 z}{\partial x\partial y}+\frac{\partial^2 z}{\partial y^2}=\sin(4x+y)$.

$$\therefore (D^2 - 5DD' + D'^2)z = \sin(4x + y).$$

For complementary function, the auxiliary equation is

$$(m^2 - 5m + 1) = 0 \Rightarrow (m-1)(m-4) = 0 \Rightarrow m = 1, 4.$$

Hence the C.F function is $CF = \phi_1(y + x) + \phi_2(y + 4x)$.

Again, for particular integral

$$P.I = \frac{1}{(D^2 - 5DD' + D'^2)} \sin(4x + y)$$

$$= \frac{1}{(D - D')(D - 4D')} \sin(4x + y)$$

$$= \frac{1}{(4-1)(D - 4D')} \sin(4x + y)$$

$$= \frac{1}{3(D - 4D')} \sin(4x + y)$$

$$\left(\because \frac{1}{\left(bD - aD'\right)^r} f(ax + by) = \frac{x^r}{b^r r!} f(ax + by) \right)$$

$$= \frac{1}{3} x \sin(4x + y)$$

Hence the general solution is

$$z = \phi_1(y + x) + \phi_2(y + 4x) - \frac{1}{3} x \sin(4x + y),$$

which is a required solution, where ϕ_1 and ϕ_2 are arbitrary functions.

Example:

Solve $\frac{\partial^3 z}{\partial x^3}-4\frac{\partial^3 z}{\partial x^2\partial y}+4\frac{\partial^3 z}{\partial x\partial y^2}=4\sin(2x+y)$

Solution:

Here, $\frac{\partial^3 z}{\partial x^3}-4\frac{\partial^3 z}{\partial x^2\partial y}+4\frac{\partial^3 z}{\partial x\partial y^2}=4\sin(2x+y)$.

$\therefore (D^3-4D^2D^{'}+4DD^{'2})z=4\sin(2x+y)$.

For complementary function, the auxiliary equation is

$(m^3-4m^2+4m)=0 \Rightarrow m(m-2)^2=0 \Rightarrow m=0,2,2$.
Hence the C.F function is

$CF=\phi_1(y)+\phi_2(y+2x)+x\phi_3(y+2x)$.

Again, for the particular integral,

$$P.I=\frac{1}{(D^3-4D^2D^{'}+4DD^{'2})}4\sin(2x+y),$$

$$=\frac{1}{D(D-2D^{'})^2}4\sin(2x+y),$$

$$\left(\because \frac{1}{\left(bD-aD^{'}\right)^r}f(ax+by)=\frac{x^r}{b^r r!}f(ax+by)\right)$$

$$=\frac{-4}{2(D-2D^{'})^2}\cos(2x+y)$$

$$= -2\frac{x^2}{2!}\cos(2x+y)$$

$$= -x^2\cos(2x+y)$$

Hence the general solution is

$$z = \phi_1(y) + \phi_2(y+2x) + x\phi_3(y+2x) - x^2\cos(2x+y)),$$

which is a required solution, where ϕ_1, ϕ_2 and ϕ_3 are arbitrary functions.

Example:

Solve $\frac{\partial^2 z}{\partial x^2} - 3\frac{\partial^2 z}{\partial x \partial y} + 2\frac{\partial^2 z}{\partial y^2} = e^{2x-y} + e^{x-y} + \cos(x+2y)$.

Solution:

Here, $\frac{\partial^2 z}{\partial x^2} - 3\frac{\partial^2 z}{\partial x \partial y} + 2\frac{\partial^2 z}{\partial y^2} = e^{2x-y} + e^{x-y} + \cos(x+2y)$.

$$\therefore (D^2 - 3DD' + 2D'^2)z = e^{2x-y} + e^{x-y} + \cos(x+2y).$$

For complementary function, the auxiliary equation is

$$(m^2 - 3m + 2) = 0 \;\Rightarrow (m-1)(m-2) = 0 \Rightarrow m = 1, 2.$$

Hence the C.F function is $CF = \phi_1(y+x) + \phi_2(y+2x)$.

Again, for particular integral

$$P.I = \frac{1}{\left(D^2 - 3DD' + 2D'^2\right)}\left(e^{2x-y} + e^{x-y} + \cos(x+2y)\right).$$

$$=\left(\begin{array}{l}\frac{1}{\left(D^2-3DD'+2D'^2\right)}e^{2x-y}+\frac{1}{\left(D-D'\right)\left(D-2D'\right)}e^{x+y} \\ +\frac{1}{\left(D^2-3DD'+2D'^2\right)}\cos(x+2y)\end{array}\right)$$

$$=\left(\frac{1}{-12}e^{2x-y}-\frac{1}{(D-D')}e^{x+y}+\frac{1}{-3}\cos\left(x+2y\right)\right)$$

$$=\frac{1}{-12}e^{2x-y}-\frac{x}{1}e^{x+y}+\frac{1}{-3}\cos(x+2y)$$

$$=\left(-\frac{1}{12}e^{2x-y}-xe^{x+y}-\frac{1}{3}\cos(x+2y)\right)$$

Hence the general solution is

$$z=\phi_1(y+x)+\phi_2(y+2x)+\left(-\frac{1}{12}e^{2x-y}-xe^{x+y}-\frac{1}{3}\cos(x+2y)\right),$$

which is a required solution, where ϕ_1 and ϕ_2 are arbitrary functions.

Example:

Solve $\frac{\partial^3 z}{\partial x^3}-7\frac{\partial^3 z}{\partial x\partial y^2}-6\frac{\partial^3 z}{\partial y^3}=e^{2x+y}+\sin(x+2y)$.

Solution:

Here, $\frac{\partial^3 z}{\partial x^3}-7\frac{\partial^3 z}{\partial x\partial y^2}-6\frac{\partial^3 z}{\partial y^3}=e^{2x+y}+\sin(x+2y)$.

$$\therefore (D^3-7DD'^2-6D'^3)z=e^{2x+y}+\sin(x+2y).$$

For complementary function, the auxiliary equation is

$(m^3-7m-6)=0 \Rightarrow (m+1)(m^2-m-6)=0 \Rightarrow (m+1)(m-3)(m+2) \Rightarrow m=3,-1,-2$.

Hence the C.F function is

$$CF=\phi_1(y+3x)+\phi_2(y-x)+\phi_3(y-2x).$$

Again, for particular integral

$$P.I=\frac{1}{(D^3-7DD'^2-6D'^3)}(e^{2x+y}+\sin(x+2y)),$$

$$=\frac{1}{(D^3-7DD'^2-6D'^3)}e^{2x+y}+\frac{1}{(D^3-7DD'^2-6D'^3)}\sin(x+2y),$$

$$=\frac{1}{(2^3-7((2)(1)^2-6(1)^3)}e^{2x+y}+\frac{1}{(1^3-7((1)(2)^2-6(2)^3)}\cos(x+2y),$$

$$=\frac{1}{-12}e^{2x+y}+\frac{1}{-75}\cos(x+2y),$$

$$=-\frac{1}{12}e^{2x+y}-\frac{1}{75}\cos(x+2y).$$

Hence the general solution is

$$z=CF=\phi_1(y+3x)+\phi_2(y-x)+\phi_3(y-2x)-\frac{1}{12}e^{2x+y}-\frac{1}{75}\cos(x+2y),$$

which is a required solution, where ϕ_1,ϕ_2 and ϕ_3 are arbitrary functions.

Example:

Solve $2\frac{\partial^2 z}{\partial x^2}-5\frac{\partial^2 z}{\partial x\partial y}-6\frac{\partial^2 z}{\partial y^3}=5\sin(2x+y)$.

Solution:

Here, $2\frac{\partial^2 z}{\partial x^2} - 5\frac{\partial^2 z}{\partial x \partial y} - 6\frac{\partial^2 z}{\partial y^3} = 5\sin(2x + y)$.

$\therefore (2D^2 - 5DD' - 6D'^3)z = 5\sin(2x + y)$.

For complementary function, the auxiliary equation is

$(2m^2 - 5m + 2) = 0 \;\; \Rightarrow m = 2, \frac{1}{2}$.

Hence the C.F function is $CF = \phi_1(y + \frac{1}{2}x) + \phi_2(y + 2x)$.

Again, for the particular integral,

$$P.I = \frac{1}{(2D^2 - 5DD' - 6D'^3)} 5(\sin(2x + y)),$$

$$= 5x\frac{1}{(4D - 5D')}(\sin(2x + y)),$$
$$= 5x\frac{1}{(4(2) - 5(1))}(-\cos(2x + y)),$$

$$= -5x\frac{1}{(8 - 5)}(\cos(2x + y)),$$

$$= -\frac{5x}{3}(\cos(2x + y)).$$

Hence the general solution is
$$z = \phi_1(y + \frac{1}{2}x) + \phi_2(y + 2x) - \frac{5x}{3}(\cos(2x + y)),$$

which is a required solution, where ϕ_1 and ϕ_2 are arbitrary functions.

3.5.4 Shortcut method when $f(x, y)$ is of the form $\sinh(ax+by)$ or $\cosh(ax+by)$

$$D^2 \sinh(ax+by+c) = a^2 \sinh(ax+by+c)$$

$$DD' \sinh(ax+by+c) = ab \sinh(ax+by+c)$$

$$D'^2 \sin(ax+by+c) = b^2 \sin(ax+by+c)$$

$$\therefore \frac{1}{F(D^2, DD', D'^2)} F(D^2, DD', D'^2) \sinh(ax+by+c)$$
$$= \frac{1}{F(D^2, DD', D'^2)} F(a^2, ab, b^2) \sinh(ax+by+c)$$

$$\therefore \sinh(ax+by+c) = F(a^2, ab, b^2) \frac{1}{F(D^2, DD', D'^2)} \sinh(ax+by+c)$$

$$\therefore \frac{1}{F(a^2, ab, b^2)} \sinh(ax+by+c) = \frac{1}{F(D^2, DD', D'^2)} \sinh(ax+by+c)$$

$$\therefore \frac{1}{F(D^2, DD', D'^2)} \sinh(ax+by+c) \, or \cosh(ax+by+c)$$
$$= \frac{1}{F(a^2, ab, b^2)} \sinh(ax+by+c) or \cosh(ax+by+c)$$

When, $F(a^2, ab, b^2) = 0$ then $bD - aD'$ must be a factor of $F(D, D')$ suppose this factor repeated r times then the value of

$$\frac{1}{\left(bD - aD'\right)^r} f(ax+by) = \frac{x^r}{b^r r!} f(ax+by)$$

3.5.5 Shortcut method when $f(x,y)$ is of the form $x^m y^n$

$P.I = \left[F(D,D')\right]^{-1} F(x,y)$, Expand $\left[F(D,D')\right]^{-1}$ in ascending powers of D and D' by binomial theorem and then operate on $F(x,y)$.

Example:

Find particular integral of $\frac{\partial^2 z}{\partial x^2} - \frac{\partial z}{\partial y} = 2y - x^2$.

Solution:

Here, $\frac{\partial^2 z}{\partial x^2} - \frac{\partial z}{\partial y} = 2y - x^2$,

$\therefore (D^2 - D')z = 2y - x^2$.

For Particular integral

$$P.I = \frac{1}{D^2 - D'}(2y - x^2)$$

$$= \frac{1}{D^2\left(1 - \frac{D'}{D^2}\right)}(2y - x^2)$$

$$= \frac{1}{D^2}\left(1 - \frac{D'}{D^2}\right)^{-1}(2y - x^2)$$

$$= \frac{1}{D^2}\left(1 + \frac{D'}{D^2} + \frac{D'^2}{D^4} + \ldots\ldots\right)(2y - x^2)$$

$$=\frac{1}{D^2}\left((2y-x^2)+\frac{D'}{D^2}(2y-x^2)+\frac{D'^2}{D^4}(2y-x^2)+......\right)$$

$$=\frac{1}{D^2}\left((2y-x^2)+\frac{1}{D^2}(2)\right)$$

$$=\frac{1}{D^2}\left((2y-x^2)+x^2\right)$$

$$=\frac{1}{D^2}(2y)$$

$$=\frac{1}{D}(2xy)$$

$$=x^2y$$

which is the required particular integral.

Example:

Solve $\frac{\partial^2 z}{\partial x^2}-\frac{\partial^2 z}{\partial y^2}=x^2y$.

Solution:

Here, $\frac{\partial^2 z}{\partial x^2}-\frac{\partial^2 z}{\partial y^2}=x^2y$.

$\therefore (D^2-D'^2)z=x^2y$.

For complementary function, the auxiliary equation is

$(m^2-1)=0 \;\; \Rightarrow (m-1)(m+1) \Rightarrow m=1,-1$.

Hence the C.F function is $CF = \phi_1(y+x)+\phi_2(y-x)$

Again, for particular integral

$$P.I = \frac{1}{(D^2 - D'^2)} x^2 y,$$

$$= \frac{1}{D^2}(1-\frac{D'^2}{D^2})^{-1} x^2 y,$$

$$= \frac{1}{D^2}\left(1+\frac{D'^2}{D^2}+\frac{D'^4}{D^4}+...\right)x^2 y,$$

$$= \frac{1}{D^2}\left(x^2 y+\frac{1}{D^2}(D'^2 x^2 y)+0\right),$$

$$= \frac{1}{D^2}\left(x^2 y+\frac{1}{D^2}(0)\right),$$

$$= \frac{1}{D^2}\left(x^2 y\right),$$

$$= \frac{x^4 y}{12}.$$

Hence the general solution is $z = \phi_1(y+x)+\phi_2(y-x)+\frac{x^4 y}{12}$,

which is a required solution, where ϕ_1 and ϕ_2 are arbitrary functions.

Example:

Solve $\frac{\partial^3 z}{\partial x^3} - 2\frac{\partial^3 z}{\partial x^2 \partial y} = x^2 y$.

Solution:

Here, $\frac{\partial^3 z}{\partial x^3}-2\frac{\partial^3 z}{\partial x^2\partial y}=x^2y$.

$\therefore (D^3-2D^2D^{'})z=x^2y$.

For complementary function, the auxiliary equation is

$(m^3-2m^2)=0 \;\Rightarrow m^2(m-2)=0 \Rightarrow m=0,0,2.$

Hence the C.F function is $CF=\phi_1(y)+x\phi_2(y)+\phi_3(y+2x)$.

Again, for particular integral

$$P.I=\frac{1}{(D^3-2D^2D^{'})}x^2y$$

$$=\frac{1}{D^3}(1-\frac{2D^{'}}{D})^{-1}x^2y$$

$$=\frac{1}{D^3}\left(1+\frac{2D^{'}}{D}+\frac{4D^{'2}}{D^2}+...\right)x^2y$$

$$=\frac{1}{D^3}\left(x^2y+\frac{2D^{'}}{D}x^2y+\frac{4D^{'2}}{D^2}x^2y+...\right)$$

$$=\frac{1}{D^3}\left(x^2y+\frac{2}{D}(D^{'}x^2y)+\frac{4}{D^2}x^2(D^{'2}y)+...\right)$$

$$=\frac{1}{D^3}\left(x^2y+\frac{2}{D}(x^2)\right)$$

$$=\frac{1}{D^3}\left(x^2y+2\frac{x^3}{3}\right)$$

$$=\frac{1}{D^3}\left(x^2y+2\frac{x^3}{3}\right)$$

$$=\frac{1}{D^2}\left(\frac{x^3}{3}y+2\frac{x^4}{12}\right)$$

$$=\frac{1}{D}\left(\frac{x^4}{12}y+2\frac{x^5}{60}\right)$$

$$=\left(\frac{x^5}{60}y+2\frac{x^6}{6(60)}\right)$$

$$=\left(\frac{x^5}{60}y+\frac{x^6}{180}\right).$$

Hence the general solution is

$$z=\phi_1(y)+x\phi_2(y)+\phi_3(y+2x)+\left(\frac{x^5}{60}y+\frac{x^6}{180}\right),$$

which is a required solution, where ϕ_1, ϕ_2 and ϕ_3 are arbitrary functions.

Example:

Find the particular integral of $\frac{\partial^2 z}{\partial x^2}-\frac{\partial^2 z}{\partial x\partial y}+2\frac{\partial z}{\partial y}-z=x^2y^2$.

Solution:

Here, $\frac{\partial^2 z}{\partial x^2} - \frac{\partial^2 z}{\partial x \partial y} + 2\frac{\partial z}{\partial y} - z = x^2 y^2$.

$\therefore (D^2 - DD' + 2D' - 1)z = x^2 y^2$.

For Particular integral

$$P.I = \frac{1}{(D^2 - DD' + 2D' - 1)} x^2 y^2$$

$$= \frac{1}{D^2(1 - \frac{D'}{D} + 2\frac{D'}{D^2} - \frac{1}{D^2})} x^2 y^2$$

$$= \frac{1}{D^2}\left(1 - \left(\frac{D'}{D} - 2\frac{D'}{D^2} + \frac{1}{D^2}\right)\right)^{-1} x^2 y^2$$

$$= \frac{1}{D^2}\left(1 + \left(\frac{D'}{D} - 2\frac{D'}{D^2} + \frac{1}{D^2}\right) + \left(\frac{D'}{D} - 2\frac{D'}{D^2} + \frac{1}{D^2}\right)^2 + \ldots\right) x^2 y^2$$

$$= \frac{1}{D^2}\left(1 + \left(\frac{D'}{D} - 2\frac{D'}{D^2} + \frac{1}{D^2}\right) + \left(\frac{D'^2}{D^2} + 4\frac{D'^2}{D^4} + \frac{1}{D^4} - 4\frac{D'^2}{D^3} - 4\frac{D'}{D^4} + 2\frac{D'}{D^3}\right) + \ldots\right) x^2 y^2$$

$$= \frac{1}{D^2}\begin{pmatrix} x^2y^2 + \left(\frac{D'}{D}x^2y^2 - 2\frac{D'}{D^2}x^2y^2 + \frac{1}{D^2}x^2y^2\right) \\ + \left(\frac{D'^2}{D^2}x^2y^2 + 4\frac{D'^2}{D^4}x^2y^2 + \frac{1}{D^4}x^2y^2 - 4\frac{D'^2}{D^3}x^2y^2 - 4\frac{D'}{D^4}x^2y^2 + 2\frac{D'}{D^3}x^2y^2\right) + \ldots \end{pmatrix}$$

$$= \frac{1}{D^2}\begin{pmatrix} x^2y^2 + \left(\frac{1}{D}2x^2y - 2\frac{1}{D^2}2x^2y + \frac{1}{D^2}x^2y^2\right) \\ + \left(\frac{1}{D^2}2x^2 + 4\frac{1}{D^4}2x^2 + \frac{1}{D^4}x^2y^2 - 4\frac{1}{D^3}2x^2 - 4\frac{1}{D^4}2x^2y + 2\frac{1}{D^3}2x^2y\right) + \ldots \end{pmatrix}$$

$$=\frac{1}{D^2}\left(\begin{array}{l}x^2y^2+\left(\frac{2}{3}x^3y-\frac{4}{12}x^4y+\frac{1}{12}x^4y^2\right)\\+\left(\frac{2}{12}x^4+\frac{8}{360}x^6+\frac{1}{360}x^6y^2-\frac{8}{60}x^5-\frac{8}{360}x^6y+\frac{4}{60}x^5y\right)\end{array}\right)$$

$$=\frac{1}{D}\left(\begin{array}{l}\frac{x^3}{3}y^2+\left(\frac{2}{12}x^4y-\frac{4}{60}x^5y+\frac{1}{60}x^5y^2\right)\\+\left(\frac{2}{60}x^5+\frac{8}{7\times 360}x^7+\frac{1}{7\times 360}x^7y^2-\frac{8}{6\times 60}x^6-\frac{8}{7\times 360}x^7y+\frac{4}{6\times 60}x^6y\right)\end{array}\right)$$

$$=\left(\begin{array}{l}\frac{x^4}{12}y^2+\left(\frac{2}{60}x^5y-\frac{4}{360}x^6y+\frac{1}{360}x^6y^2\right)\\+\left(\frac{2}{360}x^6+\frac{8}{8\times 7\times 360}x^8+\frac{1}{8\times 7\times 360}x^8y^2-\frac{8}{7\times 6\times 60}x^7-\frac{8}{8\times 7\times 360}x^8y+\frac{4}{7\times 6\times 60}x^7y\right)\end{array}\right)$$

Example:

Solve $\frac{\partial^2 z}{\partial x^2}+\frac{\partial^2 z}{\partial y^2}=x^2y^2.$

Solution:

Here, $\frac{\partial^2 z}{\partial x^2}+\frac{\partial^2 z}{\partial y^2}=x^2y^2.$

$\therefore (D^2+D'^2)z=x^2y^2.$

For complementary function, the auxiliary equation is

$(m^2+1)=0 \Rightarrow m=i,-i.$

Hence the C.F function is $CF=\phi_1(y+ix)+\phi_2(y-ix)$

Again, for particular integral

$$P.I = \frac{1}{(D^2 + D'^2)} x^2 y^2$$

$$= \frac{1}{D^2}\left(1 + \frac{D'^2}{D^2}\right)^{-1} x^2 y^2$$

$$= \frac{1}{D^2}\left(1 - \frac{D'^2}{D^2} + \frac{D'^4}{D^4} - \ldots\right) x^2 y^2$$

$$= \frac{1}{D^2}\left(x^2 y^2 - \frac{D'^2}{D^2} x^2 y^2 + \frac{D'^4}{D^4} x^2 y^2 - \ldots\right)$$

$$= \frac{1}{D^2}\left(x^2 y^2 - \frac{1}{D^2} 2x^2 + 0\right)$$

$$= \frac{1}{D^2}\left(x^2 y^2 - \frac{2}{12} x^4\right)$$

$$= \frac{1}{D}\left(\frac{x^3}{3} y^2 - \frac{2}{12}\frac{x^5}{5}\right)$$

$$= \left(\frac{x^4}{12} y^2 - \frac{1}{180} x^6\right)$$

Hence the general solution is

$$z = \phi_1(y + ix) + \phi_2(y - ix) + \left(\frac{x^4}{12} y^2 - \frac{1}{180} x^6\right),$$

which is a required solution, where ϕ_1 and ϕ_2 are arbitrary functions.

Example:

Solve $\frac{\partial^2 z}{\partial x^2} - 6\frac{\partial^2 z}{\partial x \partial y} + 9\frac{\partial^2 z}{\partial y^2} = 12x^2 + 36xy$.

Solution:

Here, $\frac{\partial^2 z}{\partial x^2} - 6\frac{\partial^2 z}{\partial x \partial y} + 9\frac{\partial^2 z}{\partial y^2} = 12x^2 + 36xy$.

$\therefore (D^2 - 6DD^{'} + 9D^{'2})z = 12x^2 + 36xy$.

For complementary function, the auxiliary equation is

$(m^2 - 6m + 9) = 0 \Rightarrow m = 3, 3.$

Hence the C.F function is $CF = \phi_1(y+3x) + x\phi_2(y+3x)$

Again, for particular integral

$$P.I = \frac{1}{(D^2 - 6DD^{'} + 9D^{'2})} 12x^2 + 36xy$$

$$= \frac{1}{D^2\left(1 - 6\frac{D^{'}}{D} + 9\frac{D^{'2}}{D^2}\right)} 12x^2 + 36xy$$

$$= \frac{1}{D^2}\left(1 - \left(6\frac{D^{'}}{D} - 9\frac{D^{'2}}{D^2}\right)\right)^{-1} 12x^2 + 36xy$$

$$= \frac{1}{D^2}\left(1 + \left(6\frac{D^{'}}{D} - 9\frac{D^{'2}}{D^2}\right) + \ldots\right) 12x^2 + 36xy$$

$$=\frac{1}{D^2}\left(12x^2+36xy+\left(6\frac{D'}{D}(12x^2+36xy)\right)\right)$$

$$=\frac{1}{D^2}\left(12x^2+36xy+\left(6\frac{1}{D}(36x)\right)\right)$$

$$=\frac{1}{D^2}\left(12x^2+36xy+\left((108)x^2\right)\right)$$

$$=\left(12\frac{x^4}{12}+36\frac{x^3}{6}y+\left((108)\frac{x^4}{12})\right)\right)$$

$$=\left(x^4+6x^3y+9x^4\right)$$

$$=\left(10x^4+6x^3y\right)$$

Hence the general solution is

$$z=\phi_1(y+3x)+x\phi_2(y+3x)+\left(10x^4+6x^3y\right),$$

which is a required solution, where ϕ_1 and ϕ_2 are arbitrary functions.

Example:

Solve $\frac{\partial^3 z}{\partial x^3}-\frac{\partial^3 z}{\partial y^3}=x^3y^3$.

Solution:

Here, $\frac{\partial^3 z}{\partial x^3}-\frac{\partial^3 z}{\partial y^3}=x^3y^3$.

$\therefore (D^3 - D'^3)z = x^3y^3$.

For complementary function, the auxiliary equation is

$$(m^3 - 1) = 0 \Rightarrow m = (m-1)(m^2+m+1) = 0 \Rightarrow m = 1,\ m = \frac{-1}{2} \pm i\frac{\sqrt{3}}{2}.$$

Hence the C.F function is

$$CF = \phi_1(y+x) + \phi_2\left(y + \left(\frac{-1}{2} + i\frac{\sqrt{3}}{2}\right)x\right) + \phi_3\left(y + \left(\frac{-1}{2} - i\frac{\sqrt{3}}{2}\right)x\right)$$

Again, for particular integral

$$P.I = \frac{1}{(D^3 - D'^3)} x^3y^3$$

$$= \frac{1}{(D^3 - D'^3)} x^3y^3$$

$$= \frac{1}{D^3}\left(1 - \frac{D'^3}{D^3}\right)^{-1} x^3y^3$$

$$= \frac{1}{D^3}(1 + \frac{D'^3}{D^3} + \ldots)^{-1} x^3y^3$$

$$= \frac{1}{D^3}(x^3y^3 + \frac{D'^3}{D^3}(x^3y^3) + 0)$$

$$= \frac{1}{D^3}(x^3y^3 + \frac{1}{D^3}(6x^3))$$

$$=\frac{1}{D^3}(x^3y^3+\frac{1}{10}(x^6))$$

$$=\left(\frac{x^6}{120}y^3+\frac{x^9}{10080}\right)$$

Hence the general solution is

$$z=\phi_1(y+x)+\phi_2\left(y+\left(\frac{-1}{2}+i\frac{\sqrt{3}}{2}\right)x\right)+\phi_3\left(y+\left(\frac{-1}{2}-i\frac{\sqrt{3}}{2}\right)x\right)+\frac{x^6}{120}y^3+\frac{1}{10080}x^9,$$

which is a required solution, where ϕ_1, ϕ_2 and ϕ_3 are arbitrary functions.

Example:

Solve $\frac{\partial^2 z}{\partial x^2}+2\frac{\partial^2 z}{\partial x\partial y}+\frac{\partial^2 z}{\partial y^2}=x^2+xy+y^2$.

Solution:

Here, $\frac{\partial^2 z}{\partial x^2}+2\frac{\partial^2 z}{\partial x\partial y}+\frac{\partial^2 z}{\partial y^2}=x^2+xy+y^2$.

$$\therefore (D^2+2DD'+D'^2)z=x^2+xy+y^2.$$

For complementary function, the auxiliary equation is

$$(m^2+2m+1)=0 \Rightarrow m=-1,-1.$$

Hence the C.F function is $CF=\phi_1(y-x)+x\phi_2(y-x)$

Again, for the particular integral,

$$P.I = \frac{1}{(D^2 + 2DD' + D'^2)} x^2 + xy + y^2,$$

$$= \frac{1}{D^2}\left(1 + \left(2\frac{D'}{D} + \frac{D'^2}{D^2}\right)\right)^{-1} x^2 + xy + y^2$$

$$= \frac{1}{D^2}\left(1 - \left(2\frac{D'}{D} + \frac{D'^2}{D^2}\right) + \left(2\frac{D'}{D} + \frac{D'^2}{D^2}\right)^2 + \ldots\right) x^2 + xy + y^2$$

$$= \frac{1}{D^2}\left(1 - \left(2\frac{D'}{D} + \frac{D'^2}{D^2}\right) + 4\frac{D'^2}{D^2} + \ldots\right) x^2 + xy + y^2$$

$$= \frac{1}{D^2}\left(1 - 2\frac{D'}{D} - \frac{D'^2}{D^2} + 4\frac{D'^2}{D^2} + \ldots\right) x^2 + xy + y^2$$

$$= \frac{1}{D^2}\left(x^2 + xy + y^2 - 2\frac{D'}{D}(x^2 + xy + y^2) + 3\frac{D'^2}{D^2}(x^2 + xy + y^2) + \ldots\right)$$

$$= \frac{1}{D^2}\left(x^2 + xy + y^2 - 2\frac{1}{D}(x + 2y) + 3\frac{1}{D^2}(2) + 0\right)$$

$$= \frac{1}{D^2}\left(x^2 + xy + y^2 - 2\left(\frac{x^2}{2} + 2yx\right) + 3x^2\right)$$

$$= \frac{1}{D^2}\left(x^2 + xy + y^2 - x^2 - 4yx + 3x^2\right)$$

$$= \frac{1}{D^2}\left(y^2 - 3yx + 3x^2\right)$$

$$=\left(\frac{x^2}{2}y^2-3y\frac{x^3}{6}+3\frac{x^4}{12}\right)$$

$$=\left(\frac{x^2}{2}y^2-y\frac{x^3}{2}+\frac{x^4}{4}\right)$$

Hence the general solution is

$$z=\phi_1(y-x)+x\phi_2(y-x)+\left(\frac{x^2}{2}y^2-y\frac{x^3}{2}+\frac{x^4}{4}\right),$$

which is a required solution, where ϕ_1 and ϕ_2 are arbitrary functions.

3.5.6 Shortcut method when $f(x,y)$ is of the form $e^{ax+by}V(x,y)$ where $V(x,y)=x^m y^n$ or $\sin(ax+by)$ or $\cos(ax+by)$

For any $g(x,y)$ by Leibnitz theorem

$D^n(e^{ax+by}g(x,y))=$

$e^{ax+by}D^n(g(x,y))+{}^nC_1 a e^{ax+by}D^{n-1}(g(x,y))+......+{}^nC_n a^n e^{ax+by}$

$$D^n(e^{ax+by}g(x,y))=e^{ax+by}(D+a)^n g(x,y)$$

Similarly,

$$D'^n(e^{ax+by}g(x,y))=e^{ax+by}(D'+b)^n g(x,y)$$

$$\therefore f(D,D')(e^{ax+by}g(x,y))=e^{ax+by}f(D+a,D'+b)g(x,y)$$

$$\therefore (e^{ax+by}g(x,y))=\frac{1}{f(D,D')}e^{ax+by}f(D+a,D'+b)g(x,y)$$

Taking $g(x,y)=\dfrac{1}{f(D+a,D'+b)}V(x,y)$

$$\therefore (e^{ax+by}\frac{1}{f(D+a,D'+b)}V(x,y)) = \frac{1}{f(D,D')}e^{ax+by}V(x,y)$$

$$\therefore \frac{1}{f(D,D')}e^{ax+by}V(x,y) = (e^{ax+by}\frac{1}{f(D+a,D'+b)}V(x,y))$$

When, $F(a,b)=0$ then $bD-aD'$ must be a factor of $F(D,D')$ suppose this factor repeated r times then the value of

$$\frac{1}{\left(bD-aD'\right)^r}f(ax+by) = \frac{x^r}{b^r r!}f(ax+by)$$

Example:

Solve $\frac{\partial^3 z}{\partial x^3}+\frac{\partial^3 z}{\partial x^2 \partial y}-\frac{\partial^3 z}{\partial x \partial y^2}-\frac{\partial^3 z}{\partial y^3}=e^x \cos 2y$.

Solution:

Here, $\frac{\partial^3 z}{\partial x^3}+\frac{\partial^3 z}{\partial x^2 \partial y}-\frac{\partial^3 z}{\partial x \partial y^2}-\frac{\partial^3 z}{\partial y^3}=e^x \cos 2y$.

$\therefore (D^3+D^2D'-DD'^2-D'^3)z=e^x \cos 2y$.

For complementary function, the auxiliary equation is

$(m^3+m^2-m-1)=0 \Rightarrow m=-1,1,-1.$

Hence the C.F function is

$CF=\phi_1(y-x)+x\phi_2(y-x)+\phi_3(y+x)$

Again, for the particular integral,

$$P.I = \frac{1}{(D^3 + D^2 D' - DD'^2 - D'^3)} e^x \cos 2y$$

$$= e^x \frac{1}{((D+1)^3 + (D+1)^2 D' - (D+1)D'^2 - D'^3)} \cos 2y$$

$$= e^x \frac{1}{(1 + D' - D'^2 - D'^3)} \cos 2y$$

$$= e^x \frac{1}{(1 + D' - (-2^2) - (-2^2)D')} \cos 2y$$

$$= e^x \frac{1}{(5 + 5D')} \cos 2y$$

$$= \frac{e^x}{5} \frac{1}{(1 + D')} \cos 2y$$

$$= \frac{e^x}{5} \frac{(1 - D')}{(1 - D'^2)} \cos 2y$$

$$= \frac{e^x}{5} \frac{(1 - D')}{(1 - (-2^2))} \cos 2y$$

$$= \frac{e^x}{25} (\cos 2y + 2 \sin 2y).$$

Hence the general solution is

$$z = \phi_1(y - x) + x\phi_2(y - x) + \phi_3(y + x) + \frac{e^x}{25} (\cos 2y + 2 \sin 2y),$$

which is a required solution, where ϕ_1, ϕ_2 and ϕ_3 are arbitrary functions.

Example:

Solve $\frac{\partial^2 z}{\partial x^2}-\frac{\partial^2 z}{\partial x \partial y}-2\frac{\partial^2 z}{\partial y^2}=e^x(y-1)$.

Solution:

Here, $\frac{\partial^2 z}{\partial x^2}-\frac{\partial^2 z}{\partial x \partial y}-2\frac{\partial^2 z}{\partial y^2}=e^x(y-1)$.

$\therefore (D^2-DD'-2D'^2)z=e^x(y-1)$.

For complementary function, the auxiliary equation is

$(m^2-m-2)=0 \Rightarrow m=-1,2.$

Hence the C.F function is $CF=\phi_1(y-x)+\phi_2(y+2x)$.

Again, for particular integral

$$P.I=\frac{1}{(D^2-DD'-2D'^2)}e^x(y-1)$$

$$=e^x\frac{1}{((D+1)^2-(D+1)D'-2D'^2)}(y-1)$$

$$=e^x\frac{1}{\left(D^2+2D+1-DD'-D'-2D'^2\right)}(y-1)$$

$$=e^x\frac{1}{\left(1+(D^2+2D-DD'-D'-2D'^2)\right)}(y-1)$$

$$= e^x \left(1+(D^2+2D-DD'-D'-2D'^2)\right)^{-1}(y-1)$$

$$= e^x \left(1-(D^2+2D-DD'-D'-2D'^2)+(D^2+2D-DD'-D'-2D'^2)^2-....\right)(y-1)$$

$$= e^x \left(y-1+D'(y-1)\right)$$

$$= e^x \left(y-1+1\right) = e^x y.$$

Hence the general solution is

$$z = \phi_1(y-x)+\phi_2(y+2x)+e^x y \ ,$$

which is a required solution, where ϕ_1 and ϕ_2 are arbitrary functions.

3.6 General method of finding the particular integral of linear homogenous equation with constant coefficients

Let the given partial differential equation:

$$F\left(D,D'\right)z = f\left(x,y\right) \tag{3.16}$$

where $F\left(D,D'\right)$ is a homogenous function of D and D' of degree *n*, so that

$$F\left(D,D'\right)=\left(D-m_1D'\right)\left(D-m_2D'\right)..........\left(D-m_nD'\right).$$

∴ Particular Integral of equation (3.16) $=\dfrac{1}{F\left(D,D'\right)}f\left(x,y\right),$

∴ Particular Integral of equation (3.16) is

$$=\frac{1}{\left(D-m_1D'\right)\left(D-m_2D'\right)...\left(D-m_nD'\right)}f\left(x,y\right) \tag{3.17}$$

To find the value of particular integral given by equation (3.17), consider a solution of the following equation $\left(D-mD'\right)z=f\left(x,y\right)$ which is a Lagrange's linear partial differential equation. The auxiliary equation associated with $\left(D-mD'\right)z=f\left(x,y\right)$ is given by

$$\frac{dx}{1}=\frac{dy}{-m}=\frac{dz}{f\left(x,y\right)}.$$

Taking fraction (a) and (b), we get

$$dy+mdx=0 \text{ so that } y=mx+c. \tag{3.18}$$

Now consider fraction (a) and (c), we get

$dz=f\left(x,y\right)dx=f\left(x,c-mx\right)dx$ (From equation (3.18)).

Integrating $z=\int f\left(x,c-mx\right)dx$.

$$\text{Thus,}\quad z=\frac{1}{D-mD'}f\left(x,y\right)=\int f\left(x,c-mx\right)dx \tag{3.19}$$

where after integration the constant c must be replaced by $y+mx$ since the particular integral does not contain any arbitrary constant. Thus, to obtain the particular integral for equation (3.16) apply the operation equation (3.19) by the factors, in succession, starting from right.

Example:

Solve $\dfrac{\partial^2 z}{\partial x^2}+\dfrac{\partial^2 z}{\partial x\partial y}-6\dfrac{\partial^2 z}{\partial y^2}=y\cos x$.

Solution:

Here, $\frac{\partial^2 z}{\partial x^2}+\frac{\partial^2 z}{\partial x\partial y}-6\frac{\partial^2 z}{\partial y^2}=y\cos x$.

$\therefore (D^2+DD^{'}-6D^{'2})z=y\cos x$.

For complementary function, the auxiliary equation is

$(m^2+m-6)=0 \Rightarrow (m-2)(m+3)\Rightarrow m=2,-3.$

Hence the C.F function is $CF=\phi_1(y+2x)+\phi_2(y-3x)$

Again, for Particular integral

$$P.I=\frac{1}{D^2+DD^{'}-6D^{'2}}y\cos x$$

$$P.I=\frac{1}{(D+3D^{'})(D-2D^{'})}y\cos x$$

$$=\frac{1}{(D+3D^{'})}\int(c-2x)\cos x dx$$

$$=\frac{1}{(D+3D^{'})}\left((c-2x)\sin x-(-2)(-\cos x)\right)$$

$$=\frac{1}{(D+3D^{'})}\left((y+2x-2x)\sin x-(-2)(-\cos x)\right)\left(\because c=y+2x\right)$$

$$=\frac{1}{(D+3D^{'})}\left(y\sin x-2\cos x\right)$$

$$=\int((c+3x)\sin x-2\cos x)dx$$

$$=(c+3x)(-\cos x)-(3)(-\sin x)-2\sin x$$

$$=-y\cos x+3\sin x-2\sin x=-y\cos x+\sin x\left(\because c=y-3x\right)$$

Hence the general solution is

$$z=\phi_1(y+2x)+\phi_2(y-3x)-y\cos x+\sin x\ ,$$

which is a required solution, where ϕ_1 and ϕ_2 are arbitrary functions.

Example:

Solve $\dfrac{\partial^2 z}{\partial x^2}+\dfrac{\partial^2 z}{\partial x\partial y}-6\dfrac{\partial^2 z}{\partial y^2}=\cos(2x+y)$.

Solution:

Here, $\dfrac{\partial^2 z}{\partial x^2}+\dfrac{\partial^2 z}{\partial x\partial y}-6\dfrac{\partial^2 z}{\partial y^2}=\cos(2x+y)$.

$$\therefore (D^2+DD'-6D'^2)z=\cos(2x+y).$$

For complementary function, the auxiliary equation is

$$(m^2+m-6)=0\ \Rightarrow (m+3)(m-2)=0 \Rightarrow m=-3,\ 2.$$

Hence the C.F function is $CF=\phi_1(y-3x)+\phi_2(y+2x)$.

Again, for the particular integral,

$$P.I=\frac{1}{(D^2+DD'-6D'^2)}\cos(2x+y)$$

$$=\frac{x}{((2)^2+(2)(1)-6(1)^2)}\cos(2x+y)$$

$$=\frac{x}{0}\cos(2x+y)$$

Hence a solution is not possible

Hence,

$$P.I = \frac{1}{(D-2D^{'})(D+3D^{'})}\cos(2x+y)$$

$$= \left(\frac{1}{(D+3D^{'})(D-2D^{'})}\right)\cos(2x+y)$$

$$= \left(\frac{1}{(D+3D^{'})}\int \cos(2x+c-2x)dx\right)(\because y = c - mx = c - 2x)$$

$$= \left(\frac{1}{(D+3D^{'})}\int \cos(c)dx\right) = \left(\frac{1}{(D+3D^{'})}x\cos(c)\right)$$

$$= \left(\frac{1}{(D+3D^{'})}x\cos(y+2x)\right)(\because c = y+2x)$$

$$= \int x\cos(c+3x+2x)dx(\because y = c - mx = c+3x)$$

$$= \int x\cos(c+5x)dx$$

$$= x\frac{\sin(c+5x)}{5} + \frac{\cos(c+5x)}{25}$$

$$= x\frac{\sin(y-3x+5x)}{5} + \frac{\cos(y-3x+5x)}{25}(\because c = y-3x)$$

$$= x\frac{\sin(y+2x)}{5} + \frac{\cos(y+2x)}{25}(\because c = y-3x)$$

Hence the general solution is

$$z=\phi_1(y-3x)+\phi_2(y+2x)+\left(\frac{x}{5}\sin(2x+y)+\frac{1}{25}\cos(2x+y)\right)$$

which is a required solution, where ϕ_1 and ϕ_2 are arbitrary functions.

Example:

Solve $2\frac{\partial^2 z}{\partial x^2}-5\frac{\partial^2 z}{\partial x\partial y}+2\frac{\partial^2 z}{\partial y^2}=5\sin(2x+y)$.

Solution:

Here, $2\frac{\partial^2 z}{\partial x^2}-5\frac{\partial^2 z}{\partial x\partial y}+2\frac{\partial^2 z}{\partial y^2}=5\sin(2x+y)$.

$$\therefore (2D^2-5DD^{'}+2D^{'2})z=5\sin(2x+y).$$

For complementary function, the auxiliary equation is

$$(2m^2-5m+2)=0 \ \Rightarrow m=2,\frac{1}{2}.$$

Hence the C.F function is $CF=\phi_1(y+\frac{1}{2}x)+\phi_2(y+2x)$.

Again, for particular integral

$$P.I=\frac{1}{(2D-D^{'})(D-2D^{'})}5\sin(2x+y)$$

$$=5\frac{1}{(2D-D^{'})(D-2D^{'})}\sin(2x+y)$$

$$=5\frac{1}{(2D-D^{'})}\int\sin(2x+c-2x)dx(\because y=c-mx=c-2x)$$

$$=5\frac{1}{(2D-D')}\int \sin c\ dx=5\frac{1}{(2D-D')}x\sin c\ dx$$

$$=5\frac{1}{(2D-D')}x\sin(y+2x)\ dx\ (\because c=y+2x)$$

$$=\frac{5}{2}\frac{1}{(D-\frac{1}{2}D')}x\sin(y+2x)\ dx$$

$$=\frac{5}{2}\int x\sin\left(2x+c-\frac{1}{2}x\right)dx(\because y=c-mx=c-\frac{1}{2}x)$$

$$=\frac{5}{2}\int x\sin\left(c+\frac{3}{2}x\right)dx$$

$$=\frac{5}{2}\left[\frac{-2x}{3}\cos\left(c+\frac{3}{2}x\right)+\frac{2}{3}\int\cos\left(c+\frac{3}{2}x\right)dx\right]$$

$$=\frac{5}{2}\left[\frac{-2x}{3}\cos\left(c+\frac{3}{2}x\right)+\frac{4}{9}\sin\left(c+\frac{3}{2}x\right)\right]$$

$$=\frac{5}{2}\left[\frac{-2x}{3}\cos\left(y+\frac{1}{2}x+\frac{3}{2}x\right)+\frac{4}{9}\sin\left(y+\frac{1}{2}x+\frac{3}{2}x\right)\right]\left(\because c=y+\frac{1}{2}x\right)$$

$$=\left[\frac{-5x}{3}\cos(y+2x)+\frac{10}{9}\sin(y+2x)\right]$$

Hence the general solution is

$$z=\phi_1(y+\frac{1}{2}x)+\phi_2(y+2x)-\frac{5x}{3}\cos(y+2x)+\frac{10}{9}\sin(y+2x),$$

which is a required solution, where ϕ_1 and ϕ_2 are arbitrary functions.

3.7 Miscellaneous Example

Example:

Solve $(D^2 + 2DD' - 8D'^2) = (2x + y)^{\frac{1}{2}}$.

Solution:

Here, $(D^2 + 2DD' - 8D'^2) = (2x + y)^{\frac{1}{2}}$.

For complementary function, the auxiliary equation is

$(m^2 + 2m - 8) = 0 \Rightarrow (m + 4)(m - 2) = -4, 2.$

Hence the C.F function is $CF = \phi_1(y - 4x) + \phi_2(y + 2x)$

Again, for particular integral

$$P.I = \frac{1}{(D^2 + 2DD' - 8D'^2)}(2x + y)^{\frac{1}{2}},$$

Integrate $z^{\frac{1}{2}}$ two times and then put $z = 2x + y$ and $D = 2, D' = 3$

$$P.I = \frac{1}{(2^2 + 2(2)(3) - 8(3)^2)} \frac{(2x + 3y)^{\frac{5}{2}}}{(3/2)(5/2)},$$

$$P.I = -\frac{1}{210}(2x + 3y)^{\frac{5}{2}}.$$

Hence the general solution is

$$z=\phi_1(y-4x)+\phi_2(y+2x)-\frac{1}{210}(2x+3y)^{\frac{5}{2}},$$

which is a required solution, where ϕ_1 and ϕ_2 are arbitrary functions.

Example:

Solve $(D^2+6DD'-D'^2)=x^2\sin(x+y)$.

Solution:

Here, $(D^2+6DD'-D'^2)=x^2\sin(x+y)$.

For complementary function, the auxiliary equation is

$(m^2+6m-1)=0 \Rightarrow (m+3)(m-2)=-3,2.$

Hence the C.F function is $CF=\phi_1(y-3x)+\phi_2(y+2x)$.

Again, for particular integral

$$P.I=\frac{1}{(D^2+6DD'-D'^2)}x^2\sin(x+y)$$

$$=\frac{1}{(D+3D')(D-2D')}x^2\sin(x+y)$$

$$=\frac{1}{(D+3D')}\int x^2\sin(x+c-2x)dx \qquad (\because y=c-2x)$$

$$=\frac{1}{(D+3D')}\int x^2\sin(c-x)dx$$

$$=\frac{1}{(D+3D')}\left(x^2\cos(c-x)-(2x)(-\sin(c-x))+(2)(-\cos(c-x))\right)$$

$$=\frac{1}{(D+3D')}\left(x^2\cos(x+y)+2x(\sin(x+y))-2\cos(x+y)\right)$$

$$=\frac{1}{(D+3D')}\left((x^2-2)\cos(x+y)+2x(\sin(x+y))\right)$$

$$=\int\left((x^2-2)\cos(x+c+3x)+2x(\sin(x+c+3x))\right)dx \quad (\because y=c+3x)$$

$$=\int\left((x^2-2)\cos(4x+c)+2x(\sin(4x+c))\right)dx$$

$$=\begin{pmatrix}(x^2-2)\frac{1}{4}\sin(4x+c)-(2x)(-\frac{1}{16})\cos(4x+c)+(2)(-\frac{1}{64})\sin(4x+c)\\ +(2x)(-\frac{1}{4})\cos(4x+c)-(2)(-\frac{1}{16})\sin(4x+c)\end{pmatrix}$$

$$=\left(\left(\frac{x^2-2}{4}-\frac{1}{32}+\frac{1}{8}\right)\sin(x+y)+\left(\frac{x}{8}-\frac{x}{2}\right)\cos(x+y)\right)$$

$$=\frac{(8x^2-13)}{32}\sin(x+y)-\frac{3}{8}x\cos(x+y).$$

Hence the general solution is

$$z=\phi_1(y-3x)+\phi_2(y+2x)-\frac{(8x^2-13)}{32}\sin(x+y)-\frac{3}{8}x\cos(x+y),$$

which is a required solution, where ϕ_1 and ϕ_2 are arbitrary functions.

Example:

Solve $(D^2-2DD'+D'^2)=2x\cos y$.

Solution:

Here, $(D^2-2DD'+D'^2)=2x\cos y$.

For complementary function, the auxiliary equation is

$(m^2-2m+1)=0 \Rightarrow (m-1)^2=1,1.$

Hence the C.F function is $CF=\phi_1(y+x)+x\phi_2(y+x)$.

Again, for particular integral

$$P.I=\frac{1}{(D-D')^2}2x\cos y$$

$$P.I=\text{Re part of}\frac{1}{(D-D')^2}2xe^{iy}$$

$$P.I=\text{Re part of}\, e^{iy}\frac{1}{(D-(D'+i))^2}2x$$

$$P.I=2\text{Re part of}\,(-e^{iy})(1+i(D-D'))^{-2}x$$
$$P.I=2\,\text{Re part of}\,(-e^{iy})(1-2iD+...)x$$

$$P.I=2\,\text{Re part of}\,(-e^{iy})(x-2i(1)+0)$$

$$P.I=-2\,\text{Re part of}\,(\cos y+i\sin y)(x-2i(1)+0)$$

$$P.I=-2\,\text{Re part of}\,(x\cos y-2i\cos y+ix\sin y+2\sin y)$$

$$P.I=-2\,\text{Re part of}\,(x\cos y+2\sin y-2i\cos y+ix\sin y)$$

$$=-2(x\cos y+2\sin y)$$

Hence the general solution is

$z=\phi_1(y+x)+x\phi_2(y+x)-2(x\cos y+2\sin y)$

which is a required solution, where ϕ_1 and ϕ_2 are arbitrary functions.

Example:

Solve $(D^2+DD^{'}-2D^{'2})z=8\log(x+5y)$.

Solution:

Here, $(D^2+DD^{'}-2D^{'2})z=8\log(x+5y)$.

For complementary function, the auxiliary equation is

$(m^2+m-2)=0 \Rightarrow (m+2)(m-1) \Rightarrow m=-2,1$.

Hence the C.F function is $CF=\phi_1(y-2x)+\phi_2(y+x)$.

Again, for particular integral

$$P.I=\frac{1}{(D^2+DD^{'}-2D^{'2})}8\log(x+5y)$$

$$P.I=\frac{1}{(D+2D^{'})(D-D^{'})}8\log(x+5y)$$

$$=8\frac{1}{(D+2D^{'})}\int\log(x+5(c-x))dx \qquad (\because y=c-x)$$

$$=8\frac{1}{(D+2D^{'})}\int\log(5c-4x)dx$$

$$=8\frac{1}{(D+2D^{'})}\left[\frac{-((5c-4x)\log(5c-4x)-(5c-4x))}{4}\right]$$

$$=8\frac{1}{(D+2D')}\left(x\log(5c-4x)-\int\left(1-\frac{5c}{5c-4x}\right)dx\right)$$

$$=8\frac{1}{(D+2D')}\left(x\log(5c-4x)-x-\frac{5c}{4}\log(5c-4x)\right)$$

$$=8\frac{1}{(D+2D')}\left(x\log(x+5y)-x-\frac{5(y+x)}{4}\log(x+5y)\right)$$

$$=8\frac{1}{(D+2D')}\left(\frac{-x-5y}{4}\log(x+5y)-x\right)$$

$$=-2\frac{1}{(D+2D')}((x+5y)\log(x+5y)+4x)$$

$$=-2\int((5c+11x)\log(5c+11x)+4x)dx \qquad (\because c=y-2x)$$

$$=-2\left[\frac{(5c+11x)^2}{22}\log(5c+11x)-\int\frac{11(5c+11x)^2}{22(5c+11x)}dx+2x^2\right]$$

$$=-\frac{1}{11}\left[(5c+11x)^2\log(5c+11x)-11\frac{(5c+11x)^2}{22}+44x^2\right]$$

$$=-\frac{1}{22}\left[2(x+5y)^2\log(x+5y)-(x+5y)^2+88x^2\right]$$

$$=-\frac{1}{22}\left[(x+5y)^2(2\log(x+5y)-1)+88x^2\right]$$

Hence, the general solution is

$$z=\phi_1(y-2x)+\phi_2(y+x)-\frac{1}{22}\left[(x+5y)^2(2\log(x+5y)-1)+88x^2\right],$$

which is a required solution, where ϕ_1 and ϕ_2 are arbitrary functions.

Example:

Solve $(4D^2 - 4DD' + D'^2)z = 16\log(x+2y)$.

Solution:

Here, $(4D^2 - 4DD' + D'^2)z = 16\log(x+2y)$.

For complementary function, the auxiliary equation is

$$(4m^2 - 4mm + 1) = 0 \Rightarrow (2m-1)^2 = 0 \Rightarrow m = \frac{1}{2}, \frac{1}{2}.$$

Hence the C.F function is $CF = \phi_1(y + \frac{x}{2}) + x\phi_2(y + \frac{x}{2})$.

Again, for particular integral

$$P.I = \frac{1}{(4D^2 - 4DD' + D'^2)} 16\log(x+2y),$$

$$P.I = \frac{1}{(2D - D')^2} 16\log(x+2y),$$

$$\left(\because \frac{1}{\left(bD - aD'\right)^r} f(ax+by) = \frac{x^r}{b^r r!} f(ax+by) \right)$$

$$P.I = \frac{x^2}{2^2 2!} 16\log(x+2y)$$

$$P.I = 2x^2 \log(x+2y)$$

Hence the general solution is

$$z = \phi_1(y+\frac{x}{2}) + x\phi_2(y+\frac{x}{2}) + 2x^2 \log(x+2y),$$

which is a required solution, where ϕ_1 and ϕ_2 are arbitrary functions.

Example:

Solve $(D^2 + DD' - 6D'^2)z = x + y$.

Solution:

Here, $(D^2 + DD' - 6D'^2)z = x + y$.

For complementary function, the auxiliary equation is

$(m^2 + m - 6) = 0 \Rightarrow (m+3)(m-2) = 0 \Rightarrow m = -3, 2$.
Hence the C.F function is $CF = \phi_1(y-3x) + x\phi_2(y+2x)$.

Again, for particular integral

$$P.I = \frac{1}{(D^2 + DD' - 6D'^2)}(x+y)$$

$$P.I = \frac{1}{(D+3D')(D-2D')}(x+y)$$

$$\because \frac{1}{\left(bD - aD'\right)^r} f(ax+by) = \frac{x^r}{b^r r!} f(ax+by)$$

$$P.I = \frac{1}{(1+3(1))(1-2(1))}\iint u\, du du \qquad (\because u = x+y)$$

$$P.I = \frac{1}{(-4)}\left(\frac{u^3}{6}\right)$$

$$P.I = \frac{1}{(-24)}(x+y)^3.$$

Hence the general solution is

$$z = \phi_1(y-3x) + x\phi_2(y+2x) - \frac{1}{(24)}(x+y)^3 .$$

which is a required solution, where ϕ_1 and ϕ_2 are arbitrary functions.

Example:

Solve $(D^2 - D'^2)z = \tan^3 x \tan y - \tan x \tan^3 y$.

Solution:

Here, $(D^2 - D'^2)z = \tan^3 x \tan y - \tan x \tan^3 y$.

For complementary function, the auxiliary equation is

$$(m^2 - 1) = 0 \Rightarrow (m+1)(m-1) = 0 \Rightarrow m = -1, 1.$$

Hence the C.F function is $CF = \phi_1(y-x) + \phi_2(y+x)$.

Again, for particular integral

$$P.I = \frac{1}{(D^2 - D'^2)}(\tan^3 x \tan y - \tan x \tan^3 y)$$

$$= \frac{1}{(D+D')(D-D')}(\tan^3 x \tan y - \tan x \tan^3 y)$$

$$=\frac{1}{(D+D')}\int(\tan^3 x\tan(c-x)-\tan x\tan^3(c-x))dx \quad (\because y=c-x)$$

$$=\frac{1}{(D+D')}\int(\tan x(\sec^2 x-1)\tan(c-x)-\tan x\tan(c-x)(\sec^2(c-x)-1)dx$$

$$=\frac{1}{(D+D')}\int(\tan x\tan(c-x)\sec^2 x-\tan x\tan(c-x)-\tan x\tan(c-x)\sec^2(c-x)+\tan x\tan(c-x))dx$$

$$=\frac{1}{(D+D')}\left(\tan(c-x)\frac{\tan^2 x}{2}+\int\sec^2(c-x)\frac{\tan^2 x}{2}dx-\tan x\left\{-\frac{1}{2}\tan^2(c-x)\right\}+\int\sec^2 x\left\{-\frac{1}{2}\tan^2(c-x)\right\}dx\right)$$

$$=\frac{1}{2(D+D')}\left(\tan(c-x)\tan^2 x+\int\sec^2(c-x)\tan^2 x dx+\tan x\tan^2(c-x)-\int\sec^2 x\tan^2(c-x)dx\right)$$

$$=\frac{1}{2(D+D')}\left(\tan(c-x)\tan^2 x+\int(1+\tan^2(c-x))\tan^2 x dx+\tan x\tan^2(c-x)-\int(1+\tan^2 x)\tan^2(c-x)dx\right)$$

$$=\frac{1}{2(D+D')}\left(\tan(c-x)\tan^2 x+\tan x\tan^2(c-x)+\int(\tan^2 x-\tan^2(c-x))dx\right)$$

$$=\frac{1}{2(D+D')}\left(\tan(c-x)\tan^2 x+\tan x\tan^2(c-x)+\int((\sec^2 x-1)-(\sec^2(c-x)-1)dx\right)$$

$$=\frac{1}{2(D+D')}\left(\tan(c-x)\tan^2 x+\tan x\tan^2(c-x)+\tan x+\tan(c-x)\right)$$

$$=\frac{1}{2(D+D')}\left(\tan y\tan^2 x+\tan x\tan^2 y+\tan x+\tan y\right)$$

$$=\frac{1}{2(D+D')}\left(\tan y(\tan^2 x+1)+\tan x(\tan^2 y+1)\right)$$

$$=\frac{1}{2(D+D')}\left(\tan y\sec^2 x+\tan x\sec^2 y\right)$$

$$=\frac{1}{2}\left(\int\tan(c+x)\sec^2 x+\tan x\sec^2(c+x)dx\right) \qquad \because y=c+x$$

$$=\frac{1}{2}\left(\tan(c+x)\tan x-\int \sec^2(c+x)\tan x\,dx+\int \tan x\sec^2(c+x)dx\right)$$

$$=\frac{1}{2}\left(\tan(c+x)\tan x\right)$$

$$=\frac{1}{2}\left(\tan y\tan x\right)$$

Hence the general solution is

$$z=\phi_1(y-x)+\phi_2(y+x)+\frac{1}{2}\left(\tan y\tan x\right)$$

which is a required solution, where ϕ_1 and ϕ_2 are arbitrary functions.

Example:

Solve $(D^2-4D'^2)z=\frac{4x}{y^2}-\frac{y}{x^2}$.

Solution:

Here, $(D^2-4D'^2)z=\frac{4x}{y^2}-\frac{y}{x^2}$.

For complementary function, the auxiliary equation is

$$(m^2-4)=0 \Rightarrow (m+2)(m-2)=0 \Rightarrow m=-2,2.$$

Hence the C.F function is $CF=\phi_1(y-2x)+\phi_2(y+2x)$.

Again, for particular integral

$$P.I=\frac{1}{(D^2-4D'^2)}\left(\frac{4x}{y^2}-\frac{y}{x^2}\right)$$

$$P.I = \frac{1}{(D+2D')(D-2D')}\left(\frac{4x}{y^2} - \frac{y}{x^2}\right)$$

Let $u = \frac{1}{D-2D'}\left(\frac{4x}{y^2} - \frac{y}{x^2}\right)$

$$\therefore (D-2D')u = \left(\frac{4x}{y^2} - \frac{y}{x^2}\right)$$

$$\therefore u = \int\left(\frac{4x}{(c-2x)^2} - \frac{(c-2x)}{x^2}\right)dx$$

$$\therefore u = \int\left(\frac{-2(c-2x)+2c}{(c-2x)^2} - \frac{c}{x^2} + \frac{2}{x}\right)dx$$

$$\therefore u = \int\left(\frac{-2}{(c-2x)} + \frac{2c}{(c-2x)^2} - \frac{c}{x^2} + \frac{2}{x}\right)dx$$

$$\therefore u = \log(c-2x) + \frac{c}{c-2x} + \frac{c}{x} + 2\log x$$

$$\therefore u = \log(2x+y-2x) + \frac{2x+y}{2x+y-2x} + \frac{2x+y}{x} + 2\log x$$

$$\therefore u = \log(y) + \frac{2x+y}{y} + 2 + \frac{y}{x} + 2\log x$$

$$P.I = \frac{1}{(D+2D')}(\log(y) + \frac{2x+y}{y} + 2 + \frac{y}{x} + 2\log x)$$

$$P.I = \int(\log(c+2x) + \frac{2x+c+2x}{c+2x} + \frac{c+2x}{x} + 2\log x)dx \quad \because y = c+2x$$

$$P.I = \int (\log(c+2x) + \frac{2x}{c+2x} + 1 + 2 + \frac{c}{x} + 2 + 2\log x)dx$$

$$P.I = \int (\log(c+2x) + \frac{2x+c-c}{c+2x} + 5 + \frac{c}{x} + 1 \cdot \log x^2)dx$$

$$P.I = \left(x\log(c+2x) - \int x\frac{1}{c+2x}2dx + 6x - \frac{c}{2}\log(c+2x) + c\log x + x\log x^2 - \int \frac{2x}{x^2}xdx \right)$$

$$P.I = \left(x\log(c+2x) - \int \frac{c+2x-c}{c+2x}dx + 6x - \frac{c}{2}\log(c+2x) + c\log x + x\log x^2 - \int 2dx \right)$$

$$P.I = \left(x\log(c+2x) - x + \frac{c}{2}\log(c+2x) + 6x - \frac{c}{2}\log(c+2x) + c\log x + x\log x^2 - 2x \right)$$

$$P.I = \left(x\log(c+2x) + 3x + c\log x + x\log x^2 \right)$$

$$P.I = \left(x\log(y-2x+2x) + 3x + (y-2x)\log x + x\log x^2 \right)$$

$$P.I = \left(x\log(y) + 3x + y\log x - x\log x^2 + x\log x^2 \right)$$

$$P.I = \left(x\log(y) + 3x + y\log x \right)$$

Hence the general solution is

$$z = \phi_1(y-2x) + \phi_2(y+2x) + \left(x\log(y) + 3x + y\log x \right)$$

which is a required solution, where ϕ_1 and ϕ_2 are arbitrary functions.

3.8 Non-Homogeneous Partial Differential Equation

Differential equation is $f(D,D')z = F(x,y)$ where $f(D,D')$ is not homogeneous (that is sum of powers of D and D' in terms may not be equal). Here the general solution is also noted as complementary function and particular integral.

Complementary function:

Factors are not repeated

When $f(D,D')$ can be factorized into linear factors of the type $D+mD'+c \ or \ D'+c$ when m and c may any constant and factors are not repeated. To find the solutions $(D+mD'+c)z=0$ that is $p+mq+cz=0$ This is Lagrange's equation.

Hence, Lagrange's auxiliary equation is

$$\frac{dx}{1}=\frac{dy}{m}=\frac{dz}{-cz}$$

By comparing the first two members $y-mx=k_1$ and the solution from first and third member is $z=k_2e^{-cx}=\phi(k_1)e^{-cx}$.

Hence the solution is $z=e^{-cx}\phi(y-mx)$

Now, we find the solutions of

$(D'+c)z=0 \ or \ q=-cz$

Lagrange's equation is

$$\frac{dx}{1}=\frac{dy}{m}=\frac{dz}{-cz}$$

The two solutions are $x=k_1, \ z=k_2e^{-ky}$. In this case the solution is $z=e^{-cy}\phi(x)$. Hence, solutions corresponding to each factor will be found if factors are not repeated. Then by superposition principle complementary function will be sum of all these solutions.

Factors are repeated

Suppose $D+mD^{'}+c$ is repeated twice

In this case the solutions are obtained as

$$(D+mD^{'}+c)^{2}z=0$$

Let, $(D+mD^{'}+c)z=u$ therefore we get,

In this case the solution is (as per non repeated case)
$u=e^{-cx}\phi(y-mx)$

Again, we have

$$(D+mD^{'}+c)z=u$$

$$\therefore(D+mD^{'}+c)z=e^{-cx}\phi(y-mx)$$

$$\therefore(D+mD^{'})=-cz+e^{-cx}\phi(y-mx)$$

Lagrange's equation is

$$\frac{dx}{1}=\frac{dy}{m}=\frac{dz}{-cz+e^{-cx}\phi(y-mx)}$$

from first two members we get $y=mx+k_1$

from first and third members

$$\frac{dx}{1}=\frac{dz}{-cz+e^{-cx}\phi(y-mx)}\Rightarrow\frac{dz}{dx}+cz=e^{-cx}\phi(k_1)$$

Hence Integrating factors is e^{cx} solution is

$$ze^{cx} = \int e^{cx}e^{-cx}\phi(k_1)dx + k_2$$

$$ze^{cx} = x\phi(k_1) + k_2$$

$$\therefore z = \left(x\phi(k_1) + k_2\right)e^{-cx}$$

$$\therefore z = \left(x\phi(y - mx) + \psi(y - mx)\right)e^{-cx}$$

Similarly, if factors are repeated three times, then the solution is

$$\therefore z = \left(\psi(y - mx) + x\phi(y - mx) + x^2\phi_1(y - mx)\right)e^{-cx}$$

Suppose $f(D, D^{'})$ cannot be factorized into linear factors. Then

$$D^r e^{hx+ky} = h^r e^{hx+ky}$$
$$D^{'r} e^{hx+ky} = k^r e^{hx+ky}$$
$$\therefore f(D, D^{'})e^{hx+ky} = f(h,k)e^{hx+ky}$$

Thus, $z = e^{hx+ky}$ will be solutions of $f(D, D^{'})z = 0$ when $f(h,k) = 0$

Hence by principle of superposition

$$C.F. = \sum_{i=1}^{\infty} c_i e^{h_i x + k_i y} \qquad where\ f(h_i, k_i) = 0$$

Operator method for finding particular integral

Consider the equation $f(D, D^{'})z = F(x, y)$ in his case the particular integral is

$$P.I. = \frac{1}{f(D, D^{'})}F(x, y)$$

Theorem: $\frac{1}{D+mD^{'}+c}F(x,y)=e^{-cx}\int e^{cx}F(x,a+mx)dx$ where we put $a=y-mx$ after integration

Proof: Let, $\frac{1}{D+mD^{'}+c}F(x,y)=G(x,y)$

$$(D+mD^{'}+c)G(x,y)=F(x,y)$$

$$\frac{\partial G}{\partial x}+m\frac{\partial G}{\partial y}=-cG+F$$

This is Lagrange's equations

The Lagrange's auxiliary equation is

$$\frac{dx}{1}=\frac{dy}{m}=\frac{dG}{-cG+F}$$

From first and second member $y=mx+c$

From first and third member

$$\frac{dG}{dx}+cG=F(x,y)=F(x,a+mx)$$

Hence, the integrating factor is I.F.= e^{cx}

Hence the solution is

$$Ge^{cx}=\int e^{cx}F(x,a+mx)dx$$

$$G=e^{-cx}\int e^{cx}F(x,a+mx)dx$$

$$\frac{1}{D+mD'+c}F(x,y)=e^{-cx}\int e^{cx}F(x,a+mx)dx$$

Example:

Find the solutions of $(3D^2+7DD'+2D'^2+7D+4D'+2)z=0$

Solution:

Here, $(3D^2+7DD'+2D'^2+7D+4D'+2)z=0$

$(D+2D'+2)(3D+D'+1)z=0$

$$(D+2D'+2)(3D+D'+1)z=0$$

$$(D+2D'+2)(D+\frac{1}{3}D'+\frac{1}{3})z=0$$

Hence the auxiliary equation is

$$(m+2)(m+\frac{1}{3})=0 \Rightarrow m=-2,-\frac{1}{3}$$

Hence the general solutions are

$$z=e^{-2x}\phi_1(y-2x)+e^{-\frac{1}{3}x}\phi_2(3y-x)$$

which is a required solution, where ϕ_1 and ϕ_2 are arbitrary functions.

Example:

Find the solutions of $(2D^2-DD'-D'^2-D+D')z=0$

Solution:

Here, $(2D^2-DD'-D'^2-D+D')z=0$

$$(D-D')(2D+D'-1)z=0$$

$$(D-D')(D+\frac{1}{2}D'-\frac{1}{2})z=0$$

Hence the auxiliary equation is

$$(m-1)(m+\frac{1}{2})=0 \Rightarrow m=1,-\frac{1}{2}$$

Hence the general solution is

$$z=\phi_1(y+x)+e^{\frac{1}{2}x}\phi_2(2y-x)$$

which is a required solution, where ϕ_1 and ϕ_2 are arbitrary functions.

Example:

Find the solutions of $(D^2-D'^2+3D'-3D)z=e^{x+2y}+xy$

Solution:

Here, $(D^2-D'^2+3D'-3D)z=e^{x+2y}+xy$

For complementary functions

$$(D^2-D'^2+3D'-3D)z=0$$

$$(D-D')(D+D'-3)z=0$$

Hence the auxiliary equation is

$$(m-1)(m+1)=0 \Rightarrow m=1,-1$$

Therefore, the complementary function is

$$C.F. = \phi_1(y+x) + e^{3x}\phi_2(y-x)$$

For particular integral

$$P.I. = \frac{1}{(D^2 - D'^2 + 3D' - 3D)}\left(e^{x+2y} + xy\right)$$

$$P.I. = \left(\frac{1}{(D^2 - D'^2 + 3D' - 3D)}e^{x+2y} + \frac{1}{(D-D')(D+D'-3)}xy\right)$$

$$P.I. = \left(x\frac{1}{(2D-3)}e^{x+2y} - \frac{1}{3D}\left(1-\frac{D'}{D}\right)^{-1}\left(1-\left(\frac{D}{3}+\frac{D'}{3}\right)\right)^{-1}(xy)\right)$$

$$P.I. = \left(x\frac{1}{(2(1)-3)}e^{x+2y} - \frac{1}{3D}\left(1+\frac{D'}{D}+\ldots\right)\left(1+\frac{D}{3}+\frac{D'}{3}+\frac{2}{9}DD'+\ldots\right)(xy)\right)$$

$$P.I. = \left(-xe^{x+2y} - \frac{1}{3D}\left(1+\frac{D}{3}+\frac{2D'}{3}+\frac{D'}{D}+\frac{2}{9}DD'+\ldots\right)(xy)\right)$$

$$P.I. = \left(-xe^{x+2y} - \frac{1}{3D}\left(xy+\frac{y}{3}+\frac{2x}{3}+\frac{1}{D}y+\frac{2}{9}\right)\right)$$

$$P.I. = \left(-xe^{x+2y} - \frac{1}{3}\left(\frac{x^2y}{2}+\frac{xy}{3}+\frac{x^2}{3}+\frac{x^3}{6}+\frac{2}{9}x\right)\right)$$

Hence the general solution is

$$z = \phi_1(y+x) + e^{3x}\phi_2(y-x) - xe^{x+2y} - \frac{1}{3}\left(\frac{x^2y}{2}+\frac{xy}{3}+\frac{x^2}{3}+\frac{x^3}{6}+\frac{2}{9}x\right)$$

which is a required solution, where ϕ_1 and ϕ_2 are arbitrary functions.

Example:

Find the solutions of $(4D^2+3DD^{'}-D^{'2}-D-D^{'})z=3e^{\frac{1}{2}(x+2y)}$

Solution:

Here, $(4D^2+3DD^{'}-D^{'2}-D-D^{'})z=3e^{\frac{1}{2}(x+2y)}$

For complementary functions

$$(4D^2+3DD^{'}-D^{'2}-D-D^{'})z=0$$

$$(4D-D^{'}-1)(D+D^{'})z=0$$

Hence the auxiliary equation is

$$(4m-1)(m+1)=0 \Rightarrow m=\frac{1}{4},-1$$

Therefore, the complementary function is

$$C.F.=\phi_1(y-x)+e^{\frac{x}{4}}\phi_2(4y+x)$$

For particular integral

$$P.I.=\frac{1}{(4D^2+3DD^{'}-D^{'2}-D-D^{'})}\left(3e^{\frac{1}{2}(x+2y)}\right)$$

$$=\frac{1}{(4D^2+3DD^{'}-D^{'2}-D-D^{'})}\left(3e^{\frac{1}{2}(x+2y)}\right)$$

$$=3x\frac{1}{(8D+3D^{'}-1)}\left(e^{\frac{1}{2}(x+2y)}\right)$$

$$= 3x\frac{1}{(4+3-1)}\left(e^{\frac{1}{2}(x+2y)}\right)$$

$$= \frac{x}{2}\left(e^{\frac{1}{2}(x+2y)}\right)$$

Hence the general solution is

$$z = \phi_1(y-x) + e^{\frac{x}{4}}\phi_2(4y+x) - \frac{x}{2}\left(e^{\frac{1}{2}(x+2y)}\right)$$

which is a required solution, where ϕ_1 and ϕ_2 are arbitrary functions.

Example:

Find the solutions of $(D^2 - D^{'})z = xe^{(ax+a^2y)}$

Solution:

Here, $(D^2 - D^{'})z = xe^{(ax+a^2y)}$

For complementary functions

$$(D^2 - D^{'})z = 0$$

But $(D^2 - D^{'})$ cannot be factorized. Hence,

$$C.F. = \sum_{i=1}^{\infty} c_i e^{h_i x + k_i y} \qquad where\ h_i^2 - k_i = 0 \Rightarrow k_i = h_i^2$$

$$C.F. = \sum_{i=1}^{\infty} c_i e^{h_i x + h_i^2 y}$$

$$C.F. = \sum_{i=1}^{\infty} c_i e^{h_i(x+h_i y)}$$

For particular integral

$$P.I. = \frac{1}{(D^2 - D')} xe^{(ax+a^2y)}$$

$$= e^{(ax+a^2y)} \frac{1}{((D+a)^2 - (D' + a^2))} x$$

$$= e^{(ax+a^2y)} \frac{1}{D^2 + 2aD} x$$

$$= e^{(ax+a^2y)} \frac{1}{2aD}\left(1 + \frac{D}{2a}\right)^{-1} x$$

$$= e^{(ax+a^2y)} \frac{1}{2aD}\left(1 - \frac{D}{2a} + \ldots\right) x$$

$$= e^{(ax+a^2y)} \frac{1}{2aD}\left(x - \frac{1}{2a}\right)$$

$$= e^{(ax+a^2y)} \frac{1}{2a}\left(\frac{x^2}{2} - \frac{x}{2a}\right)$$

Hence, the general solution is

$$z = \sum_{i=1}^{\infty} c_i e^{h_i(x+h_i y)} + e^{(ax+a^2y)} \frac{1}{2a}\left(\frac{x^2}{2} - \frac{x}{2a}\right)$$

Example:

Find the solution of
$(D^2 - DD' - 2D'^2 + 2D + 2D')z = e^{2x+3y} + \sin(2x+y) + xy$

Solution:

Here,

$$(D^2 - DD' - 2D'^2 + 2D + 2D')z = e^{2x+3y} + \sin(2x+y) + xy$$

For complementary functions

$$(D^2 - DD' - 2D'^2 + 2D + 2D')z = 0$$

$$(D - 2D' + 2)(D + D')z = 0$$

Hence the auxiliary equation is

$$(m-2)(m+1) = 0 \Rightarrow m = 2, -1$$

Therefore, the complementary function is

$$C.F. = \phi_1(y-x) + e^{-2x}\phi_2(y+2x)$$

For particular integral

$$P.I. = \frac{1}{(D^2 - DD' - 2D'^2 + 2D + 2D')}\left(e^{2x+3y} + \sin(2x+y) + xy\right)$$

$$= -\frac{1}{10}e^{2x+3y} - \frac{1}{6}\cos(2x+y) + \frac{1}{2D}\left(xy - \frac{y}{2} + \frac{3}{2}x - 1 - \frac{1}{D}x\right)$$

$$= -\frac{1}{10}e^{2x+3y} - \frac{1}{6}\cos(2x+y) + \frac{1}{24}\left(6x^2y - 6xy + 9x^2 - 12x - 2x^3\right)$$

Hence the general solution is

$$z = \phi_1(y-x) + e^{-2x}\phi_2(y+2x) - \frac{1}{10}e^{2x+3y} - \frac{1}{6}\cos(2x+y) + \frac{1}{24}\left(6x^2y - 6xy + 9x^2 - 12x - 2x^3\right)$$

which is a required solution, where ϕ_1 and ϕ_2 are arbitrary functions.

Example:

Find the solution of $\left(D+D^{'}+1\right)\left(2+D^{'}\right)z = e^{x-3y}$

Solution:

Here, $\left(D+D^{'}+1\right)\left(2+D^{'}\right)z = e^{x-3y}$

For complementary functions $\left(D+D^{'}+1\right)\left(2+D^{'}\right)z = e^{x-3y}$

Therefore, the complementary function is

$$C.F. = e^{-x}\phi_1(y-x) + e^{-2y}\phi_2(x)$$

For particular integral

$$P.I. = \frac{1}{(D+D^{'}+1)(2+D^{'})}e^{x-3y}$$

$$= \frac{1}{(2-3)(1-3+1)}e^{x-3y} = e^{x-3y}$$

Hence the general solution is

$$z = e^{-x}\phi_1(y-x) + e^{-2y}\phi_2(x) + e^{x-3y}$$

which is a required solution, where ϕ_1 and ϕ_2 are arbitrary functions.

Example:

Find the solution of $(D+D^{'}+2)z=\frac{1}{1+e^{x+y}}$

Solution:

Here, $(D+D^{'}+2)z=\frac{1}{1+e^{x+y}}$

For complementary functions $(D+D^{'}+2)z=0$

Therefore, the complementary function is given by C.F $=e^{-2x}\phi_1(y-x)$.

For particular integral

$$P.I.=\frac{1}{(D+D^{'}+2)z}\left(\frac{1}{1+e^{x+y}}\right)$$

$$=e^{-2x}\int e^{2x}\frac{1}{1+e^{x}e^{a+x}}dx, a=y-x$$

$$=e^{-2x}\int\frac{e^{2x}}{1+e^{a}e^{2x}}dx$$

$$=e^{-2x}\frac{1}{2e^{y-x}}\ln(1+e^{x+y})(\because a=y-x)$$

$$=\frac{1}{2e^{y+x}}\ln(1+e^{x+y})$$

Hence the general solution is

$$z = e^{-2x}\phi_1(y-x) + \frac{1}{2e^{y+x}}\ln(1+e^{x+y})$$

which is a required solution, where ϕ_1 is arbitrary function.

Exercise

Solve the following:

\# $\dfrac{\partial^2 z}{\partial x^2} - 4\dfrac{\partial^2 z}{\partial x \partial y} + 4\dfrac{\partial^2 z}{\partial y^2} = e^{2x+y}$

Ans:

$$z = \phi_1(y+2x) + x\phi_2(y+2x) + \frac{x^2}{2}e^{2x+y}$$

\# $\dfrac{\partial^2 z}{\partial x^2} - 5\dfrac{\partial^2 z}{\partial x \partial y} + 6\dfrac{\partial^2 z}{\partial y^2} = e^{x+y}$

Ans:

$$z = \phi_1\left(y+2x\right) + \phi_2\left(y+3x\right) + \frac{1}{2}e^{x+y}$$

\# $\dfrac{\partial^3 z}{\partial x^3} - 2\dfrac{\partial^3 z}{\partial x^2 \partial y} = 2e^{2x-y}$

Ans:

$$z = \phi_1\left(y\right) + x\phi_2\left(y\right) + \phi_3\left(y+2x\right) + \frac{1}{8}e^{2x-y}$$

\# $\dfrac{\partial^2 z}{\partial x^2} - \dfrac{\partial^2 z}{\partial y^2} - 2\dfrac{\partial z}{\partial x} + 2\dfrac{\partial z}{\partial y} = e^{2x+3y}$

Ans:

$$z = \phi_1(y+x) + e^{2x}\phi_2(y-x) - \frac{1}{3}e^{2x+3y}$$

$\frac{\partial^2 z}{\partial x^2}-5\frac{\partial^2 z}{\partial x\partial y}+6\frac{\partial^2 z}{\partial y^2}=e^{3x-2y}$

Ans:

$$z=\phi_1\left(y+2x\right)+\phi_2\left(y+3x\right)+\frac{1}{63}e^{3x-2y}$$

$\frac{\partial^2 z}{\partial x^2}-2\frac{\partial^2 z}{\partial x\partial y}+\frac{\partial^2 z}{\partial y^2}=\sin(2x+3y)$

Ans:

$$z=\phi_1\left(y-x\right)+x\phi_2\left(y-x\right)-\frac{1}{25}\sin(2x+3y)$$

$\left(D^2-DD'\right)z=\cos x\cos 2y$

Ans:

$$z=\phi_1\left(y\right)+\phi_2\left(y+x\right)+\frac{1}{2}\cos(x+2y)-\frac{1}{6}\cos(x-2y)$$

$\left(D^3-7D\left(D'\right)^2-6\left(D'\right)^3\right)z=\sin(x+2y)+e^{3x+y}$

Ans:

$$z=\phi_1\left(y-x\right)+\phi_2\left(y-2x\right)+\phi_3\left(y+3x\right)-\frac{1}{75}\cos(x+2y)+\frac{x}{20}e^{3x+y}$$

$\left(D^2-DD'\right)z=\cos 2y(\sin x+\cos x)$

Ans:

$$z=\phi_1\left(y\right)+\phi_2\left(y+x\right)+\frac{1}{2}\left[\sin(x+2y)+\cos(x+2y)\right]$$

$$-\frac{1}{6}\left[\sin(x-2y)+\cos(x-2y)\right]$$

$\frac{\partial^2 z}{\partial x^2}-3\frac{\partial^2 z}{\partial x\partial y}+2\frac{\partial^2 z}{\partial y^2}=e^{2x-y}+e^{x+y}+\cos(x+2y)$

Ans:

$$z=\phi_1(y+x)+\phi_2(y+2x)+\frac{1}{12}e^{2x-y}$$

$$-xe^{x+y}-\frac{1}{3}\cos(x+2y)$$

$\frac{\partial^2 z}{\partial x^2}+3\frac{\partial^2 z}{\partial x\partial y}-4\frac{\partial^2 z}{\partial y^2}=x+\sin y$

Ans:

$$z=\phi_1(y+x)+\phi_2(y-4x)+\frac{x^3}{6}+\frac{1}{4}\sin y$$

$\left(D^3-4D^2D'+5D(D')^2-2(D')^3\right)z=(x+y)^{\frac{1}{2}}$

$+e^{2x+y}$

Ans:

$$z=(1+x)\phi_1(y+x)+\phi_3(y+2x)-\frac{x^3}{3}(y+x)^{\frac{3}{2}}+xe^{x+y}$$

$\frac{\partial^3 z}{\partial x^3}-4\frac{\partial^3 z}{\partial x^2\partial y}+4\frac{\partial^3 z}{\partial x\partial y^2}=2\sin(3x+2y)$

Ans:

$$z=\phi_1(y)+\phi_2(2y+x)+x\phi_3(2y+x)+\frac{2}{3}\cos(3x+2y)$$

$\frac{\partial^2 z}{\partial x^2}-\frac{\partial^2 z}{\partial x\partial y}-6\frac{\partial^2 z}{\partial y^2}=xy$

Ans:

$$z=\phi_1(y-2x)+\phi_2(y+3x)+\frac{x^3 y}{6}+\frac{x^4}{24}$$

$\frac{\partial^2 z}{\partial x^2} - 3\frac{\partial z}{\partial x} + 3\frac{\partial z}{\partial y} - \frac{\partial^2 z}{\partial y^2} = xy + e^{x+2y}$

Ans:

$$z = \phi_1(y+x) + e^{3x}\phi_2(y-x) - \frac{1}{3}\left(\frac{x^2 y}{2} + \frac{x^3}{6} + \frac{x^2}{3} + \frac{xy}{3} + \frac{2x}{9}\right)$$

$$-xe^{x+2y}$$

$\frac{\partial^3 z}{\partial x^3} - \frac{\partial^3 z}{\partial y^3} = x^3 y^3$

Ans:

$$z = \phi_1(y+x) + \phi_2(y+wx) + \phi_3(y+w^2x)$$

$$+\frac{x^6 y^3}{120} + \frac{x^9}{10080}$$

Hint: A.F.is $m^3 - 1$ or $m = 1, w, w^2$ where w, w^2 are complex cube roots of unity.

$\left(D^3 - 3D^2 D'\right)z = x^2 y$

Ans:

$$z = \phi_1(y+x) + \phi_2(y-x) + \frac{1}{12}e^{2x-y} - xe^{x+y} - \frac{1}{3}\cos(x+2y)$$

$\left(D^2 - (D')^2 + D + 3D' - 2\right)z = e^{x-y} - x^2 y$

Ans:

$$z = e^{-2x}\phi_1(y+x) + e^{x}\phi_2(y-x) - \frac{1}{4}e^{x-y}$$

$$+\frac{1}{2}\left(x^2 y + xy + \frac{3x^2}{2} + \frac{3y}{2} + 3x + \frac{21}{4}\right)$$

$\left(D - D'\right)\left(D + 2D'\right)z = (x+y)e^{x}$

Ans:

$$z=\phi_1(y+x)+\phi_2(y-2x)+ye^x$$

$\left(D^2+DD'-6(D')^2\right)z=x^2\sin(x+y)$

Ans:

$$z=\phi_1(y+2x)+\phi_2(y-3x)+\frac{1}{4}\left(x^2-\frac{13}{8}\right)\sin(y+x)$$

$$-\frac{3}{8}x\cos(x+y)$$

$(D-D')(D+2D')z=(1+y)e^x$

Ans:

$$z=\phi_1(y+x)+\phi_2(y-2x)+ye^x$$

$\left(D^2-DD'-2(D')^2\right)z=\left(2x^2+xy-y^2\right)\sin xy-\cos xy$

Ans:

$$z=\phi_1(y+2x)+\phi_2(y-x)+\sin xy$$

$\left(D^2-6DD'+9(D')^2\right)z=12x^2+36xy$

Ans:

$$z=\phi_1(y+3x)+x\phi_2(y+3x)+10x^4+6x^3y$$

$\left(D^3-7D(D')^2-6(D')^3\right)z=0$

Ans:

$$z=\phi_1(y-x)+\phi_2(y-2x)+\phi_3(y+3x)$$

$25r+16t-40s=0$

Ans:

$$z=\phi_1(5y+4x)+x\phi_2(5y+4x)$$

$2\frac{\partial^2 z}{\partial x^2}+5\frac{\partial^2 z}{\partial x\partial y}+2\frac{\partial^2 z}{\partial y^2}=0$

Ans:

$$z=\phi_1(y-2x)+\phi_2(2y-x)$$

$\left(D^2-DD'-6(D')^2\right)z=0$

Ans:

$$z=\phi_1(y+3x)+\phi_2(y-2x)$$

$\left(D^3-4D'D^2+3D(D')^2\right)z=0$

Ans:

$$z=\phi_1(y)+\phi_2(y+x)+\phi_3(y+3x)$$

$\left(D^3-6D'D^2+11DD'^2-6(D')^3\right)z=0$

Ans:

$$z=\phi_1(y+2x)+\phi_2(y+x)+\phi_3(y+3x)$$

$\left(D^3-2D'D^2+D(D')^2\right)z=0$

Ans:

$$z=\phi_1(y)+\phi_2(y+x)+x\phi_3(y+x)$$

$r+9t+6s=0$

Ans:

$$z=\phi_1(y-3x)+x\phi_2(y-3x)$$

$(D+2D')(D-3D')^2 z=0$

Ans:

$$z=\phi_1(y-2x)+\phi_2(y+3x)+x\phi_3(y+3x)$$

$\frac{\partial^2 z}{\partial x^2} - 4\frac{\partial^2 z}{\partial x \partial y} + 4\frac{\partial^2 z}{\partial y^2} = 0$

Ans:

$$z = \phi_1(y+2x) + x\phi_2(y+2x)$$

$\left(D^4 - 2D'D^3 + 2D(D')^3 - (D')^4\right)z = 0$

Ans:

$$z = \phi_1(y+x) + x\phi_2(y+x) + x^2\phi_3(y+x) + \phi_4(y-x)$$

$\left(D^4 - 2D'^2D^2 + (D')^4\right)z = 0$

Ans:

$$z = \phi_1(y+x) + x\phi_2(y+x) + x\phi_3(y-x) + \phi_4(y-x)$$

$\left(D^2 + 5DD' + 6D'^2\right)z = e^{x-y}$

Ans:

$$z = \phi_1(y-2x) + \phi_2(y-3x) + \frac{1}{2}e^{x-y}$$

$\left(D^3 - 3D^2D' + 4D'^3\right)z = e^{x+2y}$

Ans:

$$z = \phi_1(y-x) + \phi_2(y+2x) + x\phi_3(y+2x) + \frac{1}{27}e^{x+2y}$$

$r + 4t - 4s = e^{2x+y}$

Ans:

$$z = \phi_1(y+2x) + x\phi_2(y+2x) + \frac{1}{2}x^2e^{2x+y}$$

$r + t + 2s = e^{2x+3y}$

Ans:

$$z=\phi_1(y-x)+x\phi_2(y-x)+\frac{1}{25}e^{2x+3y}$$

$\left(D^3-4D'D^2+5D(D')^2-2(D')^3\right)z=e^{y-2x}+e^{y+2x}$
$+e^{y+x}$

Ans:

$$z=\phi_1(y+x)+x\phi_2(y+x)+\phi_3(y+2x)-\frac{1}{36}e^{y-2x}$$
$$+xe^{y+2x}-\frac{x^2}{2}e^{y+x}$$

$\frac{\partial^2 z}{\partial x^2}+3\frac{\partial^2 z}{\partial x\partial y}+2\frac{\partial^2 z}{\partial y^2}=x+y$

Ans:

$$z=\phi_1(y-x)+\phi_2(y-2x)+\frac{x^2y}{2}-\frac{x^3}{3}$$

$\left(D^2-DD'-6D'^2\right)z=x+y$

Ans:

$$z=\phi_1(y+3x)+\phi_2(y-2x)+\frac{x^2y}{2}+\frac{x^3}{3}$$

$\left(D^2+DD'-6D'^2\right)z=x+y$

Ans:

$$z=\phi_1(y-3x)+\phi_2(y+2x)+\frac{x^2y}{2}$$

$\frac{\partial^2 z}{\partial x^2}-\frac{\partial^2 z}{\partial y^2}=x^2+y^2$

Ans: $z=\phi_1(y+x)+\phi_2(y-x)+\frac{x^2}{6}(x^2+3y^2)$

$\left(D^2+D'^2\right)z=x^2y^2$

Ans:

$$z=\phi_1(y+ix)+\phi_2(y-ix)+\frac{1}{180}\left(15x^4y^2-x^6\right)$$

$\left(D^2+2DD'+D'^2\right)z=3x+2y$

Ans:

$$z=\phi_1(y-x)+x\phi_2(y-x)+x^2y-\frac{x^3}{6}$$

$\left(4D^3-3D^2D'+D'^3\right)z=6x^2y^2$

Ans:

$$z=\phi_1(2y+x)+\phi_2(y-x)+x\phi_3(2y+x)$$

$$+\frac{1}{40}x^5y^2+\frac{1}{1120}x^7$$

$\frac{\partial^3 z}{\partial x^3}-2\frac{\partial^3 z}{\partial x^2\partial y}=2e^{2x}+3x^2y$

Ans:

$$z=\phi_1(y)+x\phi_2(y)+\phi_3(y+2x)+\frac{15e^{2x}+3x^5y+x^6}{60}$$

$\frac{\partial^2 z}{\partial x^2}-\frac{\partial^2 z}{\partial x\partial y}=\sin x\cos 2y$

Ans:

$$z=\phi_1(y+x)+\phi_2(y)+\frac{\sin(x+2y)}{2}-\frac{\sin(x-2y)}{6}$$

\# $\left(2D^2-3DD'+D'^2\right)z=\sin(x-2y)$

Ans:

$$z=\phi_1(2y+x)+\phi_2(y+x)-\frac{\sin(x-2y)}{6}$$

\# $\left(D^2+DD'-6D'^2\right)z=\cos(2x+y)$

Ans:

$$z=\phi_1(y+2x)+\phi_2(y-3x)+\frac{x\sin(2x+y)}{5}$$

\# $\frac{\partial^2 z}{\partial x^2}+\frac{\partial^2 z}{\partial y^2}=\cos mx\cos ny+30(2x+y)$

Ans:

$$z=\phi_1(y+ix)+\phi_2(y-ix)-\frac{\cos mx\cos ny}{m^2+n^2}+(2x+y)^3$$

\# $p-2q=\sin(2y+x)$

Ans:

$$z=\phi_1(y+x)+\frac{\cos(2y+x)}{3}$$

\# $\frac{\partial^3 z}{\partial x^3}-7\frac{\partial^3 z}{\partial x\partial y^2}-6\frac{\partial^3 z}{\partial y^3}=\sin(x+2y)+x^2y$

Ans:

$$z=\phi_1(y-x)+\phi_2(y+3x)+\phi_3(y-2x)+\frac{x^5y}{60}-\frac{\cos(x+2y)}{75}$$

$\left(D^2-2DD'+D'^2\right)z=\sin x$

Ans:

$$z=\phi_1\left(y+x\right)+x\phi_2\left(y+x\right)-\sin x$$

$\left(2D^2+5DD'+3D'^2\right)z=ye^x$

Ans:

$$z=\phi_1\left(2y-3x\right)+\phi_2\left(y-x\right)+\frac{(2y-5)e^x}{4}$$

$4r-4s+t=16\log(x+2y)$

Ans:

$$z=\phi_1\left(2y+x\right)+x\phi_2\left(2y+x\right)+2x^2\log(x+2y)$$

$\frac{\partial^2 y}{\partial t^2}-a^2\frac{\partial^2 y}{\partial x^2}=E\sin pt$

Ans: $y=\phi_1\left(x-at\right)+\phi_2\left(x+at\right)-\frac{E\sin pt}{p^2}$

$\left(D^2-DD'-2D'^2\right)z=e^x(y-1)$

Ans:

$$z=\phi_1\left(y-x\right)+\phi_2\left(y+2x\right)+ye^x$$

$\left(D^2-DD'\right)z=\cos x\cos 2y$

Ans: $z=\phi_1\left(y\right)+\phi_2\left(y+x\right)+\frac{\cos\left(x+2y\right)}{2}-\frac{\cos\left(x-2y\right)}{6}$

$\left(D^3 - 7DD'^2 - 6D'^3\right)z = x^2 + xy^2 + y^3 + \cos(x-y)$

Ans:

$$z = \phi_1\left(y-x\right) + \phi_2\left(y-2x\right) + \phi_3\left(y+3x\right) + \frac{x\cos(x-y)}{4} + \frac{5x^6}{72} + \frac{x^5}{60} + \frac{7x^5y}{50} + \frac{x^4y^2}{24} + \frac{x^3y^3}{6}$$

$\left(D^2 - D'^2\right)z = \tan^3 x \tan y - \tan^3 y \tan x$

Ans:

$$z = \phi_1\left(y-x\right) + \phi_2\left(y+x\right) + \frac{\tan x \tan y}{2}$$

$\frac{\partial^2 z}{\partial x^2} - \frac{\partial^2 z}{\partial y^2} = e^{x+2y}$

Ans: $z = \phi_1(y+x) + x\phi_2(y-x) - \frac{1}{3}e^{x+2y}$

$\left(D^2 - DD' + D'\right)z = x^2 + y^2$

Ans:

$$z = \phi_1\left(y\right) + e^{-x}\phi_2\left(y+x\right) + \frac{x^3}{3} - x^2 + xy^2 + 6x$$

$\left(D^2 + D + D'\right)z = 0$

Ans:

$z = ce^{ax-\left(a^2+a\right)y}$ where a and c are constants

$\left(D'^2 + 2DD' - 3D'\right)z = 3\cos(3x-2y)$

Ans:

$$z=\phi_1(x)+e^{3y}\phi_2(2y-x)$$
$$+\frac{3}{50}\left[3\sin(3x-2y)+4\cos(3x-2y)\right]$$

\# $r-s+p=1$

Ans:
$$z=\phi_1(y)+e^{-x}\phi_2(y+x)+x$$

\# $\left(D^2-D'-1\right)z=x^2y$

Ans:
$$z=c_1e^{ax+\left(a^2-1\right)y}-x^2y+x^2-2y+8$$
where a and c are constants

3.9 Partial Differential Equations with Variable Coefficients Reducible to Partial Differential Equations with Constant Coefficients

Consider the PDE of the form $f(xD,yD')z=F(x,y)$.

Let $|x|=e^X,|y|=e^Y$ $i.e\ X=\log|x|,Y=\log|y|$.

$$\frac{\partial z}{\partial x}=\frac{\partial z}{\partial X}\frac{\partial X}{\partial x}=\frac{1}{x}\frac{\partial X}{\partial x}\Rightarrow xD=D_X$$

$$\frac{\partial^2 z}{\partial x^2}=-\frac{1}{x^2}\frac{\partial z}{\partial X}+\frac{1}{x}\frac{\partial^2 z}{\partial X^2}\frac{dX}{dx}\Rightarrow\frac{\partial^2 z}{\partial x^2}=-\frac{1}{x^2}\frac{\partial z}{\partial X}+\frac{1}{x^2}\frac{\partial^2 z}{\partial X^2}$$

$$\Rightarrow x^2D^2=D_X(D_X-1)$$

Similarly, $x^3D^3=D_X(D_X-1)(D_X-2)$

$yD' = D_Y, y^2D'^2 = D_Y(D_Y - 1), y^3D'^3 = D_Y(D_Y - 1)(D_Y - 2)$ and so on.

Hence the differential equation will be reduced to partial equation with constant coefficients and be solved by methods already discussed.

Example:

Find the general solutions of the following partial differential equations:

(i) $x^2r - 3xys + 2y^2t + px + 2qy = x + 2y$

(ii) $\left(x^2D^2 - y^2D'^2 + xD - yD'\right)z = \log x$

(iii) $\frac{1}{x^2}\frac{\partial^2 z}{\partial x^2} - \frac{1}{x^3}\frac{\partial z}{\partial x} = \frac{1}{y^2}\frac{\partial^2 z}{\partial y^2} - \frac{1}{y^3}\frac{\partial z}{\partial y}$

Solution: (i)

Here differential equations in symbolic form are

$$\left(x^2D_x^2 - 3xyD_xD_y + 2y^2D_y^2 + xD_y\right)z = x + 2y$$

$$|x| = e^X, |y| = e^Y \ \ i.e\ X = \log|x|, Y = \log|y|$$

We have

$$xD_x = D_X, yD_y = D_Y, x^2D_x^2 = D_X(D_X - 1), \text{xy}\, D_xD_y = D_XD_Y, \text{y}^2\, D_y^2 = D_Y(D_Y - 1)$$

∴ The given differential equation becomes

$$\left[D_X(D_X - 1) - 3D_XD_Y + 2D_Y(D_Y - 1) + D_X + 2D_Y\right]z = e^X + 2e^Y$$

if $x > 0, y > 0$

$= -e^X + 2e^Y$ if $x < 0, y > 0$

$= -e^X - 2e^Y$ if $x < 0, y < 0$

$= e^X - 2e^Y$ if $x > 0, y < 0$

$\therefore \left[D_X^{\;2} - 3D_X D_Y + 2D_Y^{\;2}\right]z = e^X + 2e^Y \text{ if } x > 0, y > 0$

or $(D_X - D_Y)(D_X - 2D_Y)z = e^X + 2e^Y$

A.E. is $(m-1)(m-2) = 0 \Rightarrow m = 1, 2$

$\therefore \text{CF} = \phi_1(Y+2X) + \phi_2(Y+X) = \phi_1(\log|y| + 2\log|x|) + \phi_2(\log|y| + \log|x|)$

$\therefore \text{CF} = \phi_1\left(\log\left|x^2y\right|\right) + \phi_2\left(\log|xy|\right) = f_1\left(x^2y\right) + f_2(xy)$

$$\text{PI} = \frac{1}{D_X^{\;2} - 3D_X D_Y + 2D_Y^{\;2}}\left(e^X + 2e^Y\right) = \frac{1}{1-0+0}e^X + \frac{2}{0-0+2}e^Y = e^X + e^Y$$

$PI = e^X + e^Y = x + y \text{ if } x > 0, y > 0$

$= -e^X + e^Y = x + y \text{ if } x < 0, y > 0$

$= e^X - e^Y = x + y \text{ if } x > 0, y < 0$

$= -e^X - e^Y = x + y \text{ if } x < 0, y < 0$

Hence in each case $\text{PI} = x + y$

Hence, the general solution is

$z = f_1\left(x^2y\right) + f_2(xy) + x + y$

which is a required solution, where f_1 and f_2 are arbitrary functions.

Solution: (ii)

Here differential equations are

$$\left(x^2D^2 - y^2D'^2 + xD - yD'\right)z = \log x$$

put $|x| = e^X, |y| = e^Y$ $i.e\ X = \log|x|, Y = \log|y|$

We have

$$xD = D_X, yD' = D_Y, x^2D^2 = D_X(D_X - 1), xyDD' = D_XD_Y, y^2D'^2 = D_Y(D_Y - 1)$$

$\therefore$ differential equation becomes

$$\left[D_X(D_X - 1) - D_Y(D_Y - 1) + D_X - D_Y\right]z = X \text{ if } x > 0$$

or $\left(D_X{}^2 - D_Y{}^2\right)z = X$

A.E. is $(m^2 - 1) = 0 \Rightarrow m = 1, -1$

$$\therefore \text{CF} = \phi_1(Y - X) + \phi_2(Y + X) = \phi_1(\log|y| - \log|x|) + \phi_2(\log|y| + \log|x|)$$

$$\therefore \text{CF} = \phi_1\left(\log\left|\frac{y}{x}\right|\right) + \phi_2\left(\log|xy|\right) = f_1\left(\frac{y}{x}\right) + f_2(xy)$$

$$\text{PI} = \frac{1}{D_X{}^2 - D_Y{}^2}X = \frac{X^3}{2.3} = \frac{X^3}{6} = \frac{(\log x)^3}{6}$$

Hence the general solution is

$$z = f_1\left(\frac{y}{x}\right) + f_2(xy) + \frac{(\log x)^3}{6}$$

which is a required solution, where f_1 and f_2 are arbitrary functions.

Solution: (iii)

Here differential equations are

$$\frac{1}{x^2}\frac{\partial^2 z}{\partial x^2}-\frac{1}{x^3}\frac{\partial z}{\partial x}=\frac{1}{y^2}\frac{\partial^2 z}{\partial y^2}-\frac{1}{y^3}\frac{\partial z}{\partial y}$$

put $\frac{x^2}{2}=X, \frac{y^2}{2}=Y$

$$\therefore \frac{\partial z}{\partial x}=\frac{\partial z}{\partial X}\frac{dX}{dx}=x\frac{\partial z}{\partial X},$$

$$\therefore \frac{\partial^2 z}{\partial x^2}=x\frac{\partial^2 z}{\partial X^2}\frac{dX}{dx}+\frac{\partial z}{\partial X}=x^2\frac{\partial^2 z}{\partial X^2}+\frac{\partial z}{\partial X}$$

$$\therefore \frac{1}{x^2}\frac{\partial^2 z}{\partial x^2}-\frac{1}{x^3}\frac{\partial z}{\partial x}=\frac{\partial^2 z}{\partial X^2}=D_X{}^2 z$$

Similarly, $\frac{1}{y^2}\frac{\partial^2 z}{\partial y^2}-\frac{1}{y^3}\frac{\partial z}{\partial y}=D_Y{}^2 z$

$\therefore$ differential equation becomes

$$\left(D_X{}^2-D_Y{}^2\right)z=0$$

A.E. is $(m^2-1)=0 \Rightarrow m=1,-1$

$$\therefore \text{CF}=\phi_1(Y-X)+\phi_2(Y+X)=\phi_1\left(\frac{y^2-x^2}{2}\right)+\phi_2\left(\frac{y^2+x^2}{2}\right)$$

$$\therefore \text{CF}=f_1\left(x^2-y^2\right)+f_2\left(y^2+x^2\right)$$

which is a required solution, where f_1 and f_2 are arbitrary functions.

Example:

$$(D-D'-1)(D-D'-2)z=e^{2x-y}$$

Solution:

The factors of $\phi(D,D')$ in the given non-homogeneous partial differential are already of the type $a_iD+b_iD'+c_i$ .so we use method I in this case. In this case $a_1=1,b_1=-1,c_1=-1$ and $a_2=1,b_2=-1,c_2=-2$

$$\therefore \text{CF}=e^{-\frac{c_1}{a_1}}f_1(a_1y-b_1x)+e^{-\frac{c_2}{a_2}}f_2(a_2y-b_2x)=e^xf_1(x+y)+e^{2x}f_2(x+y)$$

$$\text{PI}=\frac{1}{(D-D'-1)(D-D'-2)}e^{2x-y}=\frac{1}{(2-(-1)-1)(2-(-1)-2)}e^{2x-y}=\frac{1}{2}e^{2x-y}$$

Hence the general solution is

$$z=e^xf_1(x+y)+e^{2x}f_2(x+y)+\frac{1}{2}e^{2x-y}$$

which is a required solution, where f_1 and f_2 are arbitrary functions.

Example:

Find the solutions of $\left(D^2-DD'+D'-1\right)z=\cos(x+2y)$

Solution:

We have $\phi(D,D')=D^2-DD'+D'-1=(D-1)(D-D'+1)$

In this case $a_1=1,b_1=0,c_1=-1$ and $a_2=1,b_2=-1,c_2=1$

$$\therefore \text{CF} = e^{-\frac{c_1}{a_1}} f_1(a_1 y - b_1 x) + e^{-\frac{c_2}{a_2}} f_2(a_2 y - b_2 x) = e^x f_1(y) + e^{-x} f_2(x+y)$$

$$\text{PI} = \frac{1}{D^2 - DD' + D' - 1} \cos(x+2y)$$

$$= \frac{1}{-1+2+D'-1} \cos(x+2y)$$

$$= \frac{1}{D'} \cos(x+2y) = \int \cos(x+2y)dy = \frac{1}{2}\sin(x+2y)$$

Hence the general solution is

$$z = e^x f_1(y) + e^{-x} f_2(x+y) + \frac{1}{2}\sin(x+2y)$$

which is a required solution, where f_1 and f_2 are arbitrary functions.

Example:

Find the solutions of $\left(D^3 - 3DD' + D' + 4\right)z = e^{2x+y}$

Solution:

We have $\phi\left(D, D'\right) = D^3 - 3DD' + D' + 4$ is irreducible (does not have linear factor) so we use method II in this case we have $\phi\left(a, b\right) = a^3 - 3ab + 4 = 0$

Being a cubic in a it has three values of a in terms of b.

Let these values be $f_1(b), f_2(b)$ and $f_3(b)$. Then

$$\therefore \text{CF} = c_1 e^{f_1(b)x+y} + c_2 e^{f_2(b)x+by} + c_3 e^{f_3(b)x+by}$$

$$\text{PI} = \frac{1}{D^3 - 3DD' + D' + 4} e^{2x+y} = \frac{1}{8-6+1+4} e^{2x+y} = \frac{1}{7} e^{2x+y}$$

Hence the general solution is

$$z = c_1 e^{f_1(b)x+y} + c_2 e^{f_2(b)x+by} + c_3 e^{f_3(b)x+by} + \frac{1}{7} e^{2x+y}$$

which is a required solution, where f_1, f_2 and f_3 are arbitrary functions.

Example:

$$\left(2D^4 - 3D^2D' + D'^2\right) z = 0$$

Solution:

We have

$$\phi(D,D') = 2D^4 - 3D^2D' + D'^2 = \left(2D^2 - D'\right)\left(D^2 - D'\right) = \phi_1(D,D')\phi_2(D,D')$$

Since $2D^2 - D'$ is irreducible (does not have linear factor

$$\phi_1(a,b) = 2a^2 - b = 0 \Rightarrow b = 2a^2$$

So, contribution to CF due to $\phi(D,D')z = 0$ is $c_1 e^{ax+2a^2y}$

Again since $D^2 - D'$ is irreducible

$$\phi_2(a',b') = a'^2 - b' = 0 \Rightarrow b' = a'^2$$

So, contribution to CF due to $\phi(D,D')z = 0$ is $c_2 e^{a'x+a'^2y}$

$\therefore$ general solution is

$$z = c_1 e^{ax+2a^2y} + c_2 e^{a'x+a'^2y}$$

which is a required solution, where c_1 and c_2 are arbitrary constants.

Exercise

Note: Here ϕ_1, ϕ_2, ϕ_3 and ϕ_4 are arbitrary functions.

Solve the following:

$r - t + p - q = 0$

Ans: $z = \phi_1(y+x) + e^{-x}\phi_2(y-x)$

$(D+2D'-3)(D+D'-1)z = 0$

Ans: $z = e^{x}\phi_1(y-x) + e^{3x}\phi_2(y-2x)$

$(D_x + 3D_y + 4)^2 z = 0$

Ans: $z = e^{-4x}\phi_1(y-3x) + xe^{-4x}\phi_2(y-3x)$

$(D+2D')(D+3D'+1)(D+2D'+2)^2 z = 0$

Ans:

$z = \phi_1(y-2x) + e^{-x}\phi_2(y-3x) + e^{-x}[\phi_3(y-2x) + x\phi_3(y-2x)]$

$(D^2 + DD' - D'^2 + D - D')z = 0$

Ans: $z = \sum_{n=1}^{\infty} c_n e^{a_n x + b_n y}$ where $a_n^2 + a_n b_n - b_n^2 + a_n - b_n = 0$

a_n, b_n and c_n are arbitrary constants.

$(D-D'-2)(D-D'-3)z = e^{3x-2y}$

Ans: $z = e^{2x}\phi_1(y+x) + e^{3x}\phi_2(y+x) + \frac{1}{6}e^{3x-2y}$

$(2D^2 - DD' - D'^2 + D - D')z = e^{2x+3y}$

Ans: $z = \phi_1(y+x) + e^{-\frac{x}{2}}\phi_2(2y-x) - \frac{1}{8}e^{2x+3y}$

$\left(D^2 - DD' + D' - 1\right)z = \cos(x+2y) + e^y$

Ans: $z = e^x\phi_1\left(y\right) + e^{-x}\phi_2\left(y+x\right) + \frac{1}{2}\sin(x+2y) - xe^y$

$\left(2D^2 + 3DD' + D'^2 + D + D'\right)z = x - y$

Ans: $z = \phi_1\left(y-x\right) + e^{-\frac{x}{2}}\phi_2\left(2y-x\right) + x^2 - x(y+1)$

$\left(D^2 - D'^2 + D + 3D' - 2\right)z = x^2y$

Ans:

$$z = e^{-2x}\phi_1\left(y+x\right) + e^x\phi_2\left(y-x\right) - \frac{1}{2}\left(x^2y + \frac{3}{2}x^2 + xy + \frac{3}{2}y + 3x + \frac{21}{4}\right)$$

$\left(D_x - D_y - 1\right)\left(D_x - D_y - 2\right)z = e^{2x-y} + x$

Ans: $z = e^{2x}\phi_1\left(y+x\right) + e^x\phi_2\left(y+x\right) + \frac{1}{2}e^{2x-y} + \frac{x}{2} + \frac{3}{4}$

$\left(D^3 - 3DD' + D' + 4\right)z = e^{2x+y}$

Ans:

$$z = \sum_{n=1}^{\infty} c_n e^{a_n x + b_n y} + \frac{1}{7}e^{2x+y} \text{ where } a_n^{\ 3} - 3a_n b_n + b_n + 4 = 0$$

a_n, b_n and c_n are arbitrary constants

$\left(D - 3D' - 2\right)^2 z = 2e^{2x}\sin(y+3x)$

Ans: $z = e^{2x}\phi_1\left(y+3x\right) + xe^{2x}\phi_2\left(y+3x\right) + x^2e^{2x}\sin(y+3x)$

$\left(D^2 + 2DD' + D'^2 - 2D - 2D'\right)z = \sin(x+2y)$

Ans:

$$z = \phi_1\left(y-x\right) + e^{2x}\phi_2\left(y-x\right) + \frac{\left[2\cos(x+2y) - 3\sin(x+2y)\right]}{39}$$

$(D+1)\left(D + D' - 1\right)z = \sin(x+2y)$

Ans: $z=e^{-x}\phi_1(y)+e^{x}\phi_2(y-x)-\dfrac{[\cos(x+2y)+2\sin(x+2y)]}{10}$

$(D+D'+1)z=e^{-x}\tan(x+2y)$

Ans: $z=e^{-x}\phi(y-x)+\dfrac{e^{-x}\ln|\sec(x+2y)|}{3}$.

$(x^2D^2+2xyDD'+y^2D'^2)z=x^m y^n$

Ans: $z=\phi_1\left(\dfrac{y}{x}\right)+x\phi_2\left(\dfrac{y}{x}\right)+\dfrac{x^m y^n}{(m+n)(m+n-1)}$

$(x^2D_x{}^2-y^2D_y{}^2)z=x^2y$

Ans: $z=\phi_1(xy)+x\phi_2\left(\dfrac{y}{x}\right)+\dfrac{x^2y}{2}$

$(x^2D^2-4xyDD'+4y^2D'^2+6yD')z=x^3y^4$

Ans: $z=\phi_1(x^2y)+x\phi_2(x^2y)+\dfrac{x^3y^4}{30}$

$x^2\dfrac{\partial^2 z}{\partial x^2}+2xy\dfrac{\partial^2 z}{\partial x\partial y}+y^2\dfrac{\partial^2 z}{\partial y^2}-nx\dfrac{\partial z}{\partial x}-ny\dfrac{\partial z}{\partial y}$

$+nz=x^2+y^2$

Ans: $z=x\phi_1\left(\dfrac{y}{x}\right)+x^n\phi_2\left(\dfrac{y}{x}\right)+\dfrac{x^2+y^2}{2-n}$

$(xD^3D'^2-yD^2D'^3)z=0$

Ans: $z=\phi_1(x)+\phi_2(y)+x\phi_3(y)+y\phi_4(x)+\phi_5(xy)$

$\left(x^2D^2 - 2xyDD' - 3y^2D'^2 + xD - 3yD'\right)z = x^2y$

Ans: $z = \phi_1\left(x^3y\right) + \phi_2\left(\frac{y}{x}\right) - \frac{x^2y}{3}$

Chapter 4

Solution Method for Higher Order PDE

4.1 Introduction

The variable separation approach is a versatile method that can be used to solve a wide range of PDE problems. It is especially useful for issues with certain symmetries or separability in their variables. The approach frequently simplifies the solution process by dividing a complex PDE into a set of ordinary differential equations (ODEs) or simpler PDEs. This makes it easier to solve problems and understand the behaviour of the system being modelled. The method separates the variables into independent components, allowing for a clear physical understanding of the solution. This separation is frequently related to physical processes or dimensions, allowing for a more intuitive comprehend of the situation.

4.2 Method of separation of Variable

It involves a solution which breaks up into a product of functions each of which contains only one of the variables.

Outline of the method of Separation of variables:

We are going to solve this problem in three steps.

Step-1: In the first step, we find all solutions of given equation that are of the special form $u(x,t) = X(x)T(t)$ for some function $X(x)$ that depends on x but not t and some function $T(t)$ that depends on t but not x. There is where the name "separation of variables" comes from. It is of course too much to except that all solutions of given equation is of this form. But if we find a bunch of solutions $X_i(x)T_i(t)$ of this form, then since given equation is a linear equation, $\sum_i a_i X_i(t)T_i(t)$ is also a solution for any choice of the constants a_i. If we are lucky, we will be able to choose the constants a_i so that the other conditions are also satisfied.

Step-2: We impose the boundary conditions.

Step-3: We impose the initial conditions.

Example:

Solve $2x\frac{\partial z}{\partial x} - 3y\frac{\partial z}{\partial y} = 0$ by method of separation of variables.

Solution:

Here partial differential equation is given by (4.1)

$$2x\frac{\partial z}{\partial x} - 3y\frac{\partial z}{\partial x} = 0$$

Assume that solution of Eq. (4.1) is of the form

$$z(x, y) = X(x) \, . \, Y(y), \tag{4.2}$$

where X is a function of x and Y is a function of y alone.

$$\therefore z = XY \tag{4.3}$$

Differentiating (4.3) w.r.t. x and y respectively

$$\therefore \frac{\partial z}{\partial x} = X'Y \tag{4.4}$$

$$\frac{\partial z}{\partial y} = XY' \tag{4.5}$$

where $X' = \frac{dX}{dx}$ and $Y' = \frac{dY}{dy}$. Using Eq. (4.4) and (4.5) in Eq. (4.1), we have

$$\therefore 2xX'Y - 3yXY' = 0 \Rightarrow 2xX'Y = 3yXY' \tag{4.6}$$

Separating the variables

$$\therefore \frac{2xX'}{X} = \frac{3yY'}{Y} \tag{4.7}$$

In (4.7) , L.H.S is a function of x only and R.H.S is a function of y only and here x and y are independent variables, so each side is constant. So rewriting (4.7) as

$$\therefore \frac{2xX'}{X} = \frac{3yY'}{Y} = k, \tag{4.8}$$

$$\therefore \frac{2xX'}{X} = k \Rightarrow \frac{X'}{X} = \frac{k}{2x}, \tag{4.9}$$

$$\therefore \frac{3yY'}{Y} = k \Rightarrow \frac{Y'}{Y} = \frac{k}{3y}. \tag{4.10}$$

Integrating Eq. (4.9) w.r.t. x we have

$$\ln X = \frac{k}{2}\ln x + \ln C_1, \tag{4.11}$$

where $\ln C_1$ is a constant.

$$\therefore \ln X = \ln x^{\frac{k}{2}} + \ln C_1 \Rightarrow X = C_1 x^{\frac{k}{2}} \tag{4.12}$$

Integrating Eq. (4.10) w.r.t. y ,we have

$$\ln Y = \frac{k}{3}\ln y + \ln C_2 \tag{4.13}$$

where $\ln C_2$ is a constant.

$$\therefore \ln Y = \ln y^{\frac{k}{3}} + \ln C_2 \Rightarrow Y = C_2 y^{\frac{k}{3}} \tag{4.14}$$

Now using equations (4.12) and (4.14) in (4.2), we have

$$\therefore z = \left(C_1 x^{\frac{k}{2}}\right)\left(C_2 y^{\frac{k}{3}}\right) = C_1 C_2 x^{\frac{k}{2}} y^{\frac{k}{3}} = C x^{\frac{k}{2}} y^{\frac{k}{3}}, \tag{4.15}$$

where $C = C_1 C_2$, is the required general solution of the given P.D.E equation (4.1).

Example:

Applying the method of separation of variables find the solution to the P.D.E. $3u_x + 2u_y = 0$ where $u_x = \frac{\partial u}{\partial x}, u_y = \frac{\partial u}{\partial y}$

Solution:

Here partial differential equation is given by

$$3\frac{\partial u}{\partial x} + 2\frac{\partial u}{\partial y} = 0. \tag{4.16}$$

Since u is a function of x and y.

Therefore, $u(x, y) = X(x)Y(y)$ (4.17)

Let $u(x, y) = X(x).Y(y)$ or $u = XY$ be the solution of (4.16), where X is a function of x only and Y is a function of y only. Differentiating (4.17) partially w.r.t. x and y respectively, we have

$$\frac{\partial u}{\partial x} = X'Y \tag{4.18}$$

$$\frac{\partial u}{\partial y} = XY' \tag{4.19}$$

where $X' = \frac{dX}{dx}$ and $Y' = \frac{dY}{dy}$.

Using equations (4.18) and (4.19) in (4.16), we have

$$3X'Y + 2XY' = 0 . \tag{4.20}$$

Separating the variables from Eq. (4.20), we get

$$3X'Y + 2XY' = 0 \Rightarrow 3X'Y = -2XY' \Rightarrow \frac{3X'}{X} = \frac{-2Y'}{Y} \tag{4.21}$$

In (4.21), L.H.S is a function of x only and R.H.S is a function of y only and here x and y are independent variables, so each side is constant. So rewriting (4.21) as

$$\therefore \frac{3X'}{X} = \frac{-2Y'}{Y} = k \tag{4.22}$$

From Eq. (4.22), we have $\frac{3X'}{X} = k$,

$$\Rightarrow 3X' - kX = 0 , \tag{4.23}$$

which is first order homogeneous linear differential equation. The auxiliary equation of (4.23) is given by

$$3m - k = 0 \Rightarrow m = \frac{k}{3}$$

So, the general solution of (4.23) is given by

$$\therefore X = C_1 e^{\frac{k}{3}x} \tag{4.25}$$

$\frac{-2Y'}{Y} = k$,

$$\Rightarrow 2Y' + kY = 0 , \tag{4.24}$$

which is first order homogeneous linear differential equation. The auxiliary equation of (4.24) is given by

$$2m + k = 0 \Rightarrow m = \frac{-k}{2}$$

So, the general solution of (4.24) is given by

$$\therefore Y = C_2 e^{\frac{-k}{2}y} \tag{4.26}$$

Using (4.25) and (4.26) in (4.17), we have $\therefore u = C_1 e^{\frac{k}{3}x} C_2 e^{\frac{-k}{2}y} = A e^{\frac{k}{3}x} e^{\frac{-k}{2}y} = A e^{k\left(\frac{x}{3}-\frac{y}{2}\right)}$, where $A = C_1 C_2$,which is the required general solution of the given P.D.E.

Example:

Using the method of separation of variables find the solution to the P.D.E. $\frac{\partial u}{\partial x} = 2\frac{\partial u}{\partial y} + u$ where $u(x,0) = 6e^{-3x}$.

Solution:

Here partial differential equation is given by

$$\frac{\partial u}{\partial x} = 2\frac{\partial u}{\partial y} + u. \tag{4.27}$$

Let $u(x,y) = X(x)Y(y)$, (4.28)

be the solution of (4.27) where X is a function of x only and Y is a function of y only.

Differentiating (4.28) partially w.r.t. x and y respectively, we have

$$\frac{\partial u}{\partial x} = X'Y \ , \tag{4.29}$$

$$\frac{\partial u}{\partial y} = XY', \tag{4.30}$$

where $X' = \frac{dX}{dx}$ and $Y' = \frac{dY}{dy}$. Using Eqs. (4.29) and (4.30) in (4.27), we have

$$X'Y = 2XY' + XY.$$

Separating the variables, we have

$$\therefore X'Y = 2XY' + XY \Rightarrow X'Y - XY = 2XY' \Rightarrow \frac{X' - X}{2X} = \frac{Y'}{Y} \tag{4.31}$$

In (4.31), L.H.S is a function of x only and R.H.S is a function of y only and here x and y are independent variables, therefore each side is constant. So rewriting (4.31) as

$$\frac{X'-X}{2X}=k \Rightarrow X'-X=2kX$$

$$\Rightarrow X'-X-2kX=0$$

$$\Rightarrow X'-X(1+2k)=0 \quad (4.32)$$

,

which is first order homogeneous linear differential equation.

The auxiliary equation of (4.30) is given by

$$m-1-2k=0 \Rightarrow m=2k+1.$$

So the general solution of (4.30) is given by

$$\therefore X=C_1e^{(2k+1)x} \quad (4.33)$$

$$\therefore \frac{Y'}{Y}=k \Rightarrow Y'-kY=0 \quad (4.34)$$

,

which is first order homogeneous linear differential equation. The auxiliary equation of (4.32) is given by

$$m-k=0 \Rightarrow m=k.$$

So , the general solution of (4.32) is given by

$$\therefore Y=C_2e^{ky} \quad (4.35)$$

Using equations (4.33) and (4.35) in (4.28), we have

$$u(x,y)=C_1e^{(2k+1)x}C_2e^{ky}=Ae^{(2k+1)x}e^{ky}=Ae^{(2k+1)x+ky} \quad (4.36)$$

where $A=C_1C_2$, which is the required general solution of the given P.D.E.

Putting $y=0$ in (4.36), we have

$$u(x,0)=Ae^{(2k+1)x} \quad (4.37)$$

For particular solution here it is given that
$u(x,0)=6e^{-3x}$

From (4.37) and (4.38), we have

$$\therefore 6e^{-3x}=Ae^{(2k+1)x}\Rightarrow A=6, 2k+1=-3\Rightarrow A=6, k=-2$$

Using (4.27) in (4.34) ,we have $u(x,y)=6e^{-3x-2y}\Rightarrow u(x,y)=6e^{-(3x+2y)}$, which is the required the particular solution to equation (4.27).

Example:

Using the method of separation of variables find the solution to $\frac{\partial u}{\partial x}=2\frac{\partial u}{\partial y}+u$ where $u(x,0)=3e^{-5x}-2e^{-3x}$.

Solution:

Here partial differential equation is given by

$$\frac{\partial u}{\partial x}=2\frac{\partial u}{\partial y}+u \quad . \tag{4.38}$$

Let $u(x,y)=X(x)Y(y)$ (4.39)

be the solution of (4.38), where X is a function of x only and Y is a function of y only. Differentiating partially (4.39) w.r.t. x and y respectively, we have

$$\frac{\partial u}{\partial x}=X'Y \tag{4.40}$$

$$\frac{\partial u}{\partial y}=XY' \tag{4.41}$$

where and $Y'=\frac{dY}{dy}$. Using equation (4.40) and equation (4.41) in equation (4.38), we have

$$X'Y=2XY'+XY$$

Separating the variables, we get

$$X'Y = 2XY' + XY \Rightarrow X'Y - XY = 2XY' \Rightarrow \frac{X'-X}{2X} = \frac{Y'}{Y}$$

In (4.41) , L.H.S is a function of x only and R.H.S is a function of y only and here x and y are independent variables, so each side is constant. So rewriting Eq. (4.41) as

$$\frac{X'-X}{2X} = k \,,$$

$$\Rightarrow X' - X = 2kX \,,$$

$$\Rightarrow X' - X - 2kX = 0,$$

$$\Rightarrow X' - X(1+2k) = 0 \qquad (4.42)$$

, which is first order homogeneous linear differential equation. The auxiliary equation of (4.42) is given by

$$m - 1 - 2k = 0 \Rightarrow m = 2k+1 \;.$$

So the general solution of (6) is given by

$$\therefore X = C_1 e^{(2k+1)x} \qquad (4.43)$$

$$\therefore \frac{Y'}{Y} = k \Rightarrow Y' - kY = 0 \qquad (4.44)$$

, which is first order homogeneous linear differential equation. The auxiliary equation of (4.43) is given by

$$m - k = 0 \Rightarrow m = k \,.$$

So the general solution of (4.43) is given by

$$Y = C_2 e^{ky} \qquad (4.45)$$

Using (4.43) and (4.45) in (4.39) ,we have $\therefore u(x,y) = C_1 e^{(2k+1)x} C_2 e^{ky} = C_1 C_2 e^{(2k+1)x+ky}$.

$$\therefore u(x,y) = \sum_{n=1}^{\infty} A_n e^{(2k_n+1)x+k_n y} = A_1 e^{(2k_1+1)x+k_1 y} + A_2 e^{(2k_2+1)x+k_2 y} \qquad (4.46)$$

where $A_n = C_1C_2$ and $k = k_n$,which is most general solution of the given P.D.E.

Putting $y = 0$ in (4.46) ,we have

$$\therefore u(x,0) = \sum_{n=1}^{\infty} A_n e^{(2k_n+1)x} = A_1 e^{(2k_1+1)x} + A_2 e^{(2k_2+1)x}. \tag{4.47}$$

For particular solution here given that

$$u(x,0) = 3e^{-5x} - 2e^{-3x}. \tag{4.48}$$

From (4.47) and (4.48), we have

$$\sum_{n=1}^{\infty} A_n e^{(2k_n+1)x} = 3e^{-5x} - 2e^{-3x} \Rightarrow A_1 e^{(2k_1+1)x} + A_2 e^{(2k_2+1)x} = 3e^{-5x} - 2e^{-3x} \tag{4.49}$$

$$\Rightarrow A_1 e^{(2k_1+1)x} = 3e^{-5x}, A_2 e^{(2k_2+1)x} = -2e^{-3x} 2,$$

$$\Rightarrow A_1 e^{(2k_1+1)x} = 3e^{-5x} \Rightarrow A_1 = 3, 2k_1 + 1 = -5 \Rightarrow A_1 = 3, k_1 = -3, \tag{4.50}$$

$$\Rightarrow A_2 e^{(2k_2+1)x} = -2e^{-3x} \Rightarrow A_2 = -2, 2k_2 + 1 = -3 \Rightarrow A_2 = -2, k_2 = -2 \tag{4.51}$$

Using (4.50) and (4.51) in (4.46), we have $u(x,y) = 3e^{-5x-3y} - 2e^{-3x-2y}$,which is the required particular solution.

Example:

Solve $\frac{\partial^2 z}{\partial x^2} - 2\frac{\partial z}{\partial x} + \frac{\partial z}{\partial y} = 0$ by method of separation of variables.

Solution:

Here partial differential equation is given by

$$\frac{\partial^2 z}{\partial x^2} - 2\frac{\partial z}{\partial x} + \frac{\partial z}{\partial y} = 0 \tag{4.52}$$

Let $z(x,y)=X(x)Y(y)$ (4.53)

be the solution of (4.52) where X is a function of x only and Y is a function of y only. Differentiating (4.53) partially w.r.t. x and y respectively, we have

$$\frac{\partial z}{\partial x}=X'Y, \quad (4.54)$$

$$\frac{\partial}{\partial x}\left(\frac{\partial z}{\partial x}\right)=\frac{\partial^2 z}{\partial x^2}=X''Y, \quad (4.55a)$$

$$\frac{\partial z}{\partial y}=XY', \quad (4.55b)$$

where $X'=\frac{dX'}{dx}, Y'=\frac{dY}{dy}$, and $X''=\frac{d^2X}{dx^2}$ .Using (4.54) to (4.55a&b) in (4.52) , we have

$$X''Y-2X'Y+XY'=0 . \quad (4.56)$$

Separating the variables from Eq. (4.56), we have

$$X''Y-2X'Y=-XY'\Rightarrow\frac{X''-2X'}{X}=\frac{-Y'}{Y}. \quad (4.57)$$

In (4.57), L.H.S is a function of x only and R.H.S is a function of y only and here x and y are independent variables, so each side is constant. So rewriting (4.57) as

From Eq. (4.57), we have $\frac{X''-2X'}{X}=\frac{-Y'}{Y}=k$,

$$\frac{X''-2X'}{X}=k,$$

$$X''-2X'-kX=0, \quad (4.58)$$

which is second order homogeneous linear differential equation.

$$\frac{-Y'}{Y}=k\Rightarrow Y'+kY=0 \quad (4.61)$$

,

which is first order homogeneous linear differential equation. The auxiliary equation of (4.61) is given by

$$m+k=0\Rightarrow m=-k .$$

The auxiliary equation of (4.58)) is given by

$$m^2 - 2m - k = 0 \qquad (4.59)$$

Solution of (4.59) is given by

$$m^2 - 2m - k = 0$$

$$\therefore m = \frac{2 \pm \sqrt{4 - 4(1)(-k)}}{2} \Rightarrow m = \frac{2 \pm 2\sqrt{1+k}}{2} = 1 \pm \sqrt{1+k}$$

So, the general solution of (4.58) is given by

$$\therefore X = C_1 e^{\left(1+\sqrt{1+k}\right)x} + C_2 e^{\left(1-\sqrt{1+k}\right)x} \qquad (4.60)$$

.

So the general solution of (4.60) is given by

$$Y = C_3 e^{-ky}. \qquad (4.62)$$

Using (4.60) and (4.62) in (4.53), we have

$$z = \left(C_1 e^{\left(1+\sqrt{1+k}\right)x} + C_2 e^{\left(1-\sqrt{1+k}\right)x}\right)\left(C_3 e^{-ky}\right) \Rightarrow z = C_1 C_3 e^{\left(1+\sqrt{1+k}\right)x} e^{-ky} + C_2 C_3 e^{\left(1-\sqrt{1+k}\right)x} e^{-ky}$$

.

$$\therefore z = \left(A e^{\left(1+\sqrt{1+k}\right)x} + B e^{\left(1-\sqrt{1+k}\right)x}\right) e^{-ky},$$

where $A = C_1 C_3$ and $B = C_2 C_3$, which is the required general solution of the given P.D.E.

Example:

Find a solution of the equation $\frac{\partial^2 u}{\partial x^2} = \frac{\partial u}{\partial y} + 2u$ in the form $u = X(x)Y(y)$ and solve the equation subject to the conditions $u = 0$ and $\frac{\partial u}{\partial x} = 2e^{-y}$ when $x = 0$ for all values of y.

Solution:

Here partial differential equation is given by

$$\frac{\partial^2 u}{\partial x^2} = \frac{\partial u}{\partial y} + 2u. \tag{4.63}$$

Let $u(x,y) = X(x)Y(y)$ (4.64)

be the solution of (4.63), where X is a function of x only and Y is a function of y only.

Differentiating partially (4.63) w.r.t. x and y respectively, we have

$$\frac{\partial u}{\partial x} = X'Y, \tag{4.65}$$

$$\frac{\partial^2 u}{\partial x^2} = X''Y, \tag{4.66}$$

$$\frac{\partial u}{\partial y} = XY', \tag{4.67}$$

where $X' = \frac{dX'}{dx}$, $Y' = \frac{dY}{dy}$, and $X'' = \frac{d^2 X}{dx^2}$. Using (4.66) and (4.67) in (4.63), we have

$$X''Y = XY' + 2XY.$$

Separating the variables

$$X''Y = XY' + 2XY \Rightarrow X''Y - 2XY = XY' \Rightarrow \frac{X'' - 2X}{X} = \frac{Y'}{Y} \tag{4.68}$$

In (4.68), L.H.S is a function of x only and R.H.S is a function of y only and here x and y are independent variables, so each side is constant. So rewriting (4.68) as

From (4.68), $\frac{X'' - 2X}{X} = k,$

$$\Rightarrow X'' - 2X = kX,$$

$$\Rightarrow X'' - 2X - kX = 0,$$

From (4.68), we have

$$\therefore \frac{Y'}{Y} = k \Rightarrow Y' - kY = 0 \tag{4.71}$$

$$\Rightarrow X'' - X(2+k) = 0, \quad (4.69)$$

which is second order homogeneous linear differential equation. The auxiliary equation of (4.69) is given by

$$m^2 - 2 - k = 0 \Rightarrow m = \pm\sqrt{k+2}.$$

So the general solution of (4.69) is given by

$$\therefore X = C_1 e^{(\sqrt{k+2})x} + C_2 e^{-(\sqrt{k+2})x} \quad (4.70)$$

.

which is first order homogeneous linear differential equation. The auxiliary equation of (9) is given by

$$m - k = 0 \Rightarrow m = k. \quad (4.72)$$

So, the general solution of (4.71) is given by

$$Y = C_3 e^{ky}. \quad (4.73)$$

Using (4.70) and (4.73) in (4.68), we have

$$u(x,y) = \left(C_1 e^{(\sqrt{k+2})x} + C_2 e^{-(\sqrt{k+2})x}\right) C_3 e^{ky} = C_1 C_3 e^{(\sqrt{k+2})x+ky} + C_2 C_3 e^{-(\sqrt{k+2})x+ky},$$

$$\therefore u(x,y) = A_1 e^{(\sqrt{k+2})x+ky} + A_2 e^{-(\sqrt{k+2})x+ky}, \quad (4.74)$$

where $A_1 = C_1 C_3, A_2 = C_2 C_3$, which is the required general solution of the given P.D.E.

For particular solution it is given that $u = 0$ when $x = 0$ .Putting $x = 0$ in (4.74) , we have

$$u(0,y) = A_1 e^{ky} + A_2 e^{ky} = 0 \Rightarrow A_1 = -A_2. \quad (4.75)$$

Using (4.75) in (4.74), we have

$$\therefore u(x,y) = -A_2 e^{(\sqrt{k+2})x+ky} + A_2 e^{-(\sqrt{k+2})x+ky} = A_2\left(e^{-(\sqrt{k+2})x+ky} - e^{(\sqrt{k+2})x+ky}\right) \quad (4.76)$$

Differentiating (4.76) , we have

$$\therefore \frac{\partial u}{\partial x} = -A_2(\sqrt{k+2})e^{-(\sqrt{k+2})x+ky} - A_2(\sqrt{k+2})e^{(\sqrt{k+2})x+ky} = -A_2\left((\sqrt{k+2})e^{-(\sqrt{k+2})x+ky} + (\sqrt{k+2})e^{(\sqrt{k+2})x+ky}\right) \quad (4.77)$$

It is also given that when

$$x = 0 \Rightarrow \frac{\partial u}{\partial x} = 2e^{-y}. \quad (4.78)$$

Using (4.78) in (4.77), we get $2e^{-y} = -A_2\left((\sqrt{k+2})e^{ky} + (\sqrt{k+2})e^{ky}\right) \Rightarrow 2e^{-y} = -2A_2e^{ky}\sqrt{k+2}$,

$$2e^{-y} = -2A_2e^{ky}\sqrt{k+2} \Rightarrow k = -1, A_2 = \frac{-1}{\sqrt{k+2}}, \quad (4.79)$$

$$\therefore k = -1 \Rightarrow A_2 = \frac{-1}{\sqrt{-1+2}} = -1. \quad (4.80)$$

Using (4.79), (4.80) in (4.74), we have

$$u(x,y) = (-1)\left(e^{-(\sqrt{-1+2})x-y} - e^{(\sqrt{-1+2})x-y}\right) \Rightarrow u(x,y) = (-1)\left(e^{-x-y} - e^{x-y}\right)$$

$$u(x,y) = e^{x-y} - e^{-x-y} \Rightarrow u(x,y) = e^{-y}(e^x - e^{-x})$$

$u(x,y) = 2e^{-y}\sinh x$, which is the required solution.

Example:

Using the method of separation of variables solve $\frac{\partial^2 u}{\partial x^2} - \frac{\partial u}{\partial y} = u$ subject to $u(0,y) = 0 = u(\pi, y)$.

Solution:

Here partial differential equation is given by

$$\frac{\partial^2 u}{\partial x^2} - \frac{\partial u}{\partial y} = u \quad (4.81)$$

Let $u(x,y)=X(x)Y(y)$ (4.82)

be the solution of (4.81), where X is a function of x only and Y is a function of y only.
Differentiating partially (4.82) w.r.t. x and y respectively, we have

$$\frac{\partial u}{\partial x}=X'Y \quad (4.83)$$

$$\frac{\partial^2 u}{\partial x^2}=X''Y \quad (4.84)$$

$$\frac{\partial u}{\partial y}=XY' \quad (4.85)$$

,

where $X'=\frac{dX'}{dx}$, $Y'=\frac{dY}{dy}$, and $X''=\frac{d^2X}{dx^2}$. Using (4.84) in (4.81), we have

$X''Y-XY'=XY$.

Separating the variables

$$X''Y-XY'=XY \Rightarrow X''Y=XY'+XY \Rightarrow \frac{X''}{X}=\frac{Y'+Y}{Y} \quad (4.86)$$

In (4.86), L.H.S is a function of x only and R.H.S is a function of y only and here x and y are independent variables, so each side is constant. So rewriting (4.86) as

From (4.86), $\frac{X''}{X}=-k^2$,

$\Rightarrow X''=-k^2X$,

$\Rightarrow X''+k^2X=0$ (4.87)

which is second order homogeneous linear differential

From (4.86), we have

$\frac{Y'+Y}{Y}=-k^2$,

$\Rightarrow Y'+Y+k^2Y=0$,

$\Rightarrow Y'+Y(1+k^2)=0$,(4.88)

which is first order homogeneous linear differential equation. The

equation. The auxiliary equation of (4.87) is given by	auxiliary equation of (9) is given by
$m^2+k^2=0 \Rightarrow m=\pm ik$.	$m+(1+k^2)=0 \Rightarrow m=-(1+k^2)$.
So the general solution of (4.87) is given by	So the general solution of (4.88) is given by
$X=C_1\cos(kx)+C_2\sin(kx)$ (4.88) .	$Y=C_3e^{-(1+k^2)y}$. (4.89)

Using (4.88) and (4.89) in (4.82) ,we have

$$u(x,y)=\left(C_1\cos(kx)+C_2\sin(kx)\right)C_3e^{-(1+k^2)y}=C_1C_3\cos(kx)e^{-(1+k^2)y}+C_2C_3\sin(kx)e^{-(1+k^2)y}$$

,

$$u(x,y)=A_1\cos(kx)e^{-(1+k^2)y}+A_2\sin(kx)e^{-(1+k^2)y}, \tag{4.90}$$

where $A_1=C_1C_3, A_2=C_2C_3$,is the required general solution of the given P.D.E. For particular solution here given that $u(0,y)=0$.

Putting $x=0$ in (4.90) , we have

$$u(0,y)=A_1e^{-(1+k^2)y}=0 \Rightarrow e^{-(1+k^2)y}\neq 0, A_1=0. \tag{4.91}$$

Using (4.91) in (4.90) , we have

$$u(x,y)=A_2\sin(kx)e^{-(1+k^2)y}. \tag{4.92}$$

It is also given that

$$u(\pi,y)=0. \tag{4.93}$$

Using (4.93) in (4.92), we have

$$u(\pi,y)=A_2\sin(k\pi)e^{-(1+k^2)y}=0 \Rightarrow A_2\neq 0, e^{-(1+k^2)y}\neq 0.$$

$$\therefore \sin(k\pi) = 0 \Rightarrow \sin(k\pi) = \sin(n\pi) \Rightarrow k = n, \text{where}\, n = 1,2,3,............ \qquad (4.94)$$

Using (4.94) in (4.92), we have $u(x, y) = A_2 \sin(nx)e^{-(1+n^2)y}$, which is the required particular solution.

Example:

Using the method of separation of variables solve the equation $\frac{\partial^2 v}{\partial x^2} = \frac{\partial v}{\partial t}$ given that $v = 0$ when $t \to \infty$,as well as $v = 0$ at $x = 0$ and $x = l.$

Solution:

Here partial differential equation is given by

$$\frac{\partial^2 v}{\partial x^2} = \frac{\partial v}{\partial t} \qquad (4.95)$$

Let $v(x, y) = X(x)T(t)$ (4.96)

be the solution of (4.95), where X is a function of x only and T is a function of t only.

Differentiating (4.96) partially w.r.t. x and y respectively, we have

$$\frac{\partial v}{\partial x} = X'T,$$

$$\Rightarrow \frac{\partial^2 v}{\partial x^2} = X''T \qquad (4.97)$$

$$\frac{\partial v}{\partial t} = XT', \qquad (4.98)$$

where $X' = \frac{dX}{dx}, X'' = \frac{d^2 X}{dx^2}$ and $T' = \frac{dT}{dt}$. Using (4.97) and (4.98) in (4.95), we have $X''T = XT'$.

Separating the variables

$$X''T = XT' \Rightarrow \frac{X''}{X} = \frac{T'}{T}. \qquad (4.99)$$

In (4.99), L.H.S is a function of x only and R.H.S is a function of t only and here x and t are independent variables, so each side is constant. So rewriting (4.99) as

From (4.99), we have $\frac{X''}{X} = -k^2$,

$$\Rightarrow X'' = -k^2 X,$$

$$\Rightarrow X'' + k^2 X = 0, \quad (4.100)$$

which is second order homogeneous linear differential equation. The auxiliary equation of (6) is given by

$$m^2 + k^2 = 0 \Rightarrow m = \pm ik .$$

So the general solution of (6) is given by

$$X = C_1 \cos(kx) + C_2 \sin(kx) \quad (4.101)$$

.

From (4.99), we have

$$\therefore \frac{T'}{T} = -k^2 \Rightarrow T' + k^2 T = 0$$

(4.101)

which is first order homogeneous linear differential equation. The auxiliary equation of (4.101) is given by

$$m + (k^2) = 0 \Rightarrow m = -k^2 .$$

So the general solution of (4.101) is given by

$$T = C_3 e^{-k^2 t} \quad (4.102)$$

Using (4.101) and (4.102) in (4.96), we have

$$v(x,t) = \left(C_1 \cos(kx) + C_2 \sin(kx)\right) C_3 e^{-k^2 t} = C_1 C_3 \cos(kx) e^{-k^2 t} + C_2 C_3 \sin(kx) e^{-k^2 t} ,$$

$$v(x,t) = A_1 \cos(kx) e^{-k^2 t} + A_2 \sin(kx) e^{-k^2 t} \quad (4.103)$$

where $A_1 = C_1 C_3, A_2 = C_2 C_3$, which is the required general solution of the given P.D.E.

For particular solution it is given that $x = 0, v = 0$. Putting $x = 0, v = 0.$ in (4.103) , we have

$$v(0,t) = A_1 e^{-k^2 t} = 0$$

Since , $e^{-k^2 t} \neq 0$ therefore
$\Rightarrow A_1 = 0.$ (4.104)

Using (4.104) in (4.103), we have

$$v(x,t) = A_2 \sin(kx) e^{-k^2 t} . \tag{4.105}$$

It is also given that when

$$x = l, v = 0. \tag{4.106}$$

Using (4.106) in (4.105), we have $v(l,t) = A_2 \sin(kl) e^{-k^2 t} = 0$

Since $A_2, e^{-k^2 t} \neq 0 \Rightarrow \sin(kl) = 0,$

$$\therefore \sin(kl) = 0 \Rightarrow \sin(kl) = \sin(n\pi) \Rightarrow k = \frac{n\pi}{l}, \text{where } n = 1,2,3,\ldots\ldots\ldots\ldots \tag{4.107}$$

Using (4.107) in (4.105), we have $v(x,t) = A_2 \sin\left(\frac{n\pi x}{l}\right) e^{-\left(\frac{n\pi}{l}\right)^2 t}$, which is the required particular solution.

Example:

Solve the following equation by the method of separation of variables $\frac{\partial^2 u}{\partial x \partial t} = e^{-t} \cos x$ given that $u = 0$ when $t = 0$ and $\frac{\partial u}{\partial t} = 0$ when $x = 0.$

Solution:

Here partial differential equation is given by

$$\frac{\partial^2 u}{\partial x \partial t} = e^{-t} \cos x . \tag{4.108}$$

Let $u = XT$ (4.109)

be the solution of (4.108) where X is a function of x only and T is a function of t only.

Differentiating (4.109) partially w.r.t. x and y respectively, we have

$$\frac{\partial u}{\partial x} = X'T, \quad (4.110)$$

$$\frac{\partial u}{\partial t} = XT' \quad (4.111)$$

$$\frac{\partial}{\partial x}\left(\frac{\partial u}{\partial t}\right) = \frac{\partial^2 u}{\partial x \partial t} = X'T' \quad (4.112)$$

where $X' = \frac{dX}{dx}$ and $T' = \frac{dT}{dt}$ .Using (4.112) in (4.108) ,we have

$X'T' = e^{-t}\cos x \Rightarrow X'T' = \frac{\cos x}{e^t}$.

Separating the variables, we have

$$e^t T' = \frac{\cos x}{X'} \Rightarrow \frac{\cos x}{X'} = e^t T'. \quad (4.113)$$

In (4.113) , L.H.S is a function of x only and R.H.S is a function of t only and here x and t are independent variables, so each side is constant. So, rewriting (4.113) as

From Eq. (4.113), we get

$$\frac{\cos x}{X'} = -k^2,$$

$$\Rightarrow \cos x = -k^2 X',$$

$$\Rightarrow X' + \frac{\cos x}{k^2} = 0. \quad (4.114)$$

which is first order homogeneous linear differential equation.

$$\therefore \frac{dX}{dx} + \frac{\cos x}{k^2} = 0 \Rightarrow dX = -\frac{\cos x}{k^2}dx \quad (4.115)$$

From (4.113) , we have

$$e^t T' = -k^2,$$

$$\Rightarrow T' = \frac{-k^2}{e^t},$$

$$\Rightarrow T' = -k^2 e^{-t}, \quad (4.117)$$

which is first order homogeneous linear differential equation.

Integrating Eq. (4.115) both sides, we get general solution of (4.114) as

$$X = -\frac{\sin x}{k^2} + C_1 . \qquad (4.116)$$

$$\frac{dT}{dt} = -k^2 e^{-t} ,$$

$$\Rightarrow dT = -k^2 e^{-t} dt \qquad (4.118)$$

Integrating both sides of Eq. (4.116), we get general solution of (4,116) is given by

$$T = k^2 e^{-t} + C_2 \qquad (4.119)$$

Using (4.116) and (4.119) in (4.109), we have

$$u(x,t) = \left(-\frac{\sin x}{k^2} + C_1 \right)\left(k^2 e^{-t} + C_2 \right), \qquad (4.120)$$

which is the required general solution of the given P.D.E. For particular solution it is given that $u = 0, t = 0$. Putting $u = 0, t = 0.$ in (4.120), we have

$$u(x,0) = \left(-\frac{\sin x}{k^2} + C_1 \right)\left(k^2 + C_2 \right) = 0 . \qquad (4.121)$$

But $\left(-\frac{\sin x}{k^2} + C_1 \right) = 0$ is not possible.

$$\therefore \left(k^2 + C_2 \right) = 0 \Rightarrow C_2 = -k^2 . \qquad (4.122)$$

Differentiating (4.120) w.r.t. to t we get

$$\frac{\partial u}{\partial t}(x,t) = \left(-\frac{\sin x}{k^2} + C_1 \right)\left(-k^2 e^{-t} \right). \qquad (4.123)$$

It is also given that

$$\frac{\partial u}{\partial t} = 0 \text{ when } x = 0. \qquad (4.124)$$

Using (4.124) in (4.123) , we have

$$\frac{\partial u}{\partial t}(0,t)=(0+C_1)\left(-k^2e^{-t}\right)=0\Rightarrow -k^2e^{-t}C_1=0$$

Since $-k^2e^{-t}\neq 0\Rightarrow C_1=0$ (4.125)

Using (4.122) and (4.125) in (4.120), we have

$$u(x,t)=\left(-\frac{\sin x}{k^2}\right)\left(k^2e^{-t}-k^2\right)\Rightarrow u(x,t)=\sin x(1-e^{-t}) .$$

$u(x,t)=\sin x(1-e^{-t})$, which is the required particular solution.

Example:

Solve the PDE $\frac{\partial^2 u}{\partial x^2}=\frac{\partial u}{\partial y}+2u$ by separation of variables method given that $u(0,y)=0$ and $\frac{\partial u}{\partial x}(0,y)=1+e^{-3y}$

Solution:

Here partial differential equation is given by

$$\frac{\partial^2 u}{\partial x^2}=\frac{\partial u}{\partial y}+2u . \tag{4.126}$$

Let $u=XY$ (4.127)

be the solution of (4.126), where X is a function of x only and Y is a function of y only.
Differentiating partially (4.127) w.r.t. x and y respectively, we have

$$\frac{\partial u}{\partial x}=X'Y ,$$

$$\frac{\partial^2 u}{\partial x^2}=X''Y , \tag{4.128}$$

$$\frac{\partial u}{\partial y}=XY' , \tag{4.129}$$

where $X' = \frac{dX}{dx}, X'' = \frac{d^2X}{dx^2}$ and $Y' = \frac{dY}{dy}$. Using (4.128) and (4.129) in (4.126), we have

$$X''Y = XY' + 2XY .$$

Separating the variables, we have

$$X''Y = XY' + 2XY \Rightarrow X''Y - 2XY = XY' \Rightarrow \frac{X'' - 2X}{X} = \frac{Y'}{Y} \tag{4.130}$$

In (4.130) , L.H.S is a function of x only and R.H.S is a function of y only and here x and y are independent variables, so each side is constant. So rewriting (4.130) as

$$\frac{X'' - 2X}{X} = k \Rightarrow X'' - 2X = kX,$$

$$\Rightarrow X'' - 2X - kX = 0 ,$$

$$\Rightarrow X'' - X(2+k) = 0 \tag{4.131}$$

,

which is second order homogeneous linear differential equation. The auxiliary equation of (6) is given by

$m^2 - 2 - k = 0 \Rightarrow m = \pm\sqrt{k+2}$.

So the general solution of (4.131) is given by

$$X = C_1 e^{(\sqrt{k+2})x} + C_2 e^{-(\sqrt{k+2})x} \tag{4.132}$$

From (5) we have

$$\therefore \frac{Y'}{Y} = k \Rightarrow Y' - kY = 0 \tag{4.133}$$

which is first order homogeneous linear differential equation. The auxiliary equation of (4.132) is given by

$$m - k = 0 \Rightarrow m = k \tag{14.134}$$

So the general solution of (4.132) is given by

$Y = C_3 e^{ky}$.

Using (4.132) and (4.134) in (4.127), we have

$$u(x,y) = \left(Ae^{(\sqrt{k+2})x} + Be^{-(\sqrt{k+2})x}\right)e^{ky}, \tag{4.135}$$

where $C_1C_3 = A$ and $C_2C_3 = B$,which is the required general solution of the given P.D.E.

According to the question $\frac{\partial u}{\partial x}(0, y) = 1 + e^{-3y}$ for $x = 0$ for all *y*. Therefore, in solution , we must have e^{0y} and e^{-3y} .Thus the value of *k* to be chosen are $k = 0, k = -3$ .Now taking $k = 0$ in equation (4.135), we have

$$u(x, y) = \left(Ae^{(\sqrt{2})x} + Be^{-(\sqrt{2})x}\right)e^{0y}. \tag{4.136}$$

Also, it is given that $u(0, y) = 0$ i.e. $u = 0$ when $x = 0$ .Putting $x = 0$ in (4.136), we have

$$A + B = 0 \tag{4.137}$$

Differentiating partially Eq.(4.135) w.r.t. *x*, we get

$$\frac{\partial u}{\partial x} = \sqrt{2}\left(Ae^{\sqrt{2}x} - Ae^{-\sqrt{2}x}\right). \tag{4.138}$$

Since $\frac{\partial u}{\partial x} = 1$ for $x = 0$, $\sqrt{2}\left(A - B\right) = 1$. Hence,

$$A - B = \frac{1}{\sqrt{2}} \quad . \tag{4.139}$$

Solving equations (4.137) and (4.139), we get $A = \frac{1}{2\sqrt{2}}, B = -\frac{1}{2\sqrt{2}}$. Substituting the values in Eq.(4.136), we have $u(x, y) = \left(\frac{1}{2\sqrt{2}}e^{\sqrt{2}x} + Be^{-\sqrt{2}x}\right) = \frac{1}{\sqrt{2}}\sinh\sqrt{2}x$, which is a part of the solution.

Considering equation (4.135) with $k = -3$. We have

$$u(x,y)=\left(Ae^{\sqrt{(-1)}x}+Be^{-\sqrt{(1)}x}\right)e^{-3y}, \tag{4.140}$$

$$\Rightarrow u(x,y)=\left(A^{*}e^{(-1)x}+B^{*}e^{-(1)x}\right)e^{-3y}, \tag{4.141}$$

$$u(0,y)=0\Rightarrow A^{*}=0. \tag{4.142}$$

$$u=B^{*}\sin x e^{-3y}\Rightarrow\frac{\partial u}{\partial x}=B^{*}\cos x e^{-3y}. \tag{4.143}$$

Consider, $\left(\frac{\partial u}{\partial x}\right)_{x=0}=e^{-3y}$ for all y. Therefore,

$$B^{*}\ e^{-3y}=e^{-3y} \text{ for all } y\Rightarrow B^{*}=1 \tag{4.144}$$

$$\Rightarrow u_2(x,y)=\sin xe^{-3y}$$

$\therefore u(x,y)=\frac{1}{\sqrt{2}}\sinh\sqrt{2}x+e^{-3y}\sin x$, which is the required particular solution.

Example:

Solve by the method of separation of variables $4\frac{\partial u}{\partial x}+\frac{\partial u}{\partial y}=3u$ given $u=3e^{-y}-e^{-5y}$ when $x=0$.

Solution:

Here the Partial differential equation is

$$4\frac{\partial u}{\partial x}+\frac{\partial u}{\partial y}-3u=0\ . \tag{4.145}$$

Let $u=XY$ (4.146)

be the solution of (4.145), where X is a function of x only and Y is a function of y only.

Differentiating Eq. (4,146) partially with respect to x and y we get

$$\frac{\partial u}{\partial x} = X'Y \qquad (14.147)$$

$$\frac{\partial u}{\partial y} = XY' \qquad (4.148)$$

where $X' = \frac{dX}{dx}$, and $Y' = \frac{dY}{dy}$. Using (4.147) and (4.148) in (4.145), we have

$$4X'Y + XY' - 3XY = 0.$$

Separating the variables, we have

$$\frac{4X' - 3X}{X} = \frac{-Y'}{Y}. \qquad (4.149)$$

In (4.149) , L.H.S is a function of x only and R.H.S is a function of y only and here x and y are independent variables, so each side is constant. So, rewriting (4.149) as

$$\frac{4X' - 3X}{X} = k,$$

$$\Rightarrow 4X' - 3X - kX = 0,$$

$$\Rightarrow 4X' - (3+k)X = 0,$$

$$\Rightarrow X' - \left(\frac{3+k}{4}\right)X = 0 \qquad (4.150)$$

.

The auxiliary equation of (4.150) is

$$\frac{-Y'}{Y} = k,$$

$$\Rightarrow -Y' - kY = 0,$$

$$\Rightarrow Y' + kY = 0. \qquad (4.152)$$

The auxiliary equation of (4.151) is

$$m + k = 0,$$

$$\Rightarrow m = -k$$

The solution is

$$m-\left(\frac{3+k}{4}\right)=0 \Rightarrow m=\frac{3+k}{4}.$$

Therefore, the solution is

$$X=Ae^{\left(\frac{3+k}{4}\right)x}. \tag{4.151}$$

$$Y=Be^{-ky}. \tag{4.153}$$

Substituting the values of X and Y from Eqs.(4.151) and (4.153) in Eq. (4.146), we get

$$u=Ae^{\left(\frac{3+k}{4}\right)x}.Be^{-ky}=Ce^{\left(\frac{3+k}{4}\right)x}.e^{-ky} \quad , \text{ where } C=AB. \tag{4.154}$$

It is given that when $x=0$, $u(0,y)=3e^{-y}-e^{-5y}$. Therefore, Eq. (4.154), we get

$$Ce^{-ky}=3e^{-y}-e^{-5y}$$
$$C[1-ky+...]=3[1-y+...]-[1-5y+...]$$

Equating constant terms on both the sides we get,

$$C=3-1=2. \tag{4.155}$$

Equating $'y'$ term on both sides , we get

$$-Ck=-3+5=2 \Rightarrow -2k=2 \Rightarrow k=-1. \tag{4.156}$$

Using equation (4.155) and equation (4.156), equation (4.154) becomes

$$u=2e^{\left(\frac{3-1}{2}\right)x}e^{y}=2e^{\frac{x}{2}}.e^{y}=2e^{\left(\frac{x}{2}+y\right)},$$

which is the required solution.

Example:

Using Method of separation of variables solve $\frac{\partial u}{\partial x}=4\frac{\partial u}{\partial y},u(0,y)=8e^{-3y}+4e^{-5y}$.

Solution:

The given partial differential equation is

$$\frac{\partial u}{\partial x}=4\frac{\partial u}{\partial y}\Rightarrow\frac{\partial u}{\partial x}-4\frac{\partial u}{\partial y}=0. \tag{4.157}$$

Let $u=XY$ (4.158)

be the solution of (4.158), where X is a function of x only and Y is a function of y only.

Differentiating Eq. (4.158) partially with respect to x and y, we get

$$\frac{\partial u}{\partial x}=X'Y \tag{4.159}$$

$$\frac{\partial u}{\partial y}=XY' \tag{4.160}$$

where $X'=\frac{dX}{dx}$, and $Y'=\frac{dY}{dy}$. Using (4.159) and (4.160) in (4.157), we have $X'Y-4XY'=0$.

Separating the variables, we have

$$\frac{X'}{X}=\frac{4Y'}{Y}. \tag{4.161}$$

In (4.160) , L.H.S is a function of x only and R.H.S is a function of y only and here x and y are independent variables, so each side is constant. So, rewriting (4.161) as

$$\frac{X'}{X}=k,$$

$$\Rightarrow X'-kX=0, \qquad (4.162)$$

The auxiliary equation of (4.162) is

$$m=k.$$

The solution of Eq. (4.162) is

$$X=Ae^{kx}. \qquad (4.163)$$

$$\frac{4Y'}{Y}=k, \quad . \qquad (4.164)$$

$$\Rightarrow 4Y'-kY=0.$$

The auxiliary equation of (4.164) is

$$4Y'-kY=0,$$

$$\Rightarrow m=\frac{k}{4}.$$

The solution is

$$Y=Be^{\left(\frac{k}{4}\right)y}. \qquad (4.165)$$

Substituting values from equations (4.163) and (4.165) in equation (4.158), we get

$$u(x,y)=Ae^{kx}.Be^{\left(\frac{k}{4}\right)y}=Ce^{kx}.e^{\left(\frac{k}{4}\right)y}. \qquad (4.166)$$

where $C=AB$. It is also given that

$$u(0,y)=8e^{-3y}+4e^{-5y}. \qquad (4.167)$$

Put $x=0$ in Eq. (4.167) , we get

$$u(0,y)=C.e^{\left(\frac{k}{4}\right)y}. \qquad (4.168)$$

Comparing (4.167) and (4.168), we get $Ce^{\left(\frac{k}{4}\right)y}=8e^{-3y}+4e^{-5y}$,

$$\Rightarrow C\left[1-\left(\frac{k}{4}\right)y+\ldots\right]=8[1-3y+\ldots]-4[1-5y+\ldots] \qquad (4.169)$$

Equating constants on both the sides of Eq. (4.169), we get $C = 8 + 4 \Rightarrow C = 12$.

Equating y terms on both sides of Eq. (4.169), we get

$$\frac{Ck}{4} = -24 - 20 \Rightarrow \frac{12 \times k}{4} = -44 \Rightarrow 3k = -44 \Rightarrow 3k = -44 \Rightarrow k = \frac{-44}{3}$$

.

Substituting value $C = 12$ and $k = \frac{-44}{3}$ in (4.166), we get

$$u = 12e^{\left(\frac{-44}{3}\right)x} .e^{\left(\frac{-11}{3}\right)y} = 12e^{\left(\frac{-11}{3}\right)(4x+y)},$$

which is the required solution.

Example:

Using method of separation of variables, solve $u_{xt} = e^{-t}\cos x$ with $u(x,0) = 0$ and $u(0,t) = 0$.

Solution:

The given partial differential equation is

$$u_{xt} = e^{-t}\cos x. \tag{4.170}$$

Let $u = XY$ (4.171)

be the solution of (4.171), where X is a function of x only and Y is a function of t only.

Differentiating Eq. (4.171) partially with respect to x and y, we get

$$\frac{\partial u}{\partial x} = X'Y \tag{4.172}$$

$$\frac{\partial u}{\partial t} = XY' \tag{4.173}$$

$$\frac{\partial^2 u}{\partial x \partial t} = X'Y'$$

where $X' = \frac{dX}{dx}$, and $Y' = \frac{dY}{dt}$. Using (4.172) and (4.173) in (4.170), we have $X'Y' = e^{-t}\cos x$.

Separating the variables, we have

$$\frac{X'}{\cos x} = \frac{e^{-t}}{T'}. \tag{4.174}$$

In (4.174) , L.H.S is a function of x only and R.H.S is a function of t only and here x and t are independent variables, so each side is constant. So rewriting (4.174) as

$$\frac{X'}{\cos x} = K,$$

$$\Rightarrow X' = K\cos x$$

$$\Rightarrow \frac{dX}{dx} = K\cos x,$$

$$\Rightarrow dX = K\cos x\, dx \tag{4.175}$$

Integrating (4.174), we get

$$X = K\sin x + c_1. \tag{4.176}$$

$$\frac{e^{-t}}{T'} = K,$$

$$\Rightarrow KT' = e^{-t},$$

$$\Rightarrow T' = \frac{1}{K}e^{-t},$$

$$\Rightarrow \frac{dT}{dt} = \frac{1}{K}e^{-t},$$

$$\Rightarrow dT = \frac{1}{K}e^{-t}\, dt, \tag{4.177}$$

Integrating (4.177), we get

$$T = \frac{-1}{K}e^{-t} + c_2,$$

Substituting values from equations (4.176) and (4.177) in equation (4.171), we get

$$u=\left(k\sin x+c_1\right)\left(\frac{-1}{K}e^{-t}+c_2\right). \tag{4.178}$$

Using the given condition $u\left(x,0\right)=0$ in Eq. (4.178), we get

$$0=\left(k\sin x+c_1\right)\left(\frac{-1}{K}+c_2\right),$$

$$\Rightarrow\frac{-1}{K}+c_2=0\Rightarrow c_2=\frac{1}{K}. \tag{4.179}$$

Using second condition $u\left(0,t\right)=0$ in Eq. (4.178), we get

$$0=c_1\left(\frac{-1}{K}e^{-t}+c_2\right)=c_1\left(\frac{-1}{K}e^{-t}+\frac{1}{K}\right),$$

$$\Rightarrow c_1=0. \tag{4.180}$$

Substituting values from Eq. (4.179) and (4.180) in Eq. (4.178), we get

$$u=K\sin x\left(\frac{1}{K}-\frac{1}{K}e^{-t}\right)=\sin x\left(1-e^{-t}\right).$$

Hence, the required solution is

$$u=K\sin x\left(\frac{1}{K}-\frac{1}{K}e^{-t}\right)=\sin x\left(1-e^{-t}\right).$$

Exercise

Solve the following PDE by the method of separation of variables

1. $2\frac{\partial^2 u}{\partial x^2}=\frac{\partial u}{\partial y}$. (Ans $u=\left[Ae^{\sqrt{k}x}+Be^{-\sqrt{k}x}\right]e^{2ky}$)

2. $3\frac{\partial u}{\partial x}+2\frac{\partial u}{\partial y}=0$, (Ans $u(x,y)=4e^{\frac{1}{2}(3y-2x)}$)

given

$u(x,0)=4e^{-x}$.

3. $\frac{\partial u}{\partial x}=\frac{\partial u}{\partial y}$, given (Ans $u(x,y)=8e^{-12x-3y}+4e^{-20x-5y}$)

$u(0,y)=8e^{-3y}+4e^{-5y}$

.

4. $2x\frac{\partial z}{\partial x}-3y\frac{\partial z}{\partial y}=0$. (Ans $u(x,y)=Cx^{\frac{k}{2}}y^{\frac{k}{3}}$)

5. $\frac{\partial^2 z}{\partial x^2}+4\frac{\partial^2 z}{\partial y^2}=0$. (Ans

$z=\left[C_1e^{\sqrt{k}x}+C_2e^{-\sqrt{k}x}\right]\left[C_3\sin\frac{\sqrt{k}}{2}y+C_4\cos\frac{\sqrt{k}}{2}\right.$

6. $\frac{\partial u}{\partial x}+4=\frac{\partial u}{\partial t}$ given (Ans $u(x,t)=4e^{-3x-2t}$

$u=4e^{-3x}$ when

$t=0$.

7. $4\frac{\partial u}{\partial x}+\frac{\partial u}{\partial y}=3u$ (Ans $u(x,y)=e^{2x-5y}$)

given $u=e^{-5y}$

when $x=0$.

8. $\frac{\partial u}{\partial x}+\frac{\partial u}{\partial y}=2(x+y)u$ (Ans $u(x,y)=e^{x^2+y^2+k(x-y)}$)

.

9. $2\frac{\partial u}{\partial x}-3\frac{\partial u}{\partial y}=0$ (Ans $u(x,y)=5e^{3x+2y}$)

given $u(x,0)=5e^{3x}$

.

10 $\frac{\partial^2 u}{\partial x^2}=\frac{\partial u}{\partial t}$ given (Ans $u=\sin\pi xe^{-\pi^2 t}$)

$u(x,0)=\sin\pi x$.

4.3 Finite Difference Method

The Finite Difference Method (FDM) is a numerical technique used to approximate solutions to differential equations by discretizing the domain into a finite number of points. It is useful for solving ordinary differential equations (ODEs) and partial differential equations (PDEs) in various fields such as physics, engineering, and finance.

The basic idea behind the Finite Difference Method is to replace the derivatives in the differential equation with approximations based on the differences between values at neighboring points. By discretizing the domain, the continuous problem is transformed into a discrete problem that can be solved using algebraic equations. Finite Difference Method offers several advantages, including its simplicity, versatility, and ease of implementation. However, the choice of finite difference scheme and grid resolution affect the accuracy of the solution, and for some equation types, stability problems may arise. The Finite Difference Method is a potent tool for solving differential equations numerically and offering insightful information about engineering challenges and physical phenomena.

4.4 Finite Difference Approximations to Partial Derivatives

Finite difference approximations to partial derivatives can be derived using Taylor series expansion. Let's consider a function $f(x, y)$ and aim to approximate its partial derivative with respect to x at a point (x, y). Taylor series expansion of $f(x, y)$ is given by

$$u(x+h, y) = u(x, y) + h\frac{\partial u}{\partial x} + \frac{1}{2!}h^2\frac{\partial^2 u}{\partial x^2} + \frac{1}{3!}h^3\frac{\partial^3 u}{\partial x^3} + \dots\dots\dots \quad (4.181)$$

From (4.181), neglecting h^2 and higher power of h, we get

$$u(x+h, y) \approx u(x, y) + h\frac{\partial u}{\partial x}\ .$$

$$\Rightarrow \frac{\partial u}{\partial x} = \frac{u(x+h, y) - u(x, y)}{h}$$. **(Forward Difference Formula)** (4.182)

$$u(x-h, y) = u(x, y) - h\frac{\partial u}{\partial x} + \frac{1}{2!}h^2\frac{\partial^2 u}{\partial x^2} - \frac{1}{3!}h^3\frac{\partial^2 u}{\partial x^3} + \dots\dots\dots\dots \quad (4.183)$$

From (4.183) ,neglecting h^2 and higher power of h, we get

$$u(x-h,y) \approx u(x,y) - h\frac{\partial u}{\partial x} \quad ,$$

$$\Rightarrow \frac{\partial u}{\partial x} \approx \frac{u(x,y)-u(x-h,y)}{h} . \quad (4.184)$$

(Backward Difference Formula)

Subtracting (4.183) from (4.11) and neglecting h^3 and higher power of h we get

$$u(x+h,y)-u(x-h,y) \approx 2\mathrm{h}\frac{\partial u}{\partial x}$$

$$\Rightarrow \frac{\partial u}{\partial x} \approx \frac{u(x+h,y)-u(x-h,y)}{2h} . \quad (4.185)$$

(Central Difference Formula)

Adding (4.181) and (4.183) neglecting h^5 and higher power of h , we get

$$u(x+h,y)+u(x-h,y) = 2u(x,y) + h^2\frac{\partial^2 u}{\partial x^2},$$

$$\Rightarrow \frac{\partial^2 u}{\partial x^2} \approx \frac{1}{h^2}\left[u(x+h,y)-2u(x,y)+u(x-h,y)\right]. \quad (4.186)$$

Similarly, $\frac{\partial^2 u}{\partial y^2} \approx \frac{1}{k^2}\left[u(x,y+k)-2u(x,y)+u(x,y-k)\right]$, and

$$\frac{\partial^2 u}{\partial x \partial y} \approx \frac{1}{4hk}\left[u(x+h,y+k)-2u(x-h,y+k)+u(x+h,y-k)+u(x-h,y-k)\right]$$

.

Similarly,

$$\frac{\partial u}{\partial y}=\frac{u(x,y+k)-u(x,y)}{k}$$, **(Forward Difference Formula)**

$$\frac{\partial u}{\partial y}=\frac{u(x,y)-u(x,y-k)}{k}$$ **(Backward Difference Formula)**

$$\frac{\partial u}{\partial y}=\frac{u(x,y+k)-u(x,y-k)}{2k}$$. **(Central Difference Formula)**

Let R be the given rectangular region and is divided into small rectangles of sides $\Delta x = h$ and $\Delta y = h$ as shown in Figure 4.1 .Now we write $u(x,y)=u(ih,\mathrm{jk})=u_{i,j}$. Then Eqs. (4.182), (4.184), (4.185) , and (4.186) changes to following:

Forward Difference Formula

1. $$\frac{\partial u}{\partial x}=\frac{u(ih+h,\mathrm{jk})-u(\mathrm{ih},\mathrm{jk})}{h}=\frac{u((i+1)h,\mathrm{jk})-u(\mathrm{ih},\mathrm{jk})}{h}=\frac{u_{i+1,j}-u_{i,j}}{h}$$

$$\frac{\partial u}{\partial x}=\frac{u_{i+1,j}-u_{i,j}}{h}.$$

2. $$\frac{\partial u}{\partial y}=\frac{u_{i,j+1}-u_{i,j}}{k}.$$

Backward Difference Formula

3. $$\frac{\partial u}{\partial x}=\frac{u_{i,j}-u_{i-1,j}}{h}.$$

4. $$\frac{\partial u}{\partial y}=\frac{u_{i,\mathrm{j}}-u_{i,j-1}}{k}.$$

Central Difference Formula

5. $$\frac{\partial u}{\partial x} = \frac{u_{i+1,j} - u_{i-1,j}}{2h}.$$

6. $$\frac{\partial u}{\partial y} = \frac{u_{i,i+!} - u_{i,j-1}}{2k}$$

Second Order Derivative

7. $$\frac{\partial^2 u}{\partial x^2} \approx \frac{\left[u_{i+1,j} - 2u_{i,j} + u_{i-1,j}\right]}{h^2}$$

8. $$\frac{\partial^2 u}{\partial y^2} \approx \frac{\left[u_{i,j+1} - 2u_{i,j} + u_{i,j-1}\right]}{k^2}$$

9. $$\frac{\partial^2 u}{\partial x \partial y} \approx \frac{\left[u_{i+1,j+1} - u_{i+1,j-1} - u_{i-1,j+1} + u_{i-1,j-1}\right]}{4hk}$$

4.5 Solution of First Order PDE using Finite Difference Method

In this section we discuss solution of first order PDE using finite difference method and also evaluate the given PDE using MATLAB software package. To evaluate the example following notation are used: $u_{ij} \equiv u(i\Delta x, j\Delta t), x_i = i\Delta x, t_j = j\Delta t$.

Example:

Solve the linear advection equation for

$$u(x,y) \quad , \quad \frac{\partial u(x,y)}{\partial t} = c\frac{\partial u(x,y)}{\partial x}, \qquad (4.187)$$

defined on $(-\infty\infty), t\in[0\infty)$, and with the boundary condition $u(x,0)=F(x)$ with $c=1, \Delta x=0.2,\ \Delta t=0.1$.

Solution:

For x and t, discretize the system on a two-dimensional grid. The PDE's domain is semi-finite in t and unbounded in x, per the question. The values of u are determined by the boundary condition along a single wall as shown in the Figure 4.1.

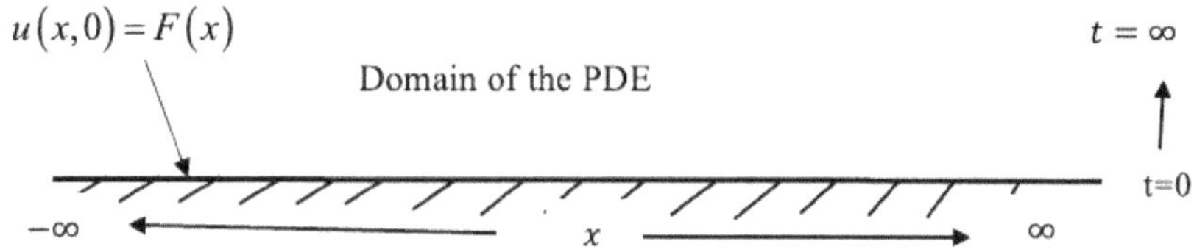

Figure 4.2: Domain of the given PDE.

More information about the grid is given in the accompanying diagram. The boundary condition $u_{i,0}=F_i$ (the discretized form of the boundary condition (I), where $F_i \equiv F(i\Delta x)$, determines the values of $u_{i,0}$ at the bottom grid points (*t* = *0* or *j* = *0,* indicated by circles). We integrate the PDE forward in t ("upward" in the following Figure) from the bottom row with known values of *u* to $t=\Delta t\,(j=1)$ produce $u_{i,1}$ or $u(x,\Delta t)$, indicated by green triangles. After that, the procedure moves on to the following level at *j* = *2* , $t=2\Delta t$, and so forth. The arrows in the diagram that join a circle to a triangle are merely for illustration purposes.

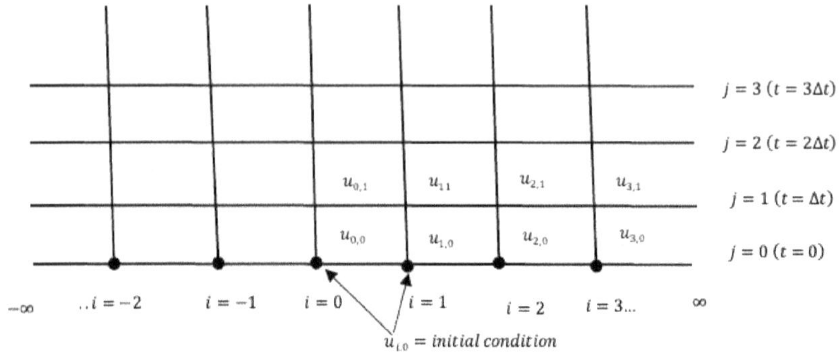

Figure 4.3 Discretized form.

In order to carry out the integration in *t*, we discretize the PDE by using the forward difference scheme and the central difference scheme to approximate $\frac{\partial u(x,y)}{\partial t}$ and $\frac{\partial u(x,y)}{\partial x}$, respectively and are mention below:

$$\frac{\partial u}{\partial t} = \frac{u_{i,j+1} - u_{i,j}}{\Delta t},$$

$$\frac{\partial u}{\partial x} = \frac{u_{i+1,j} - u_{i-1,j}}{2\Delta x}.$$

Substituting the value of $\frac{\partial u(x,y)}{\partial t}$ and $\frac{\partial u(x,y)}{\partial x}$ in Eq.(4.187), we get

$$\frac{u_{i,j+1} - u_{i,j}}{\Delta t} \approx \frac{u_{i+1,j} - u_{i-1,j}}{2\Delta x},$$

$$u_{i,j+1} = -\left(\frac{c\Delta t}{2\Delta x}\right)u_{i-1,j} + u_{i,j} + \left(\frac{c\Delta t}{2\Delta x}\right)u_{i+1,j}.$$

Assuming $\alpha = \frac{c\Delta t}{2\Delta x}$, the above equation changes to

$$u_{i,j+1} = -\alpha u_{i-1,j} + u_{i,j} + \alpha u_{i+1,j}. \tag{4.188}$$

Using the given values $c = 1, \Delta x = 0.2,\ \Delta t = 0.1$, we get $\alpha = 0.25$ with the boundary condition $u_{4,0} = 1$ and $u_{i,0} = 0$ for all $i \neq 4$.

Using the given values $c=1, \Delta x=0.2, \Delta t=0.1$, we get $\alpha=0.25$, Eq. (4.188) reduces to

$$u_{i,j+1} = -0.25u_{i-1,j} + u_{i,j} + 0.25u_{i+1,j}. \quad (4.189)$$

For $j=0$, the above equation changes to $u_{i,1} = -0.25u_{i-1,0} + u_{i,0} + 0.25u_{i+1,0}$.

Clearly $u_{i,1}$ is zero for all *i* except the following :

$$u_{3,1} = -0.25u_{2,0} + u_{3,0} + 0.25u_{4,0} = 0.25, \quad (4.190)$$

$$u_{4,1} = -0.25u_{3,0} + u_{4,0} + 0.25u_{5,0} = 1, \quad (4.191)$$

$$u_{5,1} = -0.25u_{4,0} + u_{5,0} + 0.25u_{6,0} = -0.25. \quad (4.192)$$

Eqs. (4.190)-(4.192) gives the solution of the PDE at $t=\Delta t=0.1$.The values evaluated at are further used to evaluate the value at $t=2\Delta t=0.2$, and so on. For $j=1$, Eq.(4.189) changes to

$u_{i,2} = -0.25u_{i-1,1} + u_{i,1} + 0.25u_{i+1,1}$. For different values of *i*, we have

$$u_{2,2} = -0.25u_{1,1} + u_{2,1} + 0.25u_{3,1} = 0.0625,$$
$$u_{3,2} = -0.25u_{2,1} + u_{3,1} + 0.25u_{4,1} = 0.5,$$
$$u_{4,2} = -0.25u_{3,1} + u_{4,1} + 0.25u_{5,1} = 0.875,$$
$$u_{5,2} = -0.25u_{4,1} + u_{5,1} + 0.25u_{6,1} = -0.5,$$
$$u_{6,2} = -0.25u_{5,1} + u_{6,1} + 0.25u_{7,1} = 0.0625.$$

All remaining $u_{i,2}$ with $i>6$ and $i<2$ are zero. The MATLAB code and graphical solution is mention below:

```
% Define and initialize the variables
for i = 1:10
xi(i) = i;
for j = 1:3
uij(i,j) = 0;
```

```
end
end
%The boundary condition
uij(4,1) = 1;
for j = 1:2
for i = 2:9
uij(i,j+1) = -0.25*uij(i-1,j)+uij(i,j)+0.25*uij(i+1,j);
end
end
% Graphical Representation
for i = 1:10
u0(i) = uij(i,1);
u1(i) = uij(i,2);
u2(i) = uij(i,3);
z(i) = 0;
end
subplot(3,1,1);plot(xi,z,'k--',xi,u0,'r-o');axis([1 10 -0.5 1])
subplot(3,1,2);plot(xi,z,'k--',xi,u1,'r-o');axis([1 10 -0.5 1])
subplot(3,1,3);plot(xi,z,'k--',xi,u2,'r-o');axis([1 10 -0.5 1])
```

The graph shows the initial condition and the solutions at t = Dt and t = 2Dt (i.e., $u_{i,j}$ at j = 0, 1, and 2). The first graph shows the nature of the given function at the initial condition. The second graph solution of the given function at time step level 1 and 2.

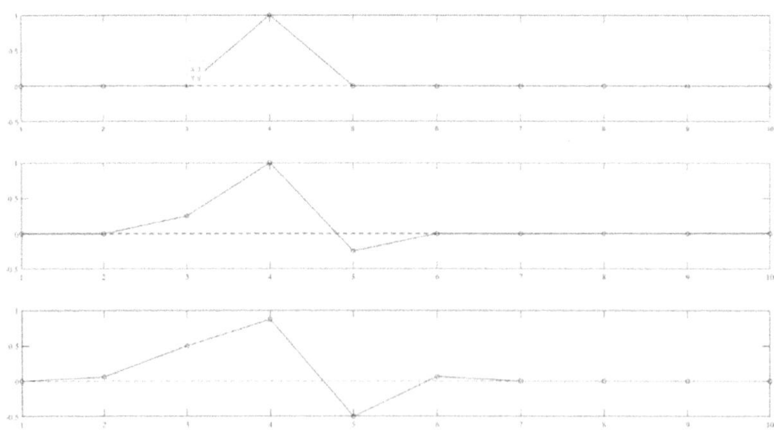

Figure 4.4 Graphical solution.

Chapter 5

Heat Equation

5.1 Introduction

The heat equation is a fundamental partial differential equation that describes how the distribution of heat (or temperature) evolves over time within a given region. It is widely used in various fields, including physics, engineering, and mathematics, to model phenomena such as heat conduction, diffusion of substances, and thermal radiation.

5.2 Derivation of one-dimensional heat flow

In this section, we shall consider the flow of heat and the accompanying variation of temperature with position and time in conducting solids. The following empirical laws are taken as the basis of investigation:

1. Heat flows from a higher to lower temperature.

2. The amount of heat required to produce a given temperature change in a body is proportional to the mass of the body and to the temperature change. This constant of proportionality is known as the specific heat (c) of the conducting material.

3. The rate at which heat flows through an area is proportional to the area and to the temperature gradient normal to the area. This constant of proportionality is known as the thermal conductivity (k) of the material.

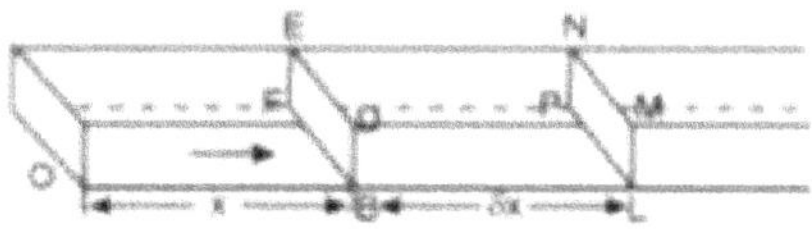

Figure 5.1: Cross sectional area of bar rod.

Consider a bar rod of homogeneous material of density $\rho\left(g/cm^3\right)$ and having a constant cross-sectional area $\left(Acm^2\right)$. We suppose that the sides of the bar are insulated and the loss of heat from the sides by conduction or radiation is negligible. Take an end of the bar as the origin and the direction of heat flow as the positive *x*-axis. Let c be the specific heat and k the thermal conductivity of the material. Consider an element between two parallel sections *BDEF* and *LMNP* at distances *x* and $x+\delta x$. From the origin O, the sections being perpendicular to the $x-$ axis. The mass of the element $= A\rho\delta x$. Let $u(x,t)$ be the temperature at a distance x at time t. By the second law stated above, the rate of increase of heat in the element $= A\rho\delta xc\frac{\partial u}{\partial t}$. If R_1 and R_2 are respectively the rates $(cal/\sec)$ of inflow and outflow, for the sections $x=x$ and $x=x+\delta x$, then $R_1=-kA\left(\frac{\partial u}{\partial x}\right)_{x=x}$ and $R_2=-kA\left(\frac{\partial u}{\partial x}\right)_{x=x+\delta x}$. The negative sign being due to the fact that heat flows from higher to lower temperature. i.e., $\frac{\partial u}{\partial x}$ is negative. Equating the rates of increase of heat from the empirical laws.

$$A\rho\delta xc\frac{\partial u}{\partial t}=R_1-R_2=-kA\left(\frac{\partial u}{\partial x}\right)_{x=x}-\left(-kA\left(\frac{\partial u}{\partial x}\right)_{x=x+\delta x}\right)$$

$$\therefore A\rho\delta xc\frac{\partial u}{\partial t}=kA\left[\left(\frac{\partial u}{\partial x}\right)_{x=x+\delta x}-kA\left(\frac{\partial u}{\partial x}\right)_{x=x}\right]$$

$$\therefore \frac{\partial u}{\partial t}=\frac{k}{\rho c}\left[\frac{\left(\frac{\partial u}{\partial x}\right)_{x=x+\delta x}-kA\left(\frac{\partial u}{\partial x}\right)_{x=x}}{\delta x}\right]$$

Taking the limit as $\delta x\to 0$ when i.e., when $x+\delta x\to x$

$$\therefore \frac{\partial u}{\partial t} = \frac{k}{\rho c} \lim_{\delta x \to 0} \left[\frac{\left(\frac{\partial u}{\partial x}\right)_{x=x+\delta x} - kA\left(\frac{\partial u}{\partial x}\right)_{x=x}}{\delta x} \right] = \frac{k}{\rho c} \frac{\partial}{\partial x}\left(\frac{\partial u}{\partial x}\right)$$

$$\therefore \frac{\partial u}{\partial t} = \frac{k}{\rho c} \frac{\partial^2 u}{\partial x^2}$$

$\frac{k}{\rho c}$ is called the diffusivity $\left(cm^2 / \sec\right)$ of the substance. If we denote it by C^2 the above equation takes the form $\frac{\partial u}{\partial t} = C^2 \frac{\partial^2 u}{\partial x^2}$. This partial differential equation is known as one-dimensional heat equation. Now we want to find the solution of a one-dimensional heat equation subject to the following initial and boundary conditions:

Boundary conditions	**Initial conditions**
BC1: $y(x=0,t)=0,\ \forall t>0$,	**IC1**: $y(x,t=0)=f(x),\ \forall x\in(0,l)$,
BC2: $y(x=l,t)=0,\ \forall t>0$,	

5.3 Solution of heat equation when both the ends of bar are kept at temperature zero and initial temperature is prescribed

The one-dimensional heat flow equation

$$\frac{\partial u}{\partial t} = C^2 \frac{\partial^2 u}{\partial x^2}. \tag{5.1}$$

with following initial and boundary condition

Boundary conditions	**Initial conditions**
BC1:	**IC1:**
$y(x=0,t)=0,\ \forall t>0$, (5.1a)	$y(x,t=0)=f(x),\ \forall x\in(0,l)$, (5.1c)
BC2:	
$y(x=l,t)=0,\ \forall t>0$, (5.1 b)	

Remark: when both the ends of bar are kept at temperature zero, the boundary conditions are given by (5.1 a) and (5.1 b) and the boundary condition are known as Dirichlet boundary condition.

Assume that a solution of equation (5.1) is of the form

$$u(x,t)=X(x)T(t) \tag{5.2}$$

where X is a function of x and T is a function of t only. Differentiate equation (5.2) partially w.r.t. x and t, we have

$$\frac{\partial u}{\partial x}=X'T \Rightarrow \frac{\partial^2 u}{\partial x^2}=X''T \tag{5.3}$$

and $\frac{\partial u}{\partial t}=XT' \Rightarrow \frac{\partial^2 u}{\partial t^2}=XT''$ (5.4)

where $X'=\frac{dX}{dx}, X''=\frac{d^2X}{dx^2}, T'=\frac{dT}{dx}, T''=\frac{d^2T}{dx^2}$. Using equations (5.3) and (5.4) in (5.2), we have

$$XT'=C^2X''T.$$

Separating the variables in the above equation, we get

$$\frac{1}{C^2}\frac{T'}{T}=\frac{X''}{X}. \tag{5.5}$$

In equation (5.5), L.H.S is a function of t only and R.H.S is a function of x only and here t and x are independent variables, so each side is constant. So rewriting equation (5.5) as

$$\frac{X''}{X}=k \Rightarrow X''=kX \Rightarrow X''-kX=0, \tag{5.6}$$

and $\frac{1}{C^2}\frac{T}{T'}=k \Rightarrow T'=kT \Rightarrow T'-C^2kT=0.$ (5.7)

equations (5.6) and (5.7) which are second and first order homogeneous linear differential equation, respectively. Now let us look at the boundary condition

BC1: $u(x=0,t)=0,\ \forall t>0$

BC2: $u(x=l,t)=0,\ \forall t>0$

The solution of equation (5.1) is given by equation (5.2) i.e., $u(x,t)=X(x)T(t)$. Using equation (5.2) and **BC1**, we get

$$u(0,t)=X(0)T(t) \Rightarrow X(0)T(t)=0 \Rightarrow X(0)=0. \tag{5.8}$$

($\because T(t)\neq 0$ as we are look for non –zero solution)

Similarly, using equation (5.2) and **BC2**, we get

$$u(l,t)=X(l)T(t) \Rightarrow X(l)T(t)=0 \Rightarrow X(l)=0. \tag{5.9}$$

($\because T(t)\neq 0$ as we are look for non –zero solution)

Now, we solve $X'' - kX = 0$ (5.10)

with boundary condition

$$X(0) = 0 \text{ and } X(l) = 0 \tag{5.11}$$

which is called an Eigen value problem and *k* is called Eigen value.

Case (i) when k is positive.

Let $k = p^2$ for some $p \in R$.

The equation (5.10) reduces to $X'' - p^2 X = 0$. The auxiliary equation associate with $X'' - p^2 X = 0$ is $m^2 - p^2 = 0$

$$\Rightarrow m = p, -p$$

Since the roots of the auxiliary equation are real and distinct, therefore

$$X(x) = Ae^{px} + Be^{-px} \tag{5.12}$$

where *A* and *B* are arbitrary constant. Now we apply the boundary condition given by equation (5.11) to find the value of *A* and *B*. Put $x = 0$ in equation (5.12), we get

$$X(0) = Ae^0 + Be^0 \Rightarrow X(0) = A + B \tag{5.13}$$

Using equation (5.11) in equation (5.13), we get

$$A = -B. \tag{5.14}$$

Put $x = l$ in equation (5.12), we get

$$X(l) = Ae^{pl} + Be^{-pl} \Rightarrow X(l) = Ae^{pl} + Be^{-pl} \quad (5.15)$$

Using equation (5.11) and (5.14) in equation (5.15), we get

$$X(l) = Ae^{-pl}\left(e^{2pl} - 1\right) = 0 \quad (5.16)$$

$$\Rightarrow A = 0 \ \left(\because e^{pl}\left(e^{2pl} - 1\right) \neq 0\right)$$

Since $A = 0$, therefore from equation (5.14), we get $B = 0$. So, equation (5.12) reduces to $X(x) = 0$. Substituting $X(x) = 0$ in equation (5.2), the required solution is $u(x,t) = 0$. So, when k is positive, there is no Eigen value, and we get a trivial solution.

Case (ii) when k is zero.

Then the auxiliary equation of equations (5.10) is $m^2 = 0 \Rightarrow m = 0, 0$. Since, the roots are real and repeated, therefore solution of equation (5.10) is $X(x) = (A + Bx)$. (5.17)

Now, we use boundary condition $X(0) = 0$ and $X(l) = 0$ to find the value of arbitrary constant.

Put $x = 0$ in equation (5.17), we get from equation (5.11)

$$X(0) = (A + B0) \Rightarrow A = 0 \quad (5.18)$$

Put $x = l$ in equation (5.17), we get $X(l) = A + Bl \Rightarrow Bl = 0$, (From equation (5.18))

$$\Rightarrow B = 0. \ (\because l \neq 0) \quad (5.19)$$

Since $A=0$ and $B=0$, therefore $X(x)=0$.Substituting $X(x)=0$ in equation (5.2), the required solution is $u(x,t)=0$. So, when k is zero, there is no Eigen value, and we get a trivial solution.

Case (iii) when k is negative

Let $k=-p^2$.

Then auxiliary equation of equations (5.10) is $m^2+p^2=0 \Rightarrow m=\pm i\,p$. Since, the roots are complex, therefore solution of equation (5.10) is

$$X(x)=(A\cos px+B\sin px) \tag{5.20}$$

where *A* and *B* are arbitrary constants. Now, we use boundary condition $X(0)=0$ and $X(l)=0$ to find the value of arbitrary constant.

Put $x=0$ in equation (5.20), we get

$$X(0)=(A\cos p0+B\sin p0) \Rightarrow A=0 \tag{5.21}$$

Put $x=l$ in equation (5.20), we get

$$X(l)=(A\cos pl+B\sin pl) \Rightarrow B\sin pl=0,$$

(From equation (5.21))

$\Rightarrow B=0$ or $\sin pl=0$,

$\Rightarrow \sin pl=0$,

($\because B\neq 0$ as $X(x)=0$)

$$\Rightarrow pl = \sin^{-1}(0)$$
$$\Rightarrow pl = n\pi,$$

$$\Rightarrow p = \frac{n\pi}{l}, n = 1, 2, 3, \ldots \quad (5.22)$$

For $n = 1, 2, 3, \ldots\ldots$, we get different value of p, so we denote p by p_n. Therefore $p = p_n = \frac{n\pi}{l}$. Since $p_n = \frac{n\pi}{l}$, therefore from equation (5.20)

$$X(x) = B\sin\left(\frac{n\pi}{l}\right)x \quad (5.23)$$

For $n = 1, 2, 3, \ldots\ldots$, we get different value of $X(x)$ using equation (5.23), so we denote $X(x)$ by $X_n(x)$. Therefore

$$X_n(x) = B_n \sin\left(\frac{n\pi}{l}\right)x. \quad (5.24)$$

Similarly, when k is negative i.e., $k = -p^2$ then equation (5.7) changes to $T' + C^2 p^2 T = 0$. Using variable separable form to solve $T' + C^2 p^2 T = 0$, we get

$$\frac{T'}{T} = -C^2 p^2\, dt,$$

Integrating both the sides, we get $\log T = -C^2 p^2 t + c$, where c is an arbitrary constant. On further simplifying $\log T = -C^2 p^2 t + c$, we get

$$T = e^{-C^2 p^2 t + c} = e^c\, e^{-C^2 p^2 t} = C' e^{-C^2 p^2 t} \quad (5.25)$$

where $C' = e^c$. Using equation (5.22), equation (5.25) changes to $T = C'e^{-C^2\left(\frac{n\pi}{l}\right)^2 t}$. For $n = 1, 2, 3, ...$, we get different value of $T(t)$, so we denote $T(t)$ by $T_n(t)$. Therefore

$$T_n(t) = C_n' e^{-C^2 p^2 t}. \tag{5.26}$$

Substituting equations (5.25) and (5.26), the solution satisfying the boundary condition is given by

$$u_n(x,t) = B_n \sin\left(\frac{n\pi x}{l}\right) C_n' e^{-C^2\left(\frac{n\pi}{l}\right)^2 t} \tag{5.27}$$

For $n = 1, 2, 3, ...$, we get different value of $u(x,t)$, so we denote $u(x,t)$ by $u_n(x,t)$. Therefore, equation (5.27) can be written as

$$u_n(x,t) = B_n \sin\left(\frac{n\pi x}{l}\right) C_n' e^{-C^2\left(\frac{n\pi}{l}\right)^2 t} \tag{5.28}$$

equation (5.28) is the solution of equation (5.1) and satisfies the boundary conditions BC1 and BC2. We know that if y_1 and y_2 are two solutions of a linear, homogenous equation satisfying linear homogenous boundary conditions, then $y_1 + y_2$ is also a solution of that equation and satisfies the same boundary condition. This is called the principle of superposition. Now the equation (5.28) with boundary conditions BC1 and BC2 are linear and homogenous. Therefore, by the principle of superposition, the series

$$u(x,t) = \sum_{n=1}^{\infty} u_n(x,t) \tag{5.29}$$

$$u(x,t)=\sum_{n=1}^{\infty} B_n \sin\left(\frac{n\pi x}{l}\right) C_n^{'} e^{-C^2\left(\frac{n\pi}{l}\right)^2 t} \tag{5.30}$$

$$u(x,t)=\sum_{n=1}^{\infty} A_n \sin\left(\frac{n\pi x}{l}\right) e^{-C^2\left(\frac{n\pi}{l}\right)^2 t} \text{ where } A_n = B_n C_n^{'} \tag{5.31}$$

converges and is solution of one-dimensional heat equation given by equation (5.1) satisfying BC1 and BC2. Now, we use initial condition **IC1** to find A_n. Put $t = 0$ in equation (5.31), we get

$$u(x,0)=\sum_{n=1}^{\infty} A_n \sin\left(\frac{n\pi}{l}\right) x.$$

From **IC1,** we get $f(x)=\sum_{n=1}^{\infty} A_n \sin\left(\frac{n\pi x}{l}\right)$, which is half range Fourier sine series and

$$A_n = \frac{2}{l}\int_0^l f(x)\sin\left(\frac{n\pi x}{l}\right)dx. \tag{5.32}$$

Therefore, the required solution to the one-dimensional heat equation given by equation (5.1) with (5.1.a)- (5.1.c) is given by

$$u(x,t)=\sum_{n=1}^{\infty} A_n \sin\left(\frac{n\pi x}{l}\right) e^{-C^2\left(\frac{n\pi}{l}\right)^2 t}, \text{ where}$$

$$A_n = \frac{2}{l}\int_0^l f(x)\sin\left(\frac{n\pi x}{l}\right)dx \tag{5.33}$$

Example:

Consider the problem

$$u_t = k u_{xx},\ 0 < x < l,\ t > 0 \tag{5.34}$$

subject to initial and boundary condition as:

$$u(x,0)=f(x)=\begin{cases} x, & ,0\le x\le \dfrac{l}{2} \\ l-x & ,\dfrac{l}{2}\le x\le l \end{cases} \tag{5.35}$$

and $u(0,t)=0=u(l,t)$. (5.36)

Solution:

Comparing the given PDE $u_t = k\,u_{xx}$ and initial and boundary condition with equation (5.1), equation (5.1a-5.1c), we get $C^2 = k$ and $f(x)=\begin{cases} x, & ,0\le x\le \dfrac{l}{2} \\ l-x & ,\dfrac{l}{2}\le x\le \pi \end{cases}$. The solution of heat equation given by equation (5.34) – (5.36) according to equation (5.33) is given by

$$u(x,t)=\sum_{n=1}^{\infty} A_n \sin\left(\frac{n\pi x}{l}\right)e^{-\frac{n^2\pi^2 kt}{l^2}} \tag{5.36}$$

where $A_n = \dfrac{2}{l}\displaystyle\int_0^l f(x)\sin\left(\frac{n\pi x}{l}\right)dx$,

$$A_n = \frac{2}{l}\int_0^{\frac{l}{2}} f(x)\sin\left(\frac{n\pi x}{l}\right)dx + \int_{\frac{l}{2}}^{l} f(x)\sin\left(\frac{n\pi x}{l}\right)dx. \tag{5.37}$$

$$A_n = \frac{2}{l}\int_0^{\frac{l}{2}} x \sin\left(\frac{n\pi x}{l}\right)dx + \frac{2}{l}\int_{\frac{l}{2}}^{l}(l-x)\sin\left(\frac{n\pi x}{l}\right)dx,$$

$$A_n = \left[\left(\frac{2x}{l}\right)\left(-\frac{\cos\left(\frac{n\pi x}{l}\right)}{\frac{n\pi}{l}}\right) - \left(\frac{2}{l}\right)\left(-\frac{\sin\left(\frac{n\pi x}{l}\right)}{\left(\frac{n\pi}{l}\right)^2}\right)\right]_0^{\frac{l}{2}}$$

$$+\left[\left(\frac{2(l-x)}{l}\right)\left(-\frac{\cos\left(\frac{n\pi x}{l}\right)}{\frac{n\pi}{l}}\right) - \left(-\frac{2}{l}\right)\left(-\frac{\sin\left(\frac{n\pi x}{l}\right)}{\left(\frac{n\pi}{l}\right)^2}\right)\right]_{\frac{l}{2}}^{l}$$

$$A_n = -\frac{l}{n\pi}\cos\left(\frac{n\pi}{2}\right) + \left(\frac{2l}{n^2\pi^2}\right)\sin\left(\frac{n\pi}{2}\right) + \frac{l}{n\pi}\cos\left(\frac{n\pi}{2}\right) + \left(\frac{2l}{n^2\pi^2}\right)\sin\left(\frac{n\pi}{2}\right),$$

$$A_n = \frac{4l}{n^2\pi^2}\sin\left(\frac{n\pi}{2}\right) = \begin{cases} 0 & ,if \;\; n = 2m \;\; and \;\; m = 1,2,3 \\ \dfrac{4l}{(2m-1)^2\pi^2} & ,if \;\; n = 2m-1 \;\; and \;\; m = 1,2,3,... \end{cases}.$$

From equation (5.36), we get

$$u(x,t) = \frac{4l}{\pi^2}\sum_{m=1}^{\infty}\frac{1}{(2m-`1)^2}\sin\left(\frac{(2m-`1)\pi x}{l}\right)e^{-\frac{(2m-`1)^2\pi^2 kt}{l^2}}$$

Example:

Consider the problem $u_t = k\,u_{xx}$, $0 < x < l$, $t > 0$, (5.38)

subject to initial and boundary condition as:

$$u(x,0)=f(x)=lx-x^2 \tag{5.39}$$

and $u(0,t)=0=u(l,t).$ (5.40)

Solution:

Comparing the given PDE $u_t = k u_{xx}$ and initial and boundary condition with equation (5.1), equation (5.1a-5.1c), we get $C^2 = k$ and $f(x)=lx-x^2$. The solution of heat equation given by equation (5.38) – (5.40) according to equation (5.33) is given by

$$u(x,t)=\sum_{n=1}^{\infty} A_n \sin\left(\frac{n\pi x}{l}\right) e^{-\frac{n^2\pi^2 kt}{l^2}} \tag{5.41}$$

where $A_n = \frac{2}{l}\int_0^l f(x)\sin\left(\frac{n\pi x}{l}\right)dx$,

$$A_n = \frac{2}{l}\int_0^l f(x)\sin\left(\frac{n\pi x}{l}\right)dx \tag{5.42}$$

$$A_n = \frac{2}{l}\int_0^l (lx-x^2)\sin\left(\frac{n\pi x}{l}\right)dx,$$

$$A_n = \left[\begin{array}{l}(lx-x^2)\left(-\frac{\cos(nx)}{\frac{n\pi}{l}}\right)-(l-2x)\left(-\frac{\sin\left(\frac{n\pi x}{l}\right)}{\left(\frac{n\pi}{l}\right)^2}\right)\\ +(-2)\left(\frac{\cos\left(\frac{n\pi x}{l}\right)}{\left(\frac{n\pi}{l}\right)^3}\right)\end{array}\right]_0^l$$

$$A_n = \frac{2}{l}\left\{-\frac{2l^3}{n^3\pi^3}\cos n\pi + \left(\frac{2l^3}{n^3\pi^3}\right)\right\},$$

$$A_n = \frac{4l^2}{n^3\pi^3}\left\{1-(-1)^n\right\} \ ,$$

$$\therefore A_n = \frac{4l^2}{n^3\pi^3}\left\{1-(-1)^n\right\} = \begin{cases} \dfrac{8l^2}{(2m-1)^3\pi^3} & ,if\ n = 2m-1\ and\ m = 1,2,3,.. \\ 0 & ,if\ n = 2m\ where\ m = 1,2,3,.... \end{cases}$$

.

From equation (5.42), we get

$$u(x,t) = \frac{8l^2}{\pi^3}\sum_{m=1}^{\infty}\frac{1}{(2m-`1)^3}\sin\left(\frac{(2m-`1)\pi x}{l}\right)e^{-\frac{(2m-`1)^2\pi^2 kt}{l^2}} \quad .$$

Example:

A homogenous rod of conducting material of length *a* has its end kept at zero temperature. The temperature at the center is T and falls uniformly to zero at the two ends. Find the temperature function $u(x,t)$.

Solution:

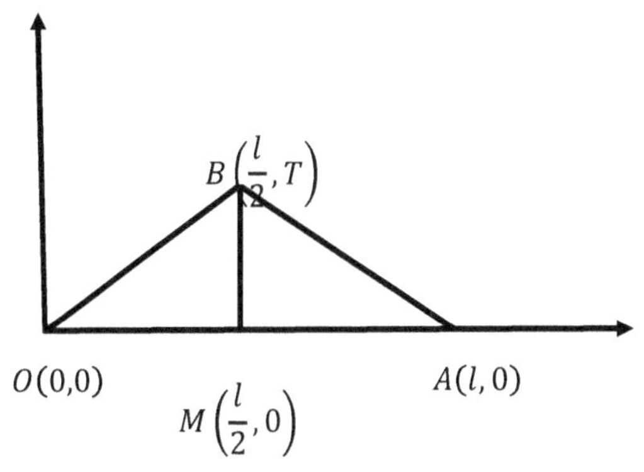

Let $u(x,t)$ be the solution of heat equation PDE $u_t = k u_{xx}$. Here the boundary conditions are $u(0,t) = 0 = u(l,t)$ for all $t \geq 0$. The given rod is represented by OA, as shown in the figure, of length l and M be its middle point. The temperature at the center is T according to given question and falls uniformly to zero at the two ends O and A of the rod. Hence the temperature distribution at initial time is shown in Figure. The equation of the line OB and BA respectively are

$$u - 0 = \frac{T-0}{\frac{l}{2}-0}(x-0) \Rightarrow u = \frac{2Tx}{l},$$

and

$$u - 0 = \frac{T-0}{\frac{l}{2}-l}(x-a) \Rightarrow u = \frac{2T}{l}(l-x).$$

Therefore, the initial condition is given by

$$u(x,0) = f(x) = \begin{cases} \frac{2Tx}{l}, & ,0 \leq x \leq \frac{l}{2} \\ \frac{2T(l-x)}{l} & ,\frac{l}{2} \leq x \leq \pi \end{cases}.$$

The solution of heat equation $u_t = k u_{xx}$ subject to boundary condition $u(0,t) = 0 = u(l,t)$ for all $t \geq 0$ and initial condition

$$u(x,0) = f(x) = \begin{cases} \frac{2Tx}{l}, & ,0 \leq x \leq \frac{l}{2} \\ \frac{2T(l-x)}{l} & ,\frac{l}{2} \leq x \leq \pi \end{cases}$$

is given by

$$u(x,t)=\sum_{n=1}^{\infty}A_n\sin\left(\frac{n\pi x}{l}\right)e^{-\frac{n^2\pi^2kt}{l^2}} \tag{5.43}$$

where $A_n=\frac{2}{l}\int_0^l f(x)\sin\left(\frac{n\pi x}{l}\right)dx$,

$$A_n=\frac{2}{l}\int_0^{\frac{l}{2}} f(x)\sin\left(\frac{n\pi x}{l}\right)dx+\int_{\frac{l}{2}}^{l} f(x)\sin\left(\frac{n\pi x}{l}\right)dx. \tag{5.44}$$

$$A_n=\frac{2}{l}\int_0^{\frac{l}{2}}\frac{2Tx}{l}\sin\left(\frac{n\pi x}{l}\right)dx+\frac{2}{l}\int_{\frac{l}{2}}^{l}\frac{2T(l-x)}{l}\sin\left(\frac{n\pi x}{l}\right)dx,$$

$$A_n=\int_0^{\frac{l}{2}}\frac{4Tx}{l^2}\sin\left(\frac{n\pi x}{l}\right)dx+\int_{\frac{l}{2}}^{l}\frac{4T(l-x)}{l^2}\sin\left(\frac{n\pi x}{l}\right)dx$$

$$A_n=\left[\left(\frac{4x}{l}\right)\left(-\frac{\cos\left(\frac{n\pi x}{l}\right)}{\frac{n\pi}{l}}\right)-\left(\frac{2}{l}\right)\left(-\frac{\sin\left(\frac{n\pi x}{l}\right)}{\left(\frac{n\pi}{l}\right)^2}\right)\right]_0^{\frac{l}{2}}$$

$$+\left[\left(\frac{2(l-x)}{l}\right)\left(-\frac{\cos\left(\frac{n\pi x}{l}\right)}{\frac{n\pi}{l}}\right)-\left(-\frac{2}{l}\right)\left(-\frac{\sin\left(\frac{n\pi x}{l}\right)}{\left(\frac{n\pi}{l}\right)^2}\right)\right]_{\frac{l}{2}}^{l}$$

$$A_n=-\frac{l}{n\pi}\cos\left(\frac{n\pi}{2}\right)+\left(\frac{2l}{n^2\pi^2}\right)\sin\left(\frac{n\pi}{2}\right)+\frac{l}{n\pi}\cos\left(\frac{n\pi}{2}\right)+\left(\frac{2l}{n^2\pi^2}\right)\sin\left(\frac{n\pi}{2}\right)$$

,

$$A_n = \frac{4l}{n^2\pi^2}\sin\left(\frac{n\pi}{2}\right) = \begin{cases} 0 & ,if \;\; n = 2m \;\; and \;\; m = 1,2,3 \\ \dfrac{4l}{(2m-1)^2\,\pi^2} & ,if \;\; n = 2m-1 \;\; and \;\; m = 1,2,3,... \end{cases}.$$

From equation (5.43), we get

$$u(x,t) = \frac{4l}{\pi^2}\sum_{m=1}^{\infty}\frac{1}{(2m-`1)^2}\sin\left(\frac{(2m-`1)\pi x}{l}\right)e^{-\frac{(2m-`1)^2\pi^2 kt}{l^2}}.$$

Example:

Consider the heat equation

$$u_t = u_{xx},\; 0 < x < 1, t > 0 \tag{5.45}$$

subject to the initial and boundary conditions

$$u(x,0) = x - x^2 \tag{5.46}$$

$$u(0,t) = u(1,t) = 0. \tag{5.47}$$

Solution:

Comparing the given PDE $u_t = k\,u_{xx}$ and initial and boundary condition with equation (5.1), equation (5.1a-5.1c), we get $C^2 = k$, $l = 1$ and $f(x) = x - x^2$. The solution of heat equation given by equation (5.45) – (5.47) according to equation (5.33) is given by

$$u(x,t) = \sum_{n=1}^{\infty} A_n \sin(n\pi x)e^{-n^2\pi^2 kt} \tag{5.48}$$

where $A_n = 2\int_0^1 f(x)\sin(n\pi x)\,dx$ (5.49)

$$A_n = 2\int_0^1 \left(x - x^2\right)\sin\left(n\pi x\right)dx,$$

$$A_n = \left[\left(x - x^2\right)\left(-\frac{\cos\left(nx\pi\right)}{n\pi}\right) - \left(1-2x\right)\left(-\frac{\sin\left(nx\pi\right)}{\left(n\pi\right)^2}\right) + \left(-2\right)\left(\frac{\cos\left(n\pi x\right)}{\left(n\pi\right)^3}\right)\right]_0^1$$

$$A_n = 2\left\{-\frac{2}{n^3\pi^3}\cos n\pi + \frac{2}{n^3\pi^3}\right\},$$

$$A_n = \frac{4}{n^3\pi^3}\left\{1-\left(-1\right)^n\right\} ,$$

$$\therefore A_n = \frac{4}{n^3\pi^3}\left\{1-\left(-1\right)^n\right\} = \begin{cases} \dfrac{8}{\left(2m-1\right)^3\pi^3} & , \mathit{if}\ n = 2m-1\ \mathit{and}\ m = 1,2,3,.. \\ 0 & , \mathit{if}\ n = 2m\ , \mathit{where}\ m = 1,2,3,.... \end{cases}$$

From equation (5.42), we get

$$u\left(x,t\right) = \frac{8}{\pi^3}\sum_{m=1}^{\infty}\frac{1}{\left(2m-`1\right)^3}\sin\left(\left(2m-1\right)\pi x\right)e^{-\left(2m-`1\right)^2\pi^2 kt} .$$

Example:

An insulated rod of length l has its end A and B maintained $0^\circ C$ and $100^\circ C$ respectively until steady state conditions prevail. If B is suddenly reduced to $0^\circ C$ and maintained at $0^\circ C$.Find the temperature at a distance x from A at time t.

Solution:

The temperature function $u\left(x,t\right)$ satisfies the one-dimensional heat equation is $u_t = a^2 u_{xx} \Rightarrow \frac{\partial u}{\partial t} = a^2\frac{\partial^2 u}{\partial x^2}$. (5.50)

From the problem, the boundary conditions are **BC1:** $u(0,t)=0$; **BC2:** $u(l,t)=0$.

According to steady state condition

$$\frac{\partial u}{\partial t}=0 . \tag{5.51}$$

Substituting the value from equation (5.51) in (5.50), we get

$$0=\frac{\partial^2 u}{\partial x^2}.$$

Integrating above equation w.r.t. x twice, we get

$$u=C_1 x+C_2 \tag{5.52}$$

where C_1 and C_2 are constants to be determined.

From first boundary condition **BC1:** $u(0,t)=0$, equation (5.52) gives at $C_2=0$.

From second boundary condition **BC2:** $u(l,t)=100$, equation (5.52) gives at

$$C_1=\frac{100}{l}.$$

Substituting values $C_1=\frac{100}{l}$ and $C_2=0$ in equation (5.52), we get

$$u(x)=\frac{100}{l}x.$$

Now the temperature at B is suddenly changed and we have again transient state. If $u(x,t)$ is the subsequent temperature function, the boundary conditions are **BC1:** $u(0,t)=0$, **BC2:** $u(l,t)=0$ and the initial condition is **IC:** $u(x,0)=\frac{100}{l}x$.

The above problem is stated as: $\frac{\partial u}{\partial t}=a^2\frac{\partial^2 u}{\partial x^2}$,

with initial and boundary condition

IC: $u(x,0)=\frac{100}{l}x$,

BC1: $u(0,t)=0$,

BC2: $u(l,t)=0$.

Comparing the given PDE $\frac{\partial u}{\partial t}=a^2\frac{\partial^2 u}{\partial x^2}$ and initial and boundary condition with equation (5.1), equation (5.1a-5.1c), we get $C^2=a^2$, $l=l$ and $f(x)=\frac{100}{l}x$. The solution of heat equation given by equation (5.38) – (5.40) according to equation (5.33) is given by

$$u(x,t)=\sum_{n=1}^{\infty}A_n\sin\left(\frac{n\pi x}{l}\right)e^{-\frac{n^2\pi^2 kt}{l^2}} \tag{5.53}$$

where $A_n=\frac{2}{l}\int_0^l f(x)\sin\left(\frac{n\pi x}{l}\right)dx$,

$$A_n=\frac{2}{l}\int_0^l \frac{100}{l}x\sin\left(\frac{n\pi x}{l}\right)dx,$$

$$A_n = \frac{200}{l^2}\left[x\left\{ \frac{\cos\frac{n\pi x}{l}}{\frac{n\pi}{l}} \right\} - 1\left\{ -\frac{\sin\frac{n\pi x}{l}}{\left(\frac{n\pi}{l}\right)^2} \right\} \right]_0^l ,$$

$$A_n = \frac{200}{l^2}\left[-\frac{l^2}{n\pi}\cos n\pi \right],$$

$$A_n = -\frac{200}{n\pi}(-1)^n ,$$

$$A_n = \frac{200}{n\pi}(-1)^{n+1} .$$

From equation (5.53), we get

$$u(x,t) = \sum_{n=1}^{\infty} \frac{200}{n\pi}(-1)^{n+1} \sin\left(\frac{n\pi x}{l}\right) e^{-\frac{c^2 n^2 \pi^2 t}{l^2}} .$$

Example:

Find the temperature in a bar of 2 whose ends kept at zero and lateral surface insulated if the initial temperature is

$$\sin\left(\frac{\pi x}{2}\right) + 3\sin\left(\frac{5\pi x}{2}\right).$$

Solution:

Here we have a one-dimensional heat flow equation is $\frac{\partial u}{\partial t} = a^2 \frac{\partial^2 u}{\partial x^2}$. According to the problem the following are the initial and boundary condition:

IC: $u(x,0)=\sin\left(\frac{\pi x}{2}\right)+3\sin\left(\frac{5\pi x}{2}\right)$,

BC1: $u(0,t)=0$,

BC2: $u(l,t)=0$.

Comparing the given problem with initial and boundary condition with equation (5.1), equation (5.1a-5.1c), we get $C^2=a^2$, $l=2$ and $f(x)=\sin\left(\frac{\pi x}{2}\right)+3\sin\left(\frac{5\pi x}{2}\right)$. The solution of heat equation is given by $u(x,t)=\sum_{n=1}^{\infty}A_n\sin\left(\frac{n\pi x}{2}\right)e^{-\frac{n^2\pi^2kt}{4}}$ (5.54)

Now to find A_n, we use the initial condition. Put $t=0$ in equation (5.54), we get

$$u(x,0)=\sum_{n=1}^{\infty}A_n\sin\left(\frac{n\pi x}{2}\right),$$

$$\sin\left(\frac{\pi x}{2}\right)+3\sin\left(\frac{5\pi x}{2}\right)=\sum_{n=1}^{\infty}A_n\sin\left(\frac{n\pi x}{2}\right), (\because \text{From IC})$$

$$\sin\left(\frac{\pi x}{2}\right)+3\sin\left(\frac{5\pi x}{2}\right)=A_1\sin\left(\frac{\pi x}{2}\right)+A_2\sin(\pi x)+A_3\sin\left(\frac{3\pi x}{2}\right)+A_4\sin(2\pi x)+A_5\sin\left(\frac{5\pi x}{2}\right)$$
$$+A_6\sin(3\pi x)+...$$

Comparing both sides of the above equation, we get

$$A_1=1, A_2=A_3=A_4=0, A_5=3, A_n=0, n\geq 6.$$

Substituting the values of $A_1=1, A_2=A_3=A_4=0, A_5=3, A_n=0, n\geq 6$ in equation (5.54), the required solution is

$$u(x,t)=\sin\left(\frac{\pi x}{2}\right)e^{-\frac{a^2\pi^2 t}{4}}+\sin\left(\frac{5\pi x}{2}\right)e^{-\frac{25a^2\pi^2 t}{4}}.$$

Example:

Determine the solution of one-dimensional heat equation $\frac{\partial u}{\partial t}=c^2\frac{\partial^2 u}{\partial x^2}$ subject to the boundary conditions $u(0,t)=0=u(l,t)$ $,t>0$ and the initial condition $u(x,0)=x$ and l being the length of the bar.

Solution:

The given heat problem is stated as:

$$\frac{\partial u}{\partial t}=c^2\frac{\partial^2 u}{\partial t^2},\ 0<x<l,t>0, \tag{5.55}$$

subject to initial and boundary condition as:

$$u(x,0)=f(x)=x \tag{5.56}$$

and $u(0,t)=0=u(l,t)$. (5.57)

Comparing the given PDE $u_t=c^2u_{xx}$ and initial and boundary condition with equation (5.1), equation (5.1a-5.1c), we get $C^2=c^2$ and $f(x)=x$. The solution of heat equation given by

$$u(x,t)=\sum_{n=1}^{\infty}A_n\sin\left(\frac{n\pi x}{l}\right)e^{-\frac{n^2\pi^2 kt}{l^2}} \tag{5.58}$$

where $A_n=\frac{2}{l}\int_0^l f(x)\sin\left(\frac{n\pi x}{l}\right)dx$,

$$A_n = \frac{2}{l}\int_0^l x\sin\left(\frac{n\pi x}{l}\right)dx,$$

$$A_n = \frac{2}{l}\left[x\left(-\frac{\cos\left(\frac{n\pi x}{l}\right)}{\frac{n\pi}{l}}\right)-\left(-\frac{\sin\left(\frac{n\pi x}{l}\right)}{\left(\frac{n\pi}{l}\right)^2}\right)\right]_0^l,$$

$$A_n = \frac{2}{l}\left[x\left(-\frac{\cos\left(\frac{n\pi x}{l}\right)}{\frac{n\pi}{l}}\right)-\left(-\frac{\sin\left(\frac{n\pi x}{l}\right)}{\left(\frac{n\pi}{l}\right)^2}\right)\right]_0^l,$$

$$A_n = \frac{2}{l}\left[x\left(-\frac{\cos\left(\frac{n\pi x}{l}\right)}{\frac{n\pi}{l}}\right)-\left(-\frac{\sin\left(\frac{n\pi x}{l}\right)}{\left(\frac{n\pi}{l}\right)^2}\right)\right]_0^l,$$

$$A_n = \frac{2}{l}\left[\left\{\left(-\frac{l^2\cos(n\pi)}{n\pi}\right)-\left(-\frac{l^2\sin(n\pi)}{(n\pi)^2}\right)\right\}-(0-0)\right]$$

$$A_n = \frac{-2l(-1)^n}{n\pi} = \frac{2l(-1)^{n+1}}{n\pi}. \tag{5.59}$$

From equation (5.58) and (5.59), we get

$$u(x,t) = \sum_{n=1}^{\infty}\frac{2l(-1)^{n+1}}{n\pi}\sin\left(\frac{n\pi x}{l}\right)e^{-\frac{n^2c^2\pi^2 t}{l^2}},$$

which is the required solution to the given problem.

Example:

Solve the equation $\frac{\partial u}{\partial t}=\frac{\partial^2 u}{\partial x^2}$ with boundary condition $u(x,0)=3\sin n\pi x,\ u(0,t)=u(1,t)=0, 0<x<1$.

Solution:

Here we have one-dimensional heat flow equation is $\frac{\partial u}{\partial t}=\frac{\partial^2 u}{\partial x^2}$,

with initial and boundary condition

$$u(x,0)=3\sin n\pi x,$$

$$u(0,t)=u(1,t)=0, 0<x<1.$$

Comparing the given above heat problem with equation (5.1), equation (5.1a-5.1c), we get $C^2=1, l=1$ and $f(x)=3\sin n\pi x$, its solution of heat equation given by

$$u(x,t)=\sum_{n=1}^{\infty} A_n \sin(n\pi x)e^{-n^2\pi^2 t} \tag{5.60}$$

We apply initial condition to equation (5.60) to find A_n. Substituting $t=0$ in equation (5.60), we get

$$u(x,0)=\sum_{n=1}^{\infty} A_n \sin(n\pi x). \tag{5.61}$$

Using the initial condition, equation (5.61) reduces to

$$3\sin n\pi x=\sum_{n=1}^{\infty} A_n \sin(n\pi x),$$

$$3\sin n\pi x = A_1\sin(\pi x) + A_2\sin(2\pi x) + A_3\sin(3\pi x) + \dots ,$$

Comparing both sides of the above equation, we get

$$A_1 = 0, A_2 = 0, A_3 = 3, A_4 = 0, A_5 = 0, \dots\dots$$

Substituting the values of $A_1 = 0, A_2 = 0, A_3 = 3, A_4 = 0, A_5 = 0, \dots..$ in equation (5.61), the required solution is

$$u(x,t) = 3\sin(n\pi x)e^{-\frac{\pi^2 t}{4}}.$$

Example:

A rod of length *l cm* with insulated lateral surface is initially at temperature u_0 at an inner point distance *x cm* from one end. If both ends are kept at zero temperature. Find the temperature function $u(x,t)$.

Solution:

The temperature function $u(x,t)$ satisfies the one-dimensional heat equation is $\frac{\partial u}{\partial t} = a^2\frac{\partial^2 u}{\partial x^2}$

where $a^2 = \frac{k}{\rho c} = \frac{\text{Thermal conductivity}}{\text{(density)(specific heat)}}$.

From the problem, the boundary and initial conditions are

(1) $u(x=0,t) = u(0,t) = 0, \forall t > 0$.

(2) $u(x=l,t) = u(l,t) = 0, \forall t > 0$.

(3) $u(x,t=0)=u(x,0)=f(x)=u_0, \forall t>0$.

Comparing the given above heat problem with equation (5.1), equations (5.1a-5.1c), we get $C^2=a^2, l=l$ and $f(x)=u_0$, its solution of heat equation given by

$$u(x,t)=\sum_{n=1}^{\infty} A_n \sin\left(\frac{n\pi x}{l}\right) e^{-\frac{n^2\pi^2 a^2 t}{l^2}} \tag{5.62}$$

where $A_n = \frac{2}{l}\int_0^l f(x)\sin\left(\frac{n\pi x}{l}\right)dx$,

$$A_n = \frac{2}{l}\int_0^l u_0 \sin\left(\frac{n\pi x}{l}\right)dx,$$

$$A_n = \frac{2}{l}\left[u_0\left(-\frac{\cos\left(\frac{n\pi x}{l}\right)}{\frac{n\pi}{l}}\right)-0\left(-\frac{\sin\left(\frac{n\pi x}{l}\right)}{\left(\frac{n\pi}{l}\right)^2}\right)\right]_0^l,$$

$$A_n = \frac{2}{l}\left[-\frac{u_0 l\cos(n\pi)}{n\pi}-\left(\frac{u_0 l}{n\pi}\right)\right]$$

$$A_n = \frac{2}{l}\left[-\frac{u_0 l(-1)^n}{n\pi}-\left(\frac{u_0 l}{n\pi}\right)\right]$$

$$A_n = \frac{2u_0\left((-1)^{n+1}-1\right)}{n\pi}$$

$$A_n = \frac{2u_0\left(\left(-1\right)^{n+1}-1\right)}{n\pi} = \begin{cases} \frac{4u_0}{n\pi} & ,if\ n\ is\ odd \\ 0 & ,if\ n\ is\ even \end{cases}. \tag{5.63}$$

Using equations (5.62) and (5.63), we get

$$u(x,t) = \sum_{m=1}^{\infty} \frac{4u_0}{(2m-1)\pi} \sin\left(\frac{(2m-1)\pi x}{l}\right) e^{-\frac{(2m-1)^2 \pi^2 a^2 t}{l^2}},$$

which is required solution.

5.4 General solution of heat equation when both the ends of a bar are insulated, and the initial temperature is prescribed

The governing one-dimensional heat flow equation is given by

$$\frac{\partial u}{\partial t} = C^2 \frac{\partial^2 u}{\partial x^2} \tag{5.64}$$

According to the physical experiment, the rate of heat flow is proportional to the gradient $\frac{\partial u}{\partial x}$ of the temperature $u(x,t)$. Hence if the ends $x=0$ and $x=l$ are insulated, no heat flows through the ends, therefore the boundary conditions are given by

$$\frac{\partial u(0,t)}{\partial x} = \frac{\partial u(l,t)}{\partial x} = 0 \tag{5.65}$$

for $\forall t$, with initial condition $u(x,0) = f(x), 0 \le x \le l$. (5.66)

Assume that a solution of equation (5.64) is of the form

$$u(x,t) = X(x)T(t) \tag{5.67}$$

where X is a function of x and T is a function of t only. Differentiate equation (5.67) partially w.r.t. x and t, we have

$$\frac{\partial u}{\partial x} = X^{'}T \Rightarrow \frac{\partial^2 u}{\partial x^2} = X^{'}T \tag{5.68}$$

and $$\frac{\partial u}{\partial t} = XT^{'} \Rightarrow \frac{\partial^2 u}{\partial t^2} = XT^{''} \tag{5.69}$$

where $X^{'} = \frac{dX}{dx}, X^{''} = \frac{d^2X}{dx^2}, T^{'} = \frac{dT}{dx}, T^{''} = \frac{d^2T}{dx^2}$. Substituting the values from equations (5.67), (5.68) and (5.69) in (5.64), we have $XT^{'} = C^2X^{''}T$.

Separating the variables in the above equation, we get

$$\frac{1}{C^2}\frac{T^{'}}{T} = \frac{X^{''}}{X}. \tag{5.70}$$

In equation (5.70), L.H.S is a function of t only and R.H.S is a function of x only and here t and x are independent variables, so each side is constant. So rewriting equation (5.71) as

$$\frac{X^{''}}{X} = k \Rightarrow X^{''} = kX \Rightarrow X^{''} - kX = 0, \tag{5.71}$$

and $$\frac{1}{C^2}\frac{T}{T^{'}} = k \Rightarrow T^{'} = kT \Rightarrow T^{'} - C^2kT = 0. \tag{5.72}$$

equations (5.71) and (5.72) which are second and first order homogeneous linear differential equation, respectively. Differentiating equation (5.67) w.r.t x, we get

$$u_x(x,t) = X^{'}(x)T(t). \tag{5.73}$$

Now using first boundary condition **BC1**: $\frac{\partial u(0,t)}{\partial x}=0,\ \forall t>0$, equation (5.73) gives

$$X^{'}(0)T(t) \Rightarrow X^{'}(0)T(t)=0 \Rightarrow X(0)=0. \tag{5.74}$$

($\because T(t)\neq 0$ as we are look for non –zero solution)

Now using second boundary condition **BC2**: $\frac{\partial u(l,t)}{\partial x}=0$, we get

$$u^{'}(l,t)=X^{'}(l)T(t) \Rightarrow X^{'}(l)T(t)=0 \Rightarrow X^{'}(l)=0 \tag{5.75}$$

($\because T(t)\neq 0$ as we are look for non –zero solution)

Now, we solve equation (5.71) i.e. $X^{''}-kX=0$, with boundary condition $X^{'}(0)=0$ and $X^{'}(l)=0$,which is called an Eigen value problem and *k* is called Eigen value.

Case (i) when k is positive.

Let $k=p^2$ for some $p\in R$.

The equation (5.71) reduces to $X^{''}-p^2X=0$. The auxiliary equation associate with $X^{''}-p^2X=0$ is $m^2-p^2=0$,

$\Rightarrow m=p,-p$.

Since the roots of the auxiliary equation are real and distinct, therefore

$$X(x)=Ae^{px}+Be^{-px} \tag{5.76}$$

where *A* and *B* are arbitrary constant. Differentiating equation (5.76), we get

$$X'(x) = Ape^{px} - Bpe^{-px}. \tag{5.77}$$

Now we apply the boundary condition given by equations (5.74) and (5.75) to find the value of *A* and *B*. Put $x = 0$ in equation (5.76), we get

$$X'(0) = Ape^{0} - Bpe^{0} \Rightarrow X(0) = A - B = 0 \Rightarrow A = B. \tag{5.78}$$

Put $x = l$ in equation (5.76), we get

$$X(l) = pAe^{pl} - pBe^{-pl} \Rightarrow X(l) = pAe^{pl} - pBe^{-pl}. \tag{5.79}$$

Using equation (5.75) and (5.79) in equation (5.79), we get $X(l) = pAe^{-pl}\left(e^{2pl} - 1\right) = 0$.

$$\Rightarrow A = 0 \quad \left(\because e^{pl}\left(e^{2pl} - 1\right) \neq 0\right)$$

Since $A = 0$, therefore from equation (5.60), we get $B = 0$. So, equation (5.67) reduces to $X(x) = 0$.

Substituting $X(x) = 0$ in equation (5.67), the required solution is $u(x,t) = 0$. So, when k is positive, there is no Eigen value, and we get a trivial solution.

Case (ii) when k is zero.

Then auxiliary equation of equations (5.71) is $m^2 = 0 \Rightarrow m = 0, 0$. Since, the roots are real and repeated, therefore solution of equation (5.71) is $X(x) = (A + Bx)$.Differentiating $X(x) = (A + Bx)$ w.r.t *x*, we get

$$X'(x) = B. \tag{5.80}$$

Now we apply the boundary condition given by equations (5.74) and (5.75) to find the value of *A* and *B*. Put $x = 0$ in equation (5.80), we get $X'(0) = B \Rightarrow B = 0$.(5.81)

Put $x = l$ in equation (5.62), we get

$$X'(l) = B \Rightarrow B = 0. \tag{5.82}$$

Since $B = 0$, therefore equation (5.62) gives $X(x) = A$. For k is zero, equation (5.53) gives $T' = 0 \Rightarrow \frac{dT}{dt} = 0 \Rightarrow T =$constant$\frac{B_0}{2A}$ (say). Substituting $T(t) = \frac{B_0}{2A}$ in equation (5.67), the required solution is

$$u(x,t) = A \times \frac{B_0}{2A} = \frac{B_0}{2}. \tag{5.83}$$

Case (iii) when k is negative

Let $k = -p^2$.

Then auxiliary equation of equations (5.71) is
$m^2 + p^2 = 0 \Rightarrow m = \pm ip$. Since, the roots are complex, therefore solution of equation (5.71) is

$$X(x) = (A\cos px + B\sin px) \tag{5.84}$$

where *A* and *B* are arbitrary constant. Differentiating equation (5.84), we get

$$X'(x) = -pA\sin px + pB\cos px. \tag{5.85}$$

Now, we use boundary condition $X'(0)=0$ and $X'(l)=0$ to find the value of arbitrary constant.

Put $x=0$ in equation (5.85), we get

$$X'(0)=-pA\sin 0+pB\cos 0 \Rightarrow B=0 \tag{5.86}$$

Put $x=l$ in equation (5.85), we get

$$X'(l)=(-pA\sin pl+pB\cos pl) \Rightarrow pA\sin pl=0,$$

(From equation (5.86))

$\Rightarrow A=0$ or $\sin pl=0$,

$\Rightarrow \sin pl=0$, ($\because B\neq 0$ as $X(x)=0$)

$\Rightarrow pl=\sin^{-1}(0)$

$\Rightarrow pl=n\pi$,

$$\Rightarrow p=\frac{n\pi}{l}, n=1,2,3,..... \tag{5.87}$$

For $n=1,2,3,...$, we get different value of p, so we denote p by p_n Therefore $p=p_n=\frac{n\pi}{l}$. Since $p_n=\frac{n\pi}{l}$, therefore from equation (5.84), we get

$$X(x)=A\cos\left(\frac{n\pi}{l}\right)x. \tag{5.88}$$

For $n = 1, 2, 3, \ldots$, we get different value of $X(x)$ using equation (5.88), so we denote $X(x)$ by $X_n(x)$. Therefore

$$X_n(x) = A_n \cos\left(\frac{n\pi}{l}\right)x. \tag{5.89}$$

Similarly, when k is negative i.e., $k = -p^2$ then equation (5.72) changes to $T' + C^2 p^2 T = 0$. Using variable separable form to solve $T' + C^2 p^2 T = 0$, we get $\frac{T'}{T} = -C^2 p^2\, dt$.

Integrating both the sides, we get $\log T = -C^2 p^2 t + c$, where c is an arbitrary constant.

On further simplifying $\log T = C^2 p^2 t + c$, we get

$$T = e^{-C^2 p^2 t + c} = e^c\, e^{-C^2 p^2 t} = C' e^{-C^2 p^2 t} \tag{5.90}$$

where $C' = e^c$. Using equation (5.87), equation (5.90) changes to

$$T = C' e^{-C^2\left(\frac{n\pi}{l}\right)^2 t}.$$

For $n = 1, 2, 3, \ldots$, we get different value of $T(t)$, so we denote $T(t)$ by $T_n(t)$. Therefore $T_n(t) = C_n' e^{-C^2 p^2 t}$. (5.91)

Substituting equations (5.89) and (5.91), the solution satisfying the boundary condition is given by

$$u(x,t) = A_n \cos\left(\frac{n\pi x}{l}\right) C_n' e^{-C^2\left(\frac{n\pi}{l}\right)^2 t}. \tag{5.92}$$

For $n=1,2,3,\ldots$, we get different value of $u(x,t)$, so we denote $u(x,t)$ by $u_n(x,t)$.

Therefore, equation (5.92) can be written as

$u_n(x,t)=A_n\cos\left(\frac{n\pi x}{l}\right)C_n^{'}e^{-C^2\left(\frac{n\pi}{l}\right)^2t}$. Let $B_n=A_nC_n^{'}$ therefore the solution changes to

$$u_n(x,t)=B_n\cos\left(\frac{n\pi x}{l}\right)e^{-C^2\left(\frac{n\pi}{l}\right)^2t} \tag{5.93}$$

Therefore, equations (5.83) and (5.93) form a set of infinite solutions of equation (5.64). Now to find a solution that satisfies the initial condition equation (5.66), consider a linear combination of these solutions. Hence the required solution is linear combination of solution given by equations (5.83) and (5.93). Therefore, the solution is given by

$$u(x,t)=\frac{B_0}{2}+\sum_{n=1}^{\infty}u_n(x,t)$$

$$u(x,t)=\frac{B_0}{2}+\sum_{n=1}^{\infty}B_n\cos\left(\frac{n\pi x}{l}\right)e^{-C^2\left(\frac{n\pi}{l}\right)^2t} \tag{5.94}$$

is solution of one-dimensional heat equation given by equation (5.64) satisfying BC1 and BC2. Now, we use the initial condition **IC1** $u(x,0)=f(x), 0\le x\le l$ to find B_n. Put $t=0$ in equation (5.94), we get $u(x,0)=\frac{B_0}{2}+\sum_{n=1}^{\infty}B_n\cos\left(\frac{n\pi x}{l}\right)$. From **IC1,** we get $f(x)=\frac{B_0}{2}+\sum_{n=1}^{\infty}B_n\cos\left(\frac{n\pi x}{l}\right)$, which is half range Fourier cosine series and

$$B_n=\frac{2}{l}\int_0^l f(x)\cos\left(\frac{n\pi x}{l}\right)dx \tag{5.95}$$

and $B_0 = \frac{2}{l}\int_0^l f(x)\, dx$. (5.96)

Therefore, the required solution to the one-dimensional heat equation given by equation (5.64) subject to boundary and initial condition given by equation (5.65)- (5.66) is given by

$$u(x,t) = \frac{B_0}{2} + \sum_{n=1}^{\infty} B_n \cos\left(\frac{n\pi x}{l}\right) e^{-C^2\left(\frac{n\pi}{l}\right)^2 t},$$

where $B_n = \frac{2}{l}\int_0^l f(x)\cos\left(\frac{n\pi x}{l}\right) dx$ and $B_0 = \frac{2}{l}\int_0^l f(x)\, dx$.

Example:

Find the solution of one-dimensional diffusion equation $\frac{\partial u}{\partial t} = C^2 \frac{\partial^2 u}{\partial x^2}$ satisfying the boundary conditions: (i) u is bounded as $t \to \infty$ (ii) $\left(\frac{\partial u}{\partial x}\right)_{x=0} = 0\ \forall t$ (iii) $\left(\frac{\partial u}{\partial x}\right)_{x=l} = 0\ \forall t$ (iv) $u(x,0) = x(l-x), x \in (0, l)$.

Solution:

The solution of the dimensional heat flow equation is given by

$$\frac{\partial u}{\partial t} = C^2 \frac{\partial^2 u}{\partial x^2} \tag{5.97}$$

with the boundary condition $\frac{\partial u(0,t)}{\partial x} = \frac{\partial u(l,t)}{\partial x} = 0$ (5.98)

for $\forall t$, and initial condition $u(x,0) = f(x), 0 \le x \le l$ (5.99) is given by $u(x,t)=\frac{B_0}{2}+\sum_{n=1}^{\infty} B_n \cos\left(\frac{n\pi x}{l}\right)e^{-C^2\left(\frac{n\pi}{l}\right)^2 t}$ (5.100)

where $B_n=\frac{2}{l}\int_0^l f(x)\cos\left(\frac{n\pi x}{l}\right)dx$ (5.101)

and $B_0=\frac{2}{l}\int_0^l f(x)\,dx$. (5.102)

Comparing the given boundary value problem with boundary value problem given by equations (5.64) and (5.66), we get

$C^2=C^2, l=l, f(x)=x(l-x)$.

Hence equation (5.101) and (5.102) reduces to

$$B_n=\frac{2}{l}\int_0^l x(l-x)\cos\left(\frac{n\pi x}{l}\right)dx$$

and $B_0=\frac{2}{l}\int_0^l x(l-x)\,dx$.

Now $B_0=\frac{2}{l}\int_0^l x(l-x)\,dx$,

$$B_0=\frac{2}{l}\int_0^l lx-x^2\,dx=\frac{2}{l}\left[\frac{lx^2}{2}-\frac{x^3}{3}\right]_0^l$$

$$B_0=\frac{2}{l}\left[\left(\frac{l^3}{2}-\frac{l^3}{3}\right)-0\right]=\frac{l^2}{3}. \quad (5.103)$$

Now we find $B_n = \frac{2}{l}\int_0^l x(l-x)\cos\left(\frac{n\pi x}{l}\right)dx$.

$$B_n = \frac{2}{l}\int_0^l \left(lx - x^2\right)\cos\left(\frac{n\pi x}{l}\right)dx,$$

$$B_n = \frac{2}{l}\left[\left(lx-x^2\right)\frac{\sin\left(\frac{n\pi x}{l}\right)}{\frac{n\pi}{l}} - (l-2x)\left(-\frac{\cos\left(\frac{n\pi x}{l}\right)}{\left(\frac{n\pi}{l}\right)^2}\right) + (-2)\left(-\frac{\sin\left(\frac{n\pi x}{l}\right)}{\left(\frac{n\pi}{l}\right)^3}\right)\right]_0^l,$$

$$B_n = \frac{2}{l}\left[\left\{0-(-l)\left(-\frac{\cos(n\pi)}{\left(\frac{n\pi}{l}\right)^2}\right)+0\right\} - \left\{0-(l)\left(-\frac{\cos(0)}{\left(\frac{n\pi}{l}\right)^2}\right)+0\right\}\right],$$

$$B_n = \frac{2}{l}\left[-\frac{l^3}{n^2\pi^2}(-1)^n - \frac{l^3}{n^2\pi^2}\right],$$

$$B_n = \frac{2l^2}{n^3\pi^3}\left[-(-1)^n - 1\right],$$

$$B_n = \begin{cases} -\frac{4l^2}{(n\pi)^2} & , n\, is\, even \\ 0 & , n\, is\, odd \end{cases}. \tag{5.104}$$

From equations (5.103) and (5.104), equations (5.100) will reduce to

$$u(x,t) = \frac{l^2}{6} - \sum_{n=even}^{\infty} \frac{4l^2}{(n\pi)^2}\cos\left(\frac{n\pi x}{l}\right)e^{-C^2\left(\frac{n\pi}{l}\right)^2 t},$$

which is a required solution.

Example:

Consider for example the heat equation $\frac{\partial u}{\partial t}=\frac{\partial^2 u}{\partial x^2}, 0<x<1, t>0$ subject to the initial and boundary conditions $u(x,0)=x-x^2, u_x(0,t)=u_x(1,t)=0.$

Solution:

The solution of the dimensional heat flow equation is given by
$\frac{\partial u}{\partial t}=C^2\frac{\partial^2 u}{\partial x^2}$.(5.105)

with the boundary condition $\frac{\partial u(0,t)}{\partial x}=\frac{\partial u(l,t)}{\partial x}=0$ (5.106)

for $\forall t$, and initial condition $u(x,0)=f(x), 0\leq x\leq l$ (5.107)

is given by $u(x,t)=\frac{B_0}{2}+\sum_{n=1}^{\infty}B_n\cos\left(\frac{n\pi x}{l}\right)e^{-C^2\left(\frac{n\pi}{l}\right)^2 t}$ (5.108)

where $B_n=\frac{2}{l}\int_0^l f(x)\cos\left(\frac{n\pi x}{l}\right)dx$ (5.109)

and $B_0=\frac{2}{l}\int_0^l f(x)\,dx$. (5.110)

Comparing the given boundary value problem with boundary value problem given by equation (5.64) -(5.66), we get

$C^2=1, l=1, f(x)=x-x^2$. Hence, equation (5.109) and (5.110) reduces to

$$B_n = 2\int_0^1 \left(x - x^2\right)\cos\left(n\pi x\right)dx$$

and $B_0 = 2\int_0^1 \left(x - x^2\right)\, dx.$

Now $B_0 = 2\int_0^1 \left(x - x^2\right)\, dx,$

$$B_0 = 2\left[\frac{x^2}{2} - \frac{x^3}{3}\right]_0^1 = \frac{1}{3}. \tag{5.111}$$

Now we find $B_n = 2\int_0^1 \left(x - x^2\right)\cos\left(n\pi x\right)dx,$

$$B_n = 2\left[\left(x - x^2\right)\frac{\sin n\pi x}{n\pi} - (1-2x)\left(-\frac{\cos n\pi x}{\left(n\pi\right)^2}\right) + (-2)\left(-\frac{\sin n\pi x}{\left(n\pi\right)^3}\right)\right]_0^1,$$

$$B_n = 2\left[\left\{0 - (-1)\left(-\frac{\cos n\pi}{\left(n\pi\right)^2}\right) + 0\right\} - \left\{0 - \left(-\frac{1}{\left(n\pi\right)^2}\right) + 0\right\}\right],$$

$$B_n = 2\left[-\frac{(-1)^n}{\left(n\pi\right)^2} - \frac{1}{\left(n\pi\right)^2}\right],$$

$$B_n = 2\left[\frac{(-1)^{n+1}}{\left(n\pi\right)^2} - \frac{1}{\left(n\pi\right)^2}\right]. \tag{5.112}$$

Substituting the values from equations (5.111) and (5.112) in equation (5.108), we get

$$u(x,t) = \frac{1}{6} + \sum_{n=1}^{\infty} 2\left[\frac{(-1)^{n+1}}{(n\pi)^2} - \frac{1}{(n\pi)^2}\right]\cos\left(\frac{n\pi x}{l}\right)e^{-C^2\left(\frac{n\pi}{l}\right)^2 t}$$

which is the required solution.

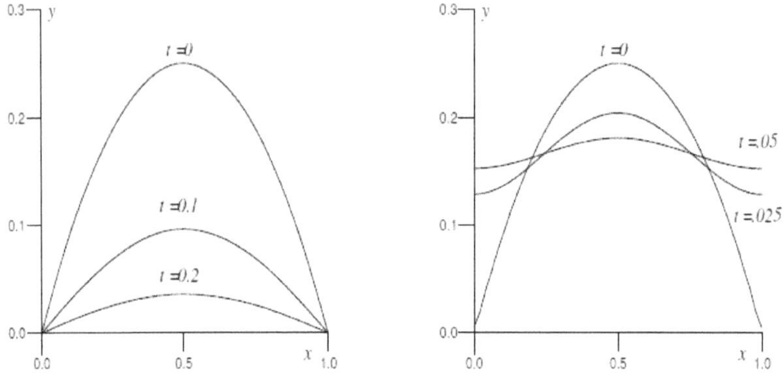

Figure 5.1: Solution of heat equation with fixed and no flux boundary condition with same initial condition.

Figure 5.1 (a) shows the solution of the heat equation for $t = 0, 0.1, 0.2$. Figure 5.1 (b) shows the solution of the heat equation given by Eq. (5.105) with initial and boundary condition given by Eq. (5.106) -(5.107) at time $t = 0, 0.25, 0.5$. It's noteworthy to observe that, despite the fact that the two problems have the same initial conditions, fixing and isolating the boundaries results in two completely distinct behaviours after $t \geq 0$.

5.5 Solution of heat equation when one end is insulated

In this section we discuss the solution of the heat equation when one end of the bar is insulated while the other end is kept at a constant temperature.

Example:

Find the temperature distribution $u(x,t)$ in a uniform bar of unit length whose one end is kept at 10 °C and the other end is insulated. It is also given that $u(x,0)=1-x,\ 0<x<1.$

Solution:

The bar is placed along x-axis with its one end whose temperature is maintained at 10 °C at the origin whereas the other end at x=1 which is insulated so that the flux is zero. The heat equation with the initial and boundary condition is mentioned as follows:

$$\frac{\partial u}{\partial t}=C^2\frac{\partial^2 u}{\partial x^2} \qquad (5.113)$$

with the boundary condition

$$u(0,t)=10,\ \frac{\partial u(1,t)}{\partial x}=0 \qquad (5.114)$$

for $\forall t$, and initial condition

$$u(x,0)=1-x,\ 0<x<1 \qquad (5.115)$$

Assume that a solution of equation (5.113) is of the form

$$u(x,t)=y(x,t)+10\Rightarrow y(x,t)=u(x,t)-10. \qquad (5.116)$$

Using equation (5.116), equation (5.113) -(5.115) changes to

$$\frac{\partial y(x,t)}{\partial t}=C^2\frac{\partial^2 y}{\partial x^2}. \qquad (5.117)$$

with boundary condition

BC1: $y(0,t)=0,$ **BC2:** $\frac{\partial y(1,t)}{\partial x}=0,$

and initial condition

IC1: $y(x,0)=u(x,0)-10=-(x+9).$

Assume that a solution of equation (5.117) is of the form

$$y(x,t)=X(x)T(t). \tag{5.118}$$

where X is a function of x and T is a function of t only. Differentiate equation (5.118) partially w.r.t. x and t, we have

$$\frac{\partial y}{\partial x}=X^{'}T \Rightarrow \frac{\partial^2 y}{\partial x^2}=X^{''}T \tag{5.119}$$

and $\frac{\partial y}{\partial t}=XT^{'} \Rightarrow \frac{\partial^2 y}{\partial t^2}=XT^{''}$ (5.120)

where $X^{'}=\frac{dX}{dx}, X^{''}=\frac{d^2X}{dx^2}, T^{'}=\frac{dT}{dx}, T^{''}=\frac{d^2T}{dx^2}.$

Using equations (5.119) and (5.120) in (5.117), we have $XT^{'}=C^2X^{''}T$. Separating the variables in the above equation, we get

$$\frac{1}{C^2}\frac{T^{'}}{T}=\frac{X^{''}}{X}. \tag{5.121}$$

In equation (5.121), L.H.S is a function of t only and R.H.S is a function of x only and here t and x are independent variables, so each side is constant. So rewriting equation (5.121) as

$$\frac{X^{''}}{X}=k \Rightarrow X^{''}=kX \Rightarrow X^{''}-kX=0, \tag{5.122}$$

and $\frac{1}{C^2}\frac{T}{T'}=k \Rightarrow T'=kT \Rightarrow T'-C^2kT=0.$ (5.123)

equations (5.122) and (5.123) are second and first order homogeneous linear differential equation, respectively. Differentiating equation (5.118) w.r.t x, we get

$$y_x(x,t)=X'(x)T(t). \quad (5.124)$$

Now using first boundary condition **BC1**: $y(0,t)=0$, equation (5.118) gives

$$X(0)T(t) \Rightarrow X(0)T(t)=0 \Rightarrow X(0)=0. \quad (5.125)$$

($\because T(t)\neq 0$ as we are look for non –zero solution)

Now using second boundary condition **BC2**: $\frac{\partial y(1,t)}{\partial x}=0$, we get from equation (5.124)

$$\frac{\partial y(1,t)}{\partial x}=X'(1)T(t) \Rightarrow X'(1)T(t)=0 \Rightarrow X'(1)=0.$$

(5.126)

($\because T(t)\neq 0$ as we are look for non –zero solution)

Now, we solve equation (5.122) $X''-kX=0$, with boundary condition

$$X(0)=0 \text{ and } X'(1)=0. \quad (5.127)$$

Case (i) when k is positive. Let $k=p^2$ for some $p\in R$.

The equation (5.122) reduces to $X''-p^2X=0$. The auxiliary equation associate with $X''-p^2X=0$ is $m^2-p^2=0 \Rightarrow m=p,-p$

Since the roots of the auxiliary equation are real and distinct, therefore

$$X(x) = Ae^{px} + Be^{-px} \tag{5.128}$$

where *A* and *B* are arbitrary constant. Differentiating equation (5.128), we get

$$X^{'}(x) = Ape^{px} - Bpe^{-px}. \tag{5.129}$$

Now we apply the boundary condition given by equation (5.127) to find the value of *A* and *B*. Put $x = 0$ in equation (5.128), we get

$$X(0) = Ae^{0} + Bpe^{0} \Rightarrow X(0) = A + B = 0 \Rightarrow A = -B. \tag{5.130}$$

Put $x = 1$ in equation (5.129), we get

$$X^{'}(1) = pAe^{p} - pBe^{-p} \Rightarrow X(1) = pAe^{p} - pBe^{-p}. \tag{5.131}$$

Using equation (5.127) and (5.130) in equation (5.131), we get $X^{'}(1) = pAe^{-pl}\left(e^{2pl} - 1\right) = 0.$

$$\Rightarrow A = 0 \ \left(\because e^{pl}\left(e^{2pl} - 1\right) \neq 0\right)$$

Since $A = 0$, therefore from equation (5.130), we get $B = 0$. So, equation (5.128) reduces to $X(x) = 0$.

Substituting $X(x) = 0$ in equation (5.118), the required solution is $y(x,t) = 0.$

Case (ii) when k is zero.

Then the auxiliary equation of equations (5.122) is $m^{2} = 0 \Rightarrow m = 0, 0$. Since, the roots are real and repeated, therefore solution of equation (5.122) is $X(x)=(A+Bx)$. Differentiating $X(x)=(A+Bx)$ w.r.t x, we get $X^{'}(x)=B$ (5.132)

Now we apply the boundary condition given by equation (5.127) and to find the value of *A* and *B*. Put $x=0$ in equation (5.132), we get

$$X(0)=A\Rightarrow A=0. \quad (5.133)$$

Put $x=1$ in equation (5.133), we get

$$X^{'}(1)=B\Rightarrow B=0. \quad (5.134)$$

Since, $A=0$ and $B=0$, therefore $X(x)=(A+Bx)$ gives $X(x)=0$.Substituting $X(x)=0$ in equation (5.118), the required solution is $u(x,t)=0$ which is a trivial solution and we are looking for non-trivial solution. So, k is zero is not possible.

Case (iii) when k is negative. Let $k=-p^2$.

Then the auxiliary equation of equations (5.122) is $m^2+p^2=0\Rightarrow m=\pm ip$. Since the roots are complex, therefore solution of equation (5.52) is

$$X(x)=(A\cos px+B\sin px) \quad (5.135)$$

where *A* and *B* are arbitrary constant. Differentiating equation (5.135), we get

$$X^{'}(x)=-pA\sin px+pB\cos px. \quad (5.136)$$

Now, we use boundary condition $X(0)=0$ and $X^{'}(1)=0$ to find the value of arbitrary constant. Put $x=0$ in equation (5.135), we get

$$X(0) = A\cos 0 + B\sin 0 \Rightarrow A = 0. \tag{5.137}$$

Put $x = 1$ in equation (5.136), we get

$$X'(1) = (-pA\sin p + pB\cos p) \Rightarrow pB\cos p = 0,$$

$$\Rightarrow B = 0 \text{ or } \cos p = 0,$$

$$\Rightarrow \cos p = 0,$$

($\because B \neq 0$ as $X(x) = 0$)

$$\Rightarrow p = \cos^{-1}(0)$$

$$\Rightarrow p = (2n-1)\frac{\pi}{2} = \frac{1}{2}(2n-1)\pi, \quad n = 1, 2, 3, ... \tag{5.138}$$

For $n = 1, 2, 3, ...$, we get different value of p, so we denote p by p_n. Therefore $p = p_n = \frac{1}{2}(2n-1)\pi$. Since $p_n = \frac{1}{2}(2n-1)\pi$, therefore from equation (5.135), we get

$$X(x) = B\sin\left(\frac{(2n-1)\pi}{2}\right)x. \tag{5.139}$$

For $n = 1, 2, 3, ...$, we get different value of $X(x)$ using equation (5.139), so we denote $X(x)$ by $X_n(x)$. Therefore

$$X_n(x) = B_n\sin\left(\frac{(2n-1)\pi}{2}\right)x. \tag{5.140}$$

Similarly, when k is negative i.e., $k=-p^2$ then equation (5.123) changes to $T^{'}+C^2\frac{(2n-1)^2\pi^2}{4}T=0$. Using variable separable form to solve $T^{'}+C^2\frac{(2n-1)^2\pi^2}{4}T=0$, we get

$$T^{'}=-C_n^2T \text{, where } C_n=-\frac{C^2(2n-1)^2\pi^2}{4}. \tag{5.141}$$

Integrating both the sides, we get $\log T=-C_n^2t+c$, where c is an arbitrary constant. On further simplifying $\log T=-C_n^2t+c$, we get

$$T=e^{-C_n^2t+c}=e^c\,e^{-C_n^2t}=De^{-C_n^2t} \tag{5.142}$$

where $D=e^c$. For $n=1,2,3,.....$, we get different value of $T(t)$, so we denote $T(t)$ by $T_n(t)$. Therefore

$$T_n(t)=D_ne^{-C_n^2t}. \tag{5.143}$$

Substituting equations (5.139) and (5.143), the solution satisfying the boundary condition is given by

$$y(x,t)=B_n\sin\left(\frac{(2n-1)\pi}{2}\right)xD_ne^{-C_n^2t}. \tag{5.144}$$

For $n=1,2,3,.....$, we get different value of $y(x,t)$, so we denote $y(x,t)$ by $y_n(x,t)$. Therefore, equation (5.144) can be written as

$$y_n(x,t)=B_n\sin\left(\frac{(2n-1)\pi x}{2}\right)D_ne^{-C_n^2t}.$$

Let $E_n=B_nD_n$, therefore the solution will change to

$$y_n(x,t) = E_n \sin\left(\frac{(2n-1)\pi}{2}\right) x e^{-C_n^2 t}. \tag{5.145}$$

Therefore, equation (5.145) forms a set of infinite solutions of equation (5.117). Now to find a solution that satisfies the initial condition **IC1:** $y(x,0) = u(x,0) - 10 = -(x+9)$), consider a linear combination of these solutions. Hence the required solution is linear combination of solution given by equation (5.145). Therefore, the solution is given by

$$y(x,t) = \sum_{n=1}^{\infty} E_n \sin\left(\frac{(2n-1)\pi}{2}\right) x e^{-C_n^2 t} \tag{5.146}$$

is solution of one-dimensional heat equation given by equation (5.113) satisfying BC1 and BC2. Now, we use the initial condition **IC1** $y(x,0) = -(x+9), 0 \le x \le 1$ to find E_n. Put $t = 0$ in equation (5.146), we get $y(x,0) = \sum_{n=1}^{\infty} E_n \sin\left(\frac{(2n-1)\pi}{2}\right) x$. From **IC1,** we get

$$-(x+9) = \sum_{n=1}^{\infty} E_n \sin\left(\frac{(2n-1)\pi}{2}\right) x,$$

Multiplying both sides of the above equation by $\sin\left(\frac{(2m-1)\pi}{2}\right) x$ and then integrating w.r.t x from 0 to 1, we get

$$-\int_0^1 (x+9)\sin\left(\frac{(2m-1)\pi x}{2}\right) dx = \sum_{n=1}^{\infty} E_n \int_0^1 \sin\left(\frac{(2n-1)\pi x}{2}\right) \sin\left(\frac{(2m-1)\pi x}{2}\right) dx \tag{5.147}$$

We have

$$\int_0^1 \sin\left(\frac{(2n-1)\pi x}{2}\right)\sin\left(\frac{(2m-1)\pi x}{2}\right)dx = \begin{cases} 0, & if\ m \neq n \\ 1, & if\ m = n \end{cases}. \tag{5.148}$$

Using equation (5.147) and (5.148), we have

$$E_m = -\int_0^1 (x+9)\sin\left(\frac{(2m-1)\pi x}{2}\right)dx.$$

Therefore, $E_n = -\int_0^1 (x+9)\sin\left(\frac{(2n-1)\pi x}{2}\right)dx$.

$$E_n = -2\left[(x+9)\left\{\frac{-\frac{\cos(2n-1)\pi x}{2}}{\frac{(2n-1)\pi}{2}}\right\} - \left\{\frac{-\frac{\sin(2n-1)\pi x}{2}}{\frac{(2n-1)^2\pi^2}{4}}\right\}\right]_0^1,$$

$$E_n = \frac{8(-1)^n}{(2n-1)^2\pi^2} - \frac{36}{(2n-1)\pi}. \tag{5.149}$$

Using equations (5.146) and (5.141), the required solution is given by

$$u(x,t) = 10 + \sum_{n=1}^{\infty} E_n \sin\left(\frac{(2n-1)\pi x}{2}\right)e^{-C_n^2 t},$$

where E_n and C_n are given by (5.149) and (5.141).

Example:

Solve $\frac{\partial u}{\partial t} = \frac{\partial^2 u}{\partial x^2}$, $0 < x < 2, t > 0$ subject to following initial and boundary condition $u(x,0) = \begin{cases} x, & 0 < x < 1 \\ 2-x, & 1 < x < 2 \end{cases}$,

$u(0,t)=u_x(2,t)=0$.

Solution:

In this problem, we have a mixture of both fixed and no flux boundary conditions.

Assume that a solution of given PDE is of the form

$$u(x,t)=X(x)T(t). \tag{5.150}$$

where X is a function of x and T is a function of t only. Differentiate equation (5.150) partially w.r.t. x and t, we have

$$\frac{\partial u}{\partial x}=X'T \Rightarrow \frac{\partial^2 u}{\partial x^2}=X'T \tag{5.151}$$

$$\text{and } \frac{\partial u}{\partial t}=XT' \Rightarrow \frac{\partial^2 u}{\partial t^2}=XT'' \tag{5.152}$$

where $X'=\frac{dX}{dx}, X''=\frac{d^2X}{dx^2}, T'=\frac{dT}{dx}, T''=\frac{d^2T}{dx^2}$. Using equations (5.150) -(5.152) in given PDE, we have $XT'=C^2X''T$. Separating the variables in the above equation, we get

$$\frac{1}{C^2}\frac{T'}{T}=\frac{X''}{X}. \tag{5.153}$$

In equation (5.153), L.H.S is a function of t only and R.H.S is a function of x only and here t and x are independent variables, so each side is constant. So rewriting equation (5.153) as

$$\frac{X''}{X}=k \Rightarrow X''=kX \Rightarrow X''-kX=0, \tag{5.154}$$

and $\frac{1}{C^2}\frac{T}{T'} = k \Rightarrow T' = kT \Rightarrow T' - C^2kT = 0$. (5.155)

equations (5.154) and (5.155) are second and first order homogeneous linear differential equation, respectively. Differentiating equation (5.150) w.r.t *x*, we get $u_x(x,t) = X'(x)T(t)$. (5.156)

Now using first boundary condition **BC1**: $u(0,t)=0$, equation (5.150) gives

$$X(0)T(t) \Rightarrow X(0)T(t) = 0 \Rightarrow X(0) = 0. \quad (5.157)$$

($\because T(t) \neq 0$ as we are look for non –zero solution)

Now using second boundary condition **BC2**: $u_x(2,t)=0$, we get

$$u_x(2,t) = X'(2)T(t) \Rightarrow X'(2)T(t) = 0 \Rightarrow X'(2) = 0. \quad (5.158)$$

($\because T(t) \neq 0$ as we are look for non –zero solution)

Now, we solve equation (5.154) $X'' - kX = 0$, with boundary condition

$$X(0) = 0 \text{ and } X'(2) = 0. \quad (5.159)$$

Case (i) when k is positive.

Let $k = p^2$ for some $p \in R$.

The equation (5.154) reduces to $X'' - p^2X = 0$. The auxiliary equation associate with $X'' - p^2X = 0$ is $m^2 - p^2 = 0 \Rightarrow m = p, -p$

Since the roots of the auxiliary equation are real and distinct, therefore

$$X(x) = Ae^{px} + Be^{-px} \tag{5.160}$$

where *A* and *B* are arbitrary constant. Differentiating equation (5.160), we get

$$X'(x) = Ape^{px} - Bpe^{-px}. \tag{5.161}$$

Now we apply the boundary condition given by equations (5.160) and (5.161) to find the value of *A* and *B*. Put $x = 0$ in equation (5.160), we get

$$X(0) = Ae^{0} + Bpe^{0} \Rightarrow X(0) = A + B = 0 \Rightarrow A = -B. \tag{5.162}$$

Put $x = 2$ in equation (5.161), we get $X'(2) = pAe^{2p} - pBe^{-2p}$,

$$\Rightarrow X(2) = pAe^{2p} - pBe^{-2p}. \tag{5.163}$$

Using equation (5.159) and (5.162) in equation (5.163), we get $X'(2) = pAe^{-2p}\left(e^{4p} - 1\right) = 0$.

$$\Rightarrow A = 0 \ \left(\because e^{-2pl}\left(e^{4p} - 1\right) \neq 0\right)$$

Since $A = 0$, therefore from equation (5.162), we get $B = 0$. So, equation (5.128) reduces to $X(x) = 0$.Substituting $X(x) = 0$ in equation (5.150), the required solution is $u(x,t) = 0$

Case (ii) when k is zero.

Then the auxiliary equation of equations (5.154) is $m^2 = 0 \Rightarrow m = 0, 0$. Since, the roots are real and repeated, therefore solution of equation (5.154) is $X(x)=(A+Bx)$ .Differentiating $X(x)=(A+Bx)$ w.r.t x, we get $X^{'}(x)=B$. (5.164)

Now we apply the boundary condition given by equations (5.157) and (5.158) to find the value of *A* and *B*. Put $x=0$ in $X(x)=(A+Bx)$, we get $X(0)=A \Rightarrow A=0$. (5.165)

Put $x=2$ in equation (5.132), we get

$$X^{'}(2)=B \Rightarrow B=0. \quad (5.166)$$

Since, $A=0$ and $B=0$, therefore $X(x)=(A+Bx)$ gives $X(x)=0$.Substituting $X(x)=0$ in equation (5.150), the required solution is $u(x,t)=0$ which is a trivial solution and we are looking for non-trivial solution. So, k zero is not possible.

Case (iii) when k is negative.

Let $k=-p^2$.

Then the auxiliary equation of equations (5.154) is
$m^2+p^2=0 \Rightarrow m=\pm ip$. Since, the roots are complex, therefore solution of equation (5.154) is

$$X(x)=(A\cos px+B\sin px) \quad (5.167)$$

where *A* and *B* are arbitrary constant. Differentiating equation (5.167), we get

$$X^{'}(x)=-pA\sin px+pB\cos px. \quad (5.168)$$

Now, we use boundary condition $X(0)=0$ and $X'(2)=0$ to find the value of arbitrary constant.

Put $x=0$ in equation (5.167), we get

$$X(0)=A\cos 0+B\sin 0 \Rightarrow A=0. \tag{5.169}$$

Put $x=2$ in equation (5.168), we get

$$X'(2)=(-pA\sin 2p+pB\cos 2p) \Rightarrow pB\cos 2p=0,$$

$$\Rightarrow B=0 \text{ or } \cos 2p=0,$$

$$\Rightarrow \cos 2p=0, (\because B\neq 0 \text{ as } X(x)=0)$$

$$\Rightarrow 2p=\cos^{-1}(0)$$

$$\Rightarrow p=(2n-1)\frac{\pi}{4}=\frac{1}{4}(2n-1)\pi,\ n=1,2,3,..... \tag{5.170}$$

For $n=1,2,3,.....$, we get different value of p, so we denote p by p_n.Therefore $p=p_n=\frac{1}{4}(2n-1)\pi$. Since $p_n=\frac{1}{4}(2n-1)\pi$, therefore from equation (5.167), we get

$$X(x)=B\sin\left(\frac{(2n-1)\pi}{4}\right)x. \tag{5.171}$$

For $n=1,2,3,.....$, we get different value of $X(x)$ using equation (5.171), so we denote $X(x)$ by $X_n(x)$. Therefore,

$$X_n(x) = B_n \sin\left(\frac{(2n-1)\pi}{4}\right)x. \tag{5.172}$$

Similarly, when k is negative i.e., $k = -p^2$ then equation (5.155) changes to $T^{'} + C^2 \frac{(2n-1)^2 \pi^2}{16} T = 0$.

Using variable separable form to solve $T^{'} + C^2 \frac{(2n-1)^2 \pi^2}{16} T = 0$

$$T^{'} = -C_n^2 T \text{, where } C_n = -\frac{C^2 (2n-1)^2 \pi^2}{16}. \tag{5.173}$$

Integrating both the sides, we get $\log T = -C_n^2 t + c$, where c is an arbitrary constant. On further simplifying $\log T = -C_n^2 t + c$, we get

$$T = e^{-C_n^2 t + c} = e^c e^{-C_n^2 t} = De^{-C_n^2 t} \tag{5.174}$$

where $D = e^c$. For $n = 1, 2, 3, ...$, we get different value of $T(t)$, so we denote $T(t)$ by $T_n(t)$.Therefore

$$T_n(t) = D_n e^{-C_n^2 t}. \tag{5.175}$$

Substituting equations (5.172) and (5.175), the solution satisfying the boundary condition is given by

$$u(x,t) = B_n \sin\left(\frac{(2n-1)\pi}{4}\right)x D_n e^{-C_n^2 t}. \tag{5.176}$$

For $n = 1, 2, 3, ...$, we get different value of $u(x,t)$, so we denote $u(x,t)$ by $u_n(x,t)$.Therefore, equation (5.176) can be written as

$$u_n(x,t) = B_n \sin\left(\frac{(2n-1)\pi x}{4}\right) D_n e^{-C_n^2 t}.$$

Let $b_n = B_n D_n$,therefore the solution changes to

$$u_n(x,t) = b_n \sin\left(\frac{(2n-1)\pi}{4}\right) x e^{-C_n^2 t}. \tag{5.177}$$

Therefore equations (5.177) form a set of infinite solutions to the given heat equation. Now to find a solution that satisfies the initial condition **IC1:** $u(x,0) = \begin{cases} x, & 0 < x < 1 \\ 2-x, & 1 < x < 2 \end{cases}$, consider a linear combination of these solutions. Hence the required solution is linear combination of solution given heat equation. Therefore, the solution is given by

$$u(x,t) = \sum_{n=1}^{\infty} b_n \sin\left(\frac{(2n-1)\pi}{4}\right) x e^{-C_n^2 t}. \tag{5.178}$$

Substituting $t = 0$ in equation (5.178), we get

$$u(x,0) = \sum_{n=1}^{\infty} b_n \sin\left(\frac{(2n-1)}{4} x\right) \tag{5.179}$$

which represents a Fourier Sine series, where

$$b_n = \frac{2}{l} \int_0^l f(x) \sin\left(\frac{n\pi x}{l}\right) dx,$$

$$\therefore b_n = 2\left[\int_0^1 x \sin\left(\frac{(2n-1)\pi x}{4}\right) dx + \int_1^2 (2-x) \sin\left(\frac{(2n-1)\pi x}{4}\right) dx\right] \tag{5.180}$$

Now $\int_0^1 x \sin\left(\frac{(2n-1)\pi x}{4}\right) dx$

$$\therefore \int_0^1 x\sin\left(\frac{(2n-1)\pi x}{4}\right)dx = \left[x\left(\frac{-\cos\left(\frac{(2n-1)\pi x}{4}\right)}{\frac{(2n-1)\pi}{4}}\right)-\left(\frac{-\sin\left(\frac{(2n-1)\pi x}{4}\right)}{\left(\frac{(2n-1)\pi}{4}\right)^2}\right)\right]_0^1,$$

$$\therefore \int_0^1 x\sin\left(\frac{(2n-1)\pi x}{4}\right)dx = \left[\left(\frac{-4x\cos\left(\frac{(2n-1)\pi x}{4}\right)}{(2n-1)\pi}\right)+\left(\frac{16\sin\left(\frac{(2n-1)\pi x}{4}\right)}{((2n-1)\pi)^2}\right)\right]_0^1,$$

$$\therefore \int_0^1 x\sin\left(\frac{(2n-1)\pi x}{4}\right)dx = \left[\left(\frac{-4\cos\left(\frac{(2n-1)\pi}{4}\right)}{(2n-1)\pi}\right)+\left(\frac{16\sin\left(\frac{(2n-1)\pi}{4}\right)}{((2n-1)\pi)^2}\right)-((0)+(0))\right]$$

,

$$\therefore \int_0^1 x\sin\left(\frac{(2n-1)\pi x}{4}\right)dx = \frac{-4\cos\left(\frac{(2n-1)\pi}{4}\right)}{(2n-1)\pi}+\frac{16\sin\left(\frac{(2n-1)\pi}{4}\right)}{((2n-1)\pi)^2}. \qquad (5.181)$$

$$\int_1^2 (2-x)\sin\left(\frac{(2n-1)\pi x}{4}\right)dx = \left[(2-x)\frac{-\cos\left(\frac{(2n-1)\pi x}{4}\right)}{\frac{(2n-1)\pi}{4}}-\left((-1)\frac{-\sin\left(\frac{(2n-1)\pi x}{4}\right)}{\left(\frac{(2n-1)\pi}{4}\right)^2}\right)\right]_1^2$$

,

$$\int_1^2 (2-x)\sin\left(\frac{(2n-1)\pi x}{4}\right)dx = \left[\frac{-(8-4x)\cos\left(\frac{(2n-1)\pi x}{4}\right)}{(2n-1)\pi}-\left(\frac{16\sin\left(\frac{(2n-1)\pi x}{4}\right)}{((2n-1)\pi)^2}\right)\right]_1^2$$

,

$$\int_1^2 (2-x)\sin\left(\frac{(2n-1)\pi x}{4}\right)dx = \left[\left(0-\left(\frac{16\sin\left(\frac{(2n-1)\pi}{2}\right)}{((2n-1)\pi)^2}\right)\right)-\left(\frac{-(4)\cos\left(\frac{(2n-1)\pi}{4}\right)}{(2n-1)\pi}-\left(\frac{16\sin\left(\frac{(2n-1)\pi}{4}\right)}{((2n-1)\pi)^2}\right.\right.\right.$$

$$\int_1^2 (2-x)\sin\left(\frac{(2n-1)\pi x}{4}\right)dx = \frac{-16\sin\left(\frac{(2n-1)\pi}{2}\right)}{((2n-1)\pi)^2}+\frac{4\cos\left(\frac{(2n-1)\pi}{4}\right)}{(2n-1)\pi}+\frac{16\sin\left(\frac{(2n-1)\pi}{4}\right)}{((2n-1)\pi)^2}. \quad (5.182)$$

Using (5.181) and (5.182) in (5.180)

$$\therefore b_n = 2\left[\frac{-4\cos\left(\frac{(2n-1)\pi}{4}\right)}{(2n-1)\pi}+\frac{16\sin\left(\frac{(2n-1)\pi}{4}\right)}{((2n-1)\pi)^2}+\frac{-16\sin\left(\frac{(2n-1)\pi}{2}\right)}{((2n-1)\pi)^2}+\frac{4\cos\left(\frac{(2n-1)\pi}{4}\right)}{(2n-1)\pi}+\frac{16\sin\left(\frac{(2n-1)\pi}{4}\right)}{((2n-1)\pi)^2}\right]$$

$$\therefore b_n = 2\left[\frac{32\sin\left(\frac{(2n-1)\pi}{4}\right)}{((2n-1)\pi)^2}-\frac{16\sin\left(\frac{(2n-1)\pi}{2}\right)}{((2n-1)\pi)^2}\right],$$

$$\therefore b_n = \left[\frac{64\sin\left(\frac{(2n-1)\pi}{4}\right)}{((2n-1)\pi)^2}-\frac{32\sin\left(n\pi-\frac{\pi}{2}\right)}{((2n-1)\pi)^2}\right],$$

$$\therefore b_n = \left[\frac{64\sin\left(\frac{(2n-1)\pi}{4}\right)}{((2n-1)\pi)^2}+\frac{32\sin\left(\frac{\pi}{2}-n\pi\right)}{((2n-1)\pi)^2}\right],$$

$$\therefore b_n = \left[\frac{64\sin\left(\frac{(2n-1)\pi}{4}\right)}{((2n-1)\pi)^2}+\frac{32\cos n\pi}{((2n-1)\pi)^2}\right]. \quad (5.183)$$

Substituting equation (5.183) in equation (5.178), we get

$$\therefore u(x,t)=\sum_{n=1}^{\infty}\left[\frac{64\sin\left(\frac{(2n-1)\pi}{4}\right)}{((2n-1)\pi)^2}+\frac{32\cos n\pi}{((2n-1)\pi)^2}\right]\sin\left(\frac{(2n-1)}{4}x\right)e^{-c^2\left(\frac{(2n-1)\pi}{4}\right)^2 t}.$$

Hence the solution of PDE is

$$\therefore u(x,t)=\frac{32}{(\pi)^2}\sum_{n=1}^{\infty}\left[\frac{2\sin\left(\frac{(2n-1)\pi}{4}\right)}{((2n-1)\pi)^2}+\frac{\cos n\pi}{((2n-1)\pi)^2}\right]\sin\left(\frac{(2n-1)}{4}x\right)e^{-c^2\left(\frac{(2n-1)\pi}{4}\right)^2 t}$$

Fig. 4.2 shows the solution of the heat equation with initial condition $u(x,0)=\begin{cases} x, & 0<x<1 \\ 2-x, & 1<x<2 \end{cases}$, and boundary condition $u(0,t)=u_x(2,t)=0$.

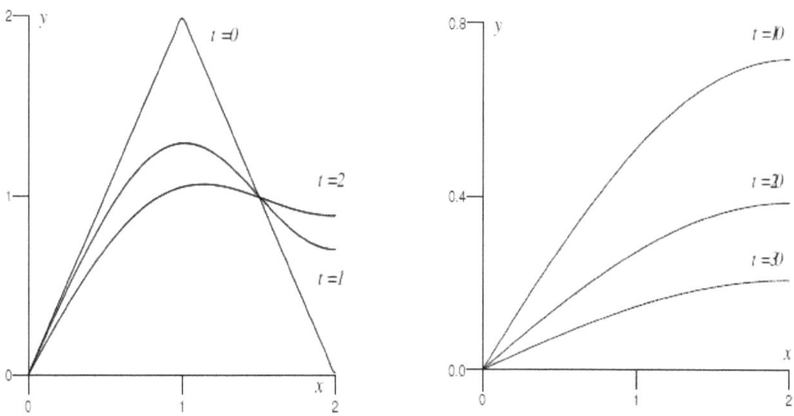

Figure 5.2: The solution's short- and long-term behaviors.

5.6 Nonhomogeneous Boundary Condition

In the above examples, the boundary conditions were either fixed, insulted, or radiating to zero. However, more often than not, we see that the boundary conditions are either nonstandard or nonhomogeneous. It is also possible that the boundary can be fixed to some constant or the flux is maintained at a certain value.

Example:

Solve $\frac{\partial u}{\partial t}=\frac{\partial^2 u}{\partial x^2},\ 0<x<3\ ,\ t>0$

(5.184)

with initial condition $u(x,0)=4x-x^2,$

(5.185)

and boundary condition $u(0,t)=0, u(3,t)=3.$ (5.186)

Solution:

To obtain a separable solution $u(x,t)=X(x)T(t)$, then from equation (5.186) we have

$$X(0)T(t)=0, \tag{5.187}$$

and $X(3)T(t)=0.$ (5.188)

So, we have a problem that equation (5.188) doesn't separate. To overcome this the transformation

$$u(x,t)=v(x,t)+ax+b. \tag{5.189}$$

Can we choose the constants *a* and *b* as to fix both boundary conditions to zero? Upon substitution of both boundary condition equations (5.186), we obtain both conditions as:

For $u(0,t)=0$ we have

$$u(0,t)=v(0,t)+a0+b \Rightarrow 0=v(0,t)+b. \tag{5.190}$$

For $u(3,t)=3$ we have

$$u(3,t)=v(3,t)+3a+b \Rightarrow 3=v(3,t)+3a+b. \tag{5.191}$$

Now to obtain the boundary value $v(0,t)=0$ and $v(3,t)=0$, which implies that we must choose $a=1$ and $b=0$ using equation (5.190) and (5.191) . Therefore, equation (5.189) reduces to

$$u(x,t)=v(x,t)+x. \tag{5.192}$$

Using the transformation given by equation (5.192), equation (5.184) change to $\frac{\partial v}{\partial t}=\frac{\partial^2 v}{\partial x^2}$,

Putting $t=0$ in $u(x,t)=v(x,t)+x$, we get

$$\begin{aligned} u(0,t)=v(0,t)+0 &\Rightarrow 4x-x^2=v(0,t)+x \\ &\Rightarrow 4x-x^2-x=v(0,t) \\ &\Rightarrow 3x-x^2=v(0,t) \end{aligned}$$

Therefore, initial condition given by equation (5.185) transform to

$$v(0,t)=3x-x^2.$$

Putting $x=0$ in $u(x,t)=v(x,t)+x$, we get

$$\begin{aligned} u(0,t)=v(0,t)+0 &\Rightarrow 0=v(0,t)+0 \\ &\Rightarrow v(0,t)=0 \end{aligned}$$

Putting $x=3$ in $u(x,t)=v(x,t)+x$, we get

$$u(3,t)=v(3,t)+0 \Rightarrow 3=v(3,t)+3$$
$$\Rightarrow v(3,t)=0$$

Therefore, the boundary condition given by equation (5.185) transform to $v(0,t)=0$ and $v(3,t)=0$.

Hence, the new problem using the transformation equation (5.192), we have

$$\frac{\partial v}{\partial t}=\frac{\partial^2 v}{\partial x^2},\ 0<x<3\ ,\ t>0 \tag{5.193}$$

IC1: $v(0,t)=3x-x^2$ (5.194)

and **BC1:** $v(0,t)=0$, **BC2:** $v(3,t)=0$. (5.195)

Comparing PDE given by equation (5.193)-(5.195) and with equation (5.1), equation (5.1a-5.1c), we get $C^2=1\ l=1$ and $f(x)=3x-x^2$. Therefore, the solution of heat equation (5.193) with initial and boundary condition (equation (5.194) -(5.195)) is given by

$$v(x,t)=\sum_{n=1}^{\infty} A_n \sin\left(\frac{n\pi x}{3}\right) e^{-\frac{n^2\pi^2 t}{9}} \tag{5.196}$$

which is a half range sine series where $A_n=\frac{2}{3}\int_0^3 f(x)\sin\left(\frac{n\pi x}{3}\right)dx$

,

$$A_n=\frac{2}{3}\int_0^3 \left(3x-x^2\right)\sin\left(\frac{n\pi x}{3}\right)dx,$$

$$A_n=\frac{2}{3}\left[\left(3x-x^2\right)\left(-\frac{\cos\left(\frac{n\pi x}{3}\right)}{\frac{n\pi}{3}}\right)-\left((3-2x)\left(\frac{-\sin\left(\frac{n\pi x}{3}\right)}{\left(\frac{n\pi}{3}\right)^3}\right)\right)+\left((-2)\left(\frac{\cos\left(\frac{n\pi x}{3}\right)}{\left(\frac{n\pi}{3}\right)^3}\right)\right)\right]_0^3,$$

$$A_n=\frac{2}{3}\left[\frac{-3\left(3x-x^2\right)\cos\left(\frac{n\pi x}{3}\right)}{n\pi}+\frac{9(3-2x)\sin\left(\frac{n\pi x}{3}\right)}{(n\pi)^2}-\frac{54\cos\left(\frac{n\pi x}{3}\right)}{(n\pi)^3}\right]_0^3,$$

$$A_n=\frac{2}{3}\left[\left(0+0-\frac{54\cos n\pi}{(n\pi)^3}\right)-\left(0+0-\frac{54}{(n\pi)^3}\right)\right],$$

$$A_n=\frac{2}{3}\left[\left(-\frac{54(-1)^n}{(n\pi)^3}\right)+\frac{54}{(n\pi)^3}\right],$$

$$A_n=\frac{36\left(1-(-1)^n\right)}{(n\pi)^3} \tag{5.197}$$

Substituting the value from equation (5.197) in equation (5.196), we get

$$v(x,t)=\sum_{n=1}^{\infty}\frac{36\left(1-(-1)^n\right)}{(n\pi)^3}\sin\left(\frac{n\pi x}{3}\right)e^{-\left(\frac{n\pi}{3}\right)^2 t},$$

$$v(x,t)=\frac{36}{\pi^3}\sum_{n=1}^{\infty}\frac{\left(1-(-1)^n\right)}{(n)^3}\sin\left(\frac{n\pi x}{3}\right)e^{-\left(\frac{n\pi}{3}\right)^2 t}. \tag{5.198}$$

Substituting value of $v(x,t)$ from equation (5.198) in equation (5.192), we have

$$u(x,t)=x+\frac{36}{\pi^{3}}\sum_{n=1}^{\infty}\frac{\left(1-(-1)^{n}\right)}{(n)^{3}}\sin\left(\frac{n\pi x}{3}\right)e^{-\left(\frac{n\pi}{3}\right)^{2}t},$$

which is the required solution. Fig. 4.3 represents the solution of the heat equation with non-homogenous boundary condition given by equations (5.185) -(5.186).

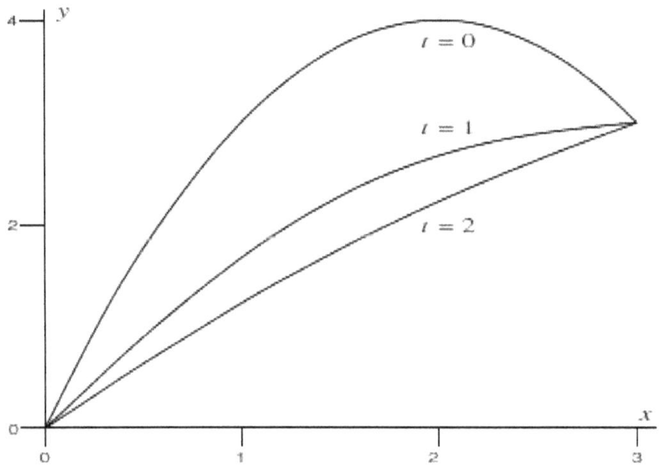

Figure 5.3: Solution of heat equation with non-homogenous boundary condition given by equations (5.185) -(5.186) at time $t=0,\ 1,\ 2$.

Example:

Consider the heat equation $\frac{\partial u}{\partial t}=\frac{\partial^{2}u}{\partial x^{2}}, 0<x<1, t>0$ subject to the initial and boundary conditions

$u(x,0)=0, u_{x}(0,t)=-1, u_{x}(1,t)=0.$

Solution:

The transformation used in the previous example $u(x,t)=v(x,t)+ax+b$ doesn't work since $u_x=v_x+a$ choosing a to fix the right boundary to zero only makes the left boundary non-zero. To overcome this, we use the transformation

$$u(x,t)=v(x,t)+a\left(x^2+2t\right)+bx \tag{5.199}$$

Can we choose the constants *a* and *b* as to fix both boundary conditions to zero? Upon substitution of both boundary condition equations (5.199), we obtain both conditions as:

For $u_x(0,t)=-1$ we have

$$u_x(x,0)=v_x(x,0)+2ax+b \Rightarrow -1=v_x(x,0)+2ax+b,$$

$$\Rightarrow v_x(x,0)=-1-2ax-b. \tag{5.200}$$

For $u_x(1,t)=0$ we have

$$u_x(1,t)=v_x(1,t)+2ax+b \Rightarrow 0=v_x(1,t)+2ax+b.$$

$$\Rightarrow v_x(1,t)+2ax+b=0. \tag{5.201}$$

Now to obtain the boundary value $v_x(0,t)=0$ and $v_x(1,t)=0$, which implies that we must choose $a=\frac{1}{2}$ and $b=-1$ using equation (5.200) and (5.201). Therefore, equation (5.199) reduces to

$$u(x,t)=v(x,t)+\frac{1}{2}\left(x^2+2t\right)-x. \tag{5.202}$$

Using the transformation given by equation (5.202), $\frac{\partial u}{\partial t}=\frac{\partial^2 u}{\partial x^2}, 0<x<1, t>0$ change to $\frac{\partial v}{\partial t}=\frac{\partial^2 v}{\partial x^2}, 0<x<1,\ t>0$. Putting $t=0$ in $u(x,t)=v(x,t)+\frac{1}{2}(x^2+2t)-x$, we get

$$u(x,0)=v(x,0)+\frac{1}{2}(x^2+0)-x,$$

$$\Rightarrow 0=v(x,0)+\frac{x^2}{2}-x,$$

$$\Rightarrow x-\frac{x^2}{2}=v(x,0).$$

Therefore, initial condition transforms to $v(x,0)=x-\frac{x^2}{2}$.

Putting $x=0$ in $u_x(x,t)=v_x(x,t)+x-1$, we get

$$u_x(x,t)=v_x(x,t)+x-1 \Rightarrow u_x(0,t)=v_x(0,t)+0-1$$

$-1=v(0,t)-1 \Rightarrow 0=v_x(0,t)$, (From given boundary condition $u_x(0,t)=-1$)

$$\Rightarrow v_x(0,t)=0.$$

Putting $x=1$ in $u_x(x,t)=v_x(x,t)+x-1$, we get

$$u_x(1,t)=v_x(1,t)+1-1$$

$\Rightarrow 0=v_x(1,t)$ (From given boundary condition $u_x(1,t)=0$)

Therefore, the boundary condition transforms to $v_x(0,t)=0$ and $v_x(1,t)=0$.

Hence, the new problem using the transformation equation (5.202), we have

$$\frac{\partial v}{\partial t}=\frac{\partial^2 v}{\partial x^2},\ 0<x<3\ ,\ t>0 \tag{5.203}$$

IC1: $v(x,0)=x-\dfrac{x^2}{2}$ (5.204)

and **BC1:** $v_x(0,t)=0$, **BC2:** $v_x(1,t)=0$. (5.205)

Comparing PDE given by equation (5.203)-(5.205) and with equation (5.64)-(5.66) we get $C^2=1\ l=1$ and $f(x)=x-\dfrac{x^2}{2}$. Therefore, the required solution to the one-dimensional heat equation given by equation (5.203) subject to boundary and initial condition given by equation (5.204)- (5.205) is given by

$$v(x,t)=\frac{B_0}{2}+\sum_{n=1}^{\infty}B_n\cos(n\pi x)e^{-(n\pi)^2 t} \tag{5.206}$$

is a half range cosine series, where

$$B_n=2\int_0^l\left(x-\frac{x^2}{2}\right)\cos(n\pi x)\,dx \tag{5.207}$$

and $B_0=2\displaystyle\int_0^l\left(x-\frac{x^2}{2}\right)dx$. (5.208)

Now, $B_0=2\displaystyle\int_0^1\left(x-\frac{x^2}{2}\right)dx$,

$$B_0 = 2\left[\frac{x^2}{2} - \frac{x^3}{6}\right]_0^1 = 2\left[\left(\frac{1}{2} - \frac{1}{6}\right) - (0+0)\right],$$

$$B_0 = \frac{2}{3}.$$

Now, $B_n = 2\int_0^l \left(x - \frac{x^2}{2}\right)\cos(n\pi x)\,dx$

$$B_n = 2\left[\begin{array}{l}\left(x - \frac{x^2}{2}\right)\left(\frac{\sin n\pi x}{n\pi}\right) - (1-x)\left(-\frac{\cos n\pi x}{(n\pi)^2}\right) \\ +(-1)\left(-\frac{\sin n\pi x}{(n\pi)^3}\right)\end{array}\right]_0^1,$$

$$B_n = 2\left[(0-0+0) - \left(0 - \left(-\frac{1}{(n\pi)^2}\right) + 0\right)\right],$$

$$B_n = -\frac{2}{(n\pi)^2}.$$

Substituting the value of $B_0 = \frac{2}{3}$ and $B_n = -\frac{2}{(n\pi)^2}$ in equation (5.206), we get

$$v(x,t) = \frac{\left(\frac{2}{3}\right)}{2} + \sum_{n=1}^{\infty} -\frac{2}{(n\pi)^2}\cos(n\pi x)e^{-(n\pi)^2 t},$$

$$\therefore v(x,t)=\frac{1}{3}-\frac{2}{\pi^2}\sum_{n=1}^{\infty}\frac{\cos(n\pi x)e^{-(n\pi)^2 t}}{n^2}. \tag{5.209}$$

For required solution substituting the value from equation (5.209) in equation (5.202), we get

$$u(x,t)=\frac{1}{3}-\frac{2}{\pi^2}\sum_{n=1}^{\infty}\frac{\cos(n\pi x)e^{-(n\pi)^2 t}}{n^2}+\frac{1}{2}\left(x^2+2t\right)-x.$$

Figure 5.4 depicts graphs at time *t* = *0.01, 0.5, 1.0*, and *1.5*. At the left boundary, $u_x=-1$, and the flux $\phi=-ku_x$ indicates that $\phi=k>0$, indicating that the flow is position-dependent and heat is being added. As a result, the left boundary has a higher profile while the right boundary remains insulated (no flux).

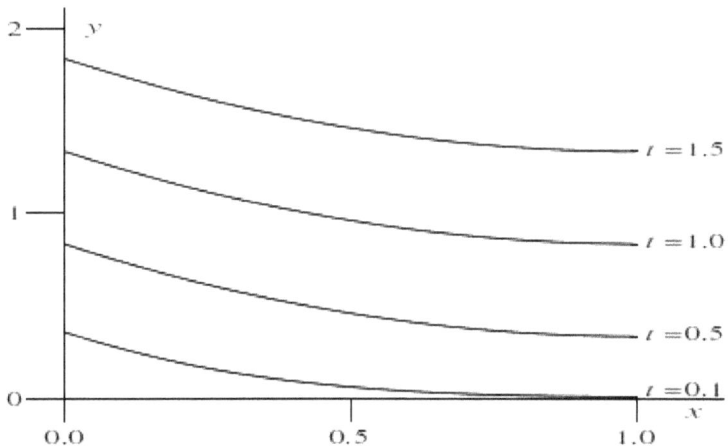

Figure 5.4: Temperature profile at time *t* = *0.01, 0.5, 1.0*, and *1.5* for $\frac{\partial u}{\partial t}=\frac{\partial^2 u}{\partial x^2}, 0<x<1, t>0$ subject to the initial and boundary conditions $u(x,0)=0, u_x(0,t)=-1, u_x(1,t)=0.$

Note: A natural question is, "Can we transform

$$u_t = u_{xx},\ 0 < x < L,\ t > 0,$$

$$u(x,0) = f(x),\ \ u(0,t) = p(t),\ u(L,t) = q(t)$$

to a problem with standard boundary conditions?" The answer is yes. We require a transformation of the form

$$u = v + A(t)x + B(t)$$

as to transform the nonstandard boundary conditions to standard boundary condition .On substitution of the u and v boundary conditions, we obtain

$$p(t) = 0 + A(t)0 + B(t),\ q(t) = 0 + A(t)L + B(t)$$

and solving for $A(t)$ and $B(t)$ gives

$$A(t) = \frac{q(t) - p(t)}{L}, B(t) = q(t),$$

which results in the transformation

$$u = v + \frac{q(t) - p(t)}{L}x + p(t)$$

$$= v + q(t)\frac{x}{L} + p(t)\frac{(L - x)}{L}.$$

However, in doing so, we change not only the original equation but also the initial condition. Now the problem changes to

$$v_t = v_{xx} - q'(t)\frac{(L - x)}{L},$$

$$v(x,0) = f(x) - q(0)\frac{x}{L} - p(0)\frac{(L-x)}{L}.$$

5.7 Nonhomogeneous equations

In this section we focus on solving the heat equation with a source term

$$u_t = u_{xx} + Q(x),\ 0 < x < L,\ t > 0 \tag{5.210}$$

$$u(0,t) = 0,\ u(L,t) = 0,\ u(x,0) = f(x). \tag{5.211}$$

To investigate this problem, we will consider a particular example where $L = 2,\ f(x) = 2x - x^2$ and $Q(x) = 1 - |x - 1|$. If we were to consider this problem without a source term, we would have solutions of the form

$$u(x,t) = \sum_{n=1}^{\infty} T_n(t) sin\frac{n\pi x}{2}, \tag{5.212}$$

where $T_n(t) = \dfrac{16\left(1-(-1)^n\right)}{n^3\pi^3} e^{-\frac{n^2\pi^2}{4}t}$ (5.213)

We note that even if this T_n wasn't known, we could find it as substitution of equation (5.212) into the heat equation and isolating coefficients of $sin\dfrac{n\pi x}{2}$, would lead to

$$T_n^{'}(t) = -\frac{n^2\pi^2}{4}T_n(t),$$

leading to the solution equation (5.213). For the problem with a source term we look for solutions of the same form i.e., Eq.5.212. However, in order that this technique works, it is necessary to expand the source term also in terms of a Fourier sine series, *i.e.*

$$Q(x)=\sum_{n=1}^{\infty} q_n \; sin\frac{n\pi x}{2} \, . \tag{5.214}$$

For $Q(x)=\begin{cases} x & \text{if } 0<x<1, \\ 2-x & \text{if } 1<x<2, \end{cases}$ (5.215)

where

$$q_n = \int_0^2 Q(x) sin\frac{n\pi x}{2}\, dx.$$

$$=\int_0^1 x\, sin\frac{n\pi x}{2}\, dx + \int_1^2 (2-x)\, sin\frac{n\pi x}{2}\, dx,$$

$$=\left[\frac{4}{n^2\pi^2} sin\frac{n\pi x}{2} - 2xn\pi \cos\frac{n\pi x}{2}\right]_0^1 + \left[-\frac{4}{n^2\pi^2} sin\frac{n\pi x}{2} + (2x-4)\cos\frac{n\pi x}{2}\right]_1^2$$

$$=\left[\frac{4}{n^2\pi^2} sin\frac{n\pi x}{2} - 2xn\pi \cos\frac{n\pi x}{2}\right]_0^1 + \left[-\frac{4}{n^2\pi^2} sin\frac{n\pi x}{2} + (2x-4)\cos\frac{n\pi x}{2}\right]_1^2$$

$$=\frac{8}{n^2\pi^2} sin\frac{n\pi x}{2} \, . \tag{5.216}$$

Substituting both (5.212) and (5.214) into (5.210) gives

$$\sum_{n=1}^{\infty} T_n^{'}(t)\, sin\frac{n\pi x}{2} = \sum_{n=1}^{\infty} -\left(\frac{n\pi x}{2}\right)^2 T_n(t)\, sin\frac{n\pi x}{2} + \sum_{n=1}^{\infty} q_n\, sin\frac{n\pi x}{2},$$

and re-grouping and isolating the coefficients of $sin\frac{n\pi x}{2}$ gives

$$T_n^{'}(t) + \frac{n^2\pi^2}{4} T_n(t) = q_n, \tag{5.217}$$

$$T_n^{'}(t)+\sum_{n=1}^{\infty}T_n^{'}(t)\sin\frac{n\pi x}{2}=\sum_{n=1}^{\infty}-\left(\frac{n\pi x}{2}\right)^2 T_n(t)\sin\frac{n\pi x}{2}+\sum_{n=1}^{\infty}q_n\sin\frac{n\pi x}{2}, \tag{5.218}$$

a linear ODE in $Tn(t)$!

On solving equation (5.218) we obtain

$$T_n(t)=\frac{4}{n^2\pi^2}q_n+b_n e^{-\left(\frac{n\pi}{2}\right)^2 t},$$

giving the final solution

$$u(x,t)=\sum_{n=1}^{\infty}\left(\frac{4}{n^2\pi^2}q_n+b_n e^{-\left(\frac{n\pi}{2}\right)^2 t}\right)\sin\frac{n\pi x}{2}.$$

Imposing the initial condition gives

$$2x-x^2=\sum_{n=1}^{\infty}\left(\frac{4}{n^2\pi^2}q_n+b_n\right)\sin\frac{n\pi x}{2}.$$

If we set, $c_n=\frac{4}{n^2\pi^2}q_n+b_n$, then we have $2x-x^2=\sum_{n=1}^{\infty}c_n\sin\frac{n\pi x}{2}$.

a regular Fourier series. Therefore,

$$c_n=\int_0^2(2x-x^2)\sin\frac{n\pi x}{2}\,dx$$

$$=\frac{16}{n^3\pi^3}\left(1-\cos\frac{n\pi}{2}\right), \tag{5.219}$$

which in turn, gives

$$b_n = c_n - \frac{4}{n^2\pi^2} q_n, \tag{5.220}$$

and finally, the solution as

$$u(x,t) = \sum_{n=1}^{\infty} \left(\frac{4}{n^2\pi^2} q_n + \left(c_n - \frac{4}{n^2\pi^2} q_n \right) e^{-\left(\frac{n\pi}{2}\right)^2 t} \right) sin \frac{n\pi x}{2}$$

where b_n and c_n are given in (5.219) and (5.220), respectively. Typical plots are given in figure 7 at times t = 0, 1, 2 and 3.

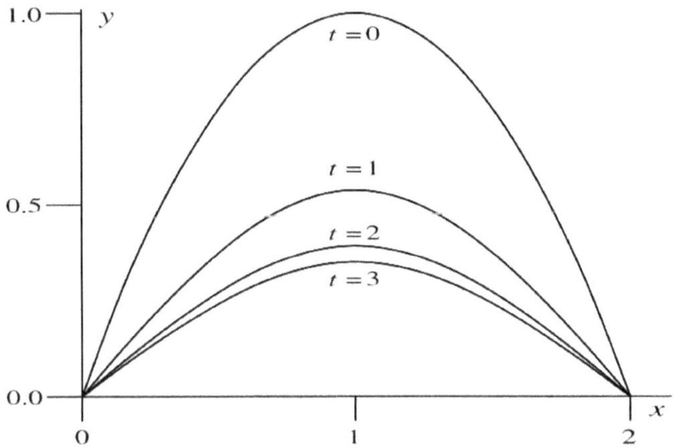

Figure 5.5: The solution equation (5.220) at time $t = 0, 1, 2$ and 3.

It is interesting to note that if we let $t \to \infty$ the solution approaches the same curve at $t = 3$. This is what is called steady state (no changes in time). It is natural to ask, "Can we find this steady state solution.?" The answer is yes. For the steady state, $u_t \to 0$ as $t \to \infty$ and the original PDE becomes $u_{xx} + Q(x) = 0$. (5.221)

Integrating twice with $Q(x)$ given in equation (5.215) gives

$$u=\begin{cases}-\dfrac{x^3}{6}+c_1x+c_2 & \text{if } 0<x<1,\\ \dfrac{x^3}{6}-x^2+k_1x+k_2 & \text{if } 1<x<2,\end{cases} \tag{5.222}$$

where c_1, c_2, k_1 and k_2 are constant of integration. Imposing that the solution and its first derivative are continuous at $x=1$ and that the solution is zero at the end points gives

$$c_1-c_1+1=0,\ c_1+c_2-k_1-k_2+\frac{2}{3}=0,$$

$$c_2=0,\ 2k_1+k_2-\frac{8}{3}=0,$$

which gives, upon solving

$$c_1=\frac{1}{2},\ c_2=0,\ k_1=\frac{3}{2},\ k_2=-\frac{1}{3}. \tag{5.223}$$

This, in turn, gives the steady state solution as

$$u=\begin{cases}-\dfrac{x^3}{6}+\dfrac{x}{2} & \text{if } 0<x<1,\\ \dfrac{x^3}{6}-x^2+\dfrac{3x}{2}-\dfrac{1}{3} & \text{if } 1<x<2.\end{cases} \tag{5.224}$$

5.8 Equations with a solution dependent source term

In this section we discuss the solution of heat equation with a solution dependent source term. Let us assume that the source term is linear for simplicity's sake. For example, $u_t = u_{xx}+\alpha u,\ 0<x<1,\ t>0,$

$$u(0,t)=0,\ u(1,t)=0,\ u(x,0)=x-x^2, \tag{5.225}$$

where α is some constant. We could try a separation of variables to obtain solutions for this problem, but for more complicated source terms like $Q(x,t)u$, a separation of variables is unsuccessful. Therefore, we try a different technique. Here we will try and transform the PDE into one that has no source term. In attempting to do so, we seek a transformation of the form

$$u\left(x,t\right)=A\left(x,t\right)v. \tag{5.226}$$

and ask, "Is possible to find A such that the source term in equation (5.226) can be removed?" Substituting of equation (5.226) in equation (5.225) gives

$$Av_t+A_tv=Av_{xx}+2A_xv_x+A_{xx}v+\alpha Av,$$

and dividing by *A* and expanding and regrouping gives

$$v_t=v_{xx}+2\frac{A_x}{A}v_x+\left(\frac{A_{xx}v}{A}-\frac{A_t}{A}+\alpha\right)v.$$

In order to target the standard heat equation, we choose

$$A_x=0,\frac{A_{xx}v}{A}-\frac{A_t}{A}+\alpha=0. \tag{5.227}$$

From the first we obtain that *A* = *A*(*t*) and from the second we obtain $A'=\alpha A$ which has the solution $A(t)=A_0e^{\alpha t}$ for some constant A_0. The boundary conditions become

$$u(0,t)=0\Rightarrow A_0e^{\alpha t}v(0,t)=0\Rightarrow v(0,t)=0,$$

$$v(1,t)\;=0\Rightarrow A_0e^{\alpha t}v(1,t)=0\Rightarrow v(1,t)=0,$$

So, the boundary conditions become unchanged. Next, we consider the initial condition, so

$$u(x,0)=x-x^2\Rightarrow A_0e^{\alpha 0}v(x,0)=0\Rightarrow A_0v(x,0)=x-x^2,$$

and so as to leave the initial condition unchanged, we choose $A_0 = 1$ Thus, under the transformation

$$u = e^{\alpha t} v . \tag{5.228}$$

the problem given by equation (5.225) changes to

$$v_t = v_{xx}, \; 0 < x < 1, t > 0, \tag{5.229}$$

$$v(x,\ 0) = 0, v(1,t) = 0,\ v(x,\ 0) = x - x^2, \tag{5.230}$$

The solution of equation (5.229) subject to initial and boundary condition given by equations (5.229) and (5.230) are

$$v(x,t) = \frac{4}{\pi^3} \sum_{n=1}^{\infty} \frac{1-(-1)^n}{n^3} e^{-n^2\pi^2 t} sin\, n\pi x$$

and so, from equation (5.228) we obtain the solution of equation (5.225) as

$$u(x,t) = \frac{4}{\pi^3} e^{\alpha t} \sum_{n=1}^{\infty} \frac{1-(-1)^n}{n^3} e^{-n^2\pi^2 t} sin\, n\pi x. \tag{5.231}$$

Figure 5. represent plots at times $t = 0, 0.1$ and 0.2 when $\alpha = 5$ and 12. It is important to note that in the case where $\alpha = 5$, the diffusion is slower in comparison with no source term (*i.e.* $\alpha = 0$ see Figure 5.6) and there is no diffusion at all when $\alpha = 12$.

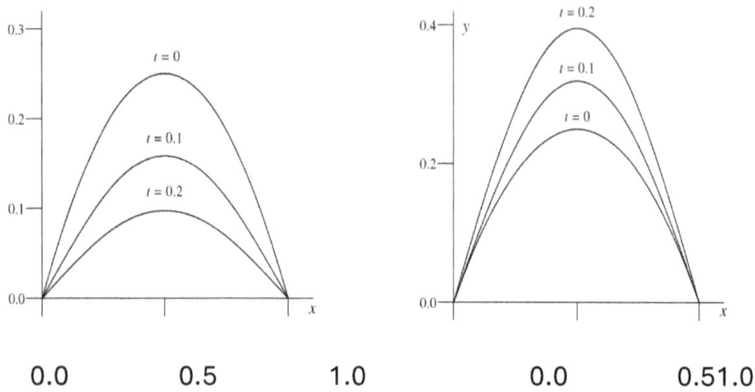

Figure 5.6: The solution equation (5.231) of the heat equation with a source $u_t = u_{xx} + \alpha u,\ 0 < x < 1,\ t > 0$, with $\alpha = 5$ and $\alpha = 12$.

Naturally, one would wonder at what alpha value a steady state solution can be reached. For that examine the first few terms of the solution in Equation (5.231) to get an answer.

$$u = \frac{8}{\pi^3} e^{\alpha t}\left(e^{-\pi^2 t} \sin \pi x + \frac{1}{27} e^{-9\pi^2 t} \sin 3\pi x + \ldots\ldots \right). \tag{5.232}$$

The first term in $(\cdot)$ will decay at the slowest rate as the other exponential components in the expression clearly approach zero. Therefore, whether or not the solution decays to zero depends on the balance between $e^{\alpha t}$ and $e^{-\pi^2 t}$. The steady state solution is provided by equality $\alpha = \pi^2$.

Example:

Solve $u_t = u_{xx} + \alpha u,\ 0 < x < 2,\ \ t > 0$ (5.233)

$$u(x,\ 0) = 4x - x^3\ u_x(0,t) = 0,\ u_x(2,t) = 0. \tag{5.234}$$

The transformation $u = e^{\alpha t} v$ will transform the equation to the heat equation and will leave the initial condition unchanged. It is now necessary to determine what happens to the boundary conditions. Using $u = e^{\alpha t} v$, equation (5.233) and (5.234) changes to

$$u_x(0,t) = 0 \Rightarrow A_0 e^{\alpha t} v_x(0,t) = 0 \Rightarrow v_x(0,t) = 0, \tag{5.235}$$

$$v_x(2,t) = 0 \Rightarrow A_0 e^{\alpha t} v_x(2,t) = 0 \Rightarrow v_x(2,t) \ = 0, \tag{5.236}$$

and so insulated boundary conditions also remain insulated. Thus, problem given by equations (5.233) – (5.234) changes to

$$v_t = v_{xx},\ 0 < x < 2, t > 0$$

$$v(x,0) = 4x - x^2,\ v_x(0,t) \ = 0, v_x(2,t) \ = 0. \tag{5.237}$$

Using a separation of variables and imposing the boundary conditions gives

$$v = \frac{1}{2} a_0 + \sum_{n=1}^{\infty} a_n e^{-\frac{n^2\pi^2}{4} t} \cos \frac{n\pi}{2} x, \tag{5.238}$$

where

$$a_0 = \frac{2}{2} \int_0^2 \left(4x - x^3\right) dx = \left[2x^2 - \frac{x^4}{4} \right]_0^2 = 4,$$

$$a_n = \frac{2}{2} \int_0^2 \left(4x - x^3\right) \cos \frac{n\pi}{2} x \, dx \tag{5.239}$$

$$= \left[\left(\frac{2\left(4x - x^3\right)}{n\pi} + \frac{48}{n^3\pi^3} \right) \sin \frac{n\pi}{2} x + \left(\frac{4\left(4 - 3x^2\right)}{n^2\pi^2} + \frac{96}{n^4\pi^4} \right) \cos \frac{n\pi}{2} x \, dx \right]_0^2$$

$$= \frac{-16}{n^2\pi^2} - \frac{96}{n^4\pi^4} + \left(\frac{6}{n\pi} + \frac{48}{n^3\pi^3} \right) \sin \frac{n\pi}{2} + \left(\frac{4}{n^2\pi^2} + \frac{96}{n^4\pi^4} \right) \cos \frac{n\pi}{2}$$

Together with the transformation $u = e^{\alpha t} v$, the solution of equations (5.233) -(5.234) is

$$u = 2e^{\alpha t} + e^{\alpha t} \sum_{n=1}^{\infty} a_n e^{-\frac{n^2 \pi^2}{4} t} \cos \frac{n\pi}{2} x. \quad (5.240)$$

where a_n is given in equation (5.239). Plots for $t = 0, 0.2, 0.4$ and 0.6 when $\alpha = -2$ and 2 are displayed in Figure 5.7. It is noteworthy to observe that the solution's exponential growth or decay depends on the sign of α.

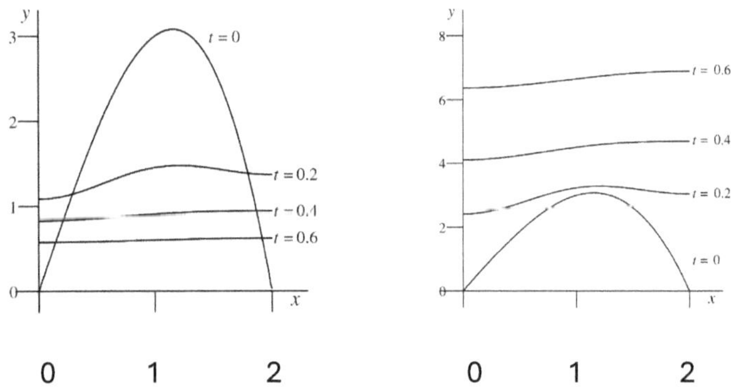

Figure 5.7: The solution equation (5.240) of the heat equation with a source with no flux boundary condition with $\alpha = -2$ and 2.

5.9 Equations with a solution dependent convective term

Consider the heat equation with a solution dependent linear convection term, *i.e.*

$$u_t = u_{xx} + \beta\, u_x,\ 0 < x < 1, t > 0 \quad (5.241)$$

$$u(0,t) = 0, u(1,t) = 0, u(x,0) = x - x^2, \quad (5.242)$$

where β is some constant. In order to compare the two problems, the same initial and boundary conditions as in the preceding section should be taken into consideration. Once more, we may attempt a separation of variables to find solutions for this issue; however, a separation of variables fails for more complex convection terms such as. Therefore, we try a different technique again. We will try and transform the PDE into one that has no convection term. In attempting to do so, we seek a transformation of the form

$$u(x,t) = A(x,t)v \tag{5.243}$$

and ask, "Is there a way to identify A so that the convection term is eliminated?" What results from substituting equations (5.243) in equations (5.241) and (5.242) are

$$Av_t + A_t v = Av_{xx} + 2A_x v_x + A_{xx} v + \beta(Av_x + A_x v),$$

$$Av_t + A_t v = Av_{xx} + 2A_x v_x + A_{xx} v + \beta(Av_x + A_x v),$$

and dividing by *A* and expanding and regrouping gives

$$v_t = v_{xx} + \frac{2A_x + \beta A}{A} vx + \frac{A_{xx} - A_t + \beta A_x}{A} v\ . \tag{5.244}$$

In order to target the standard heat equation, we choose

$$2A_x + \beta A = 0,\ \ A_{xx} - A_t + \beta A_x = 0. \tag{5.245}$$

From the first we obtain that $A(x,t) = C(t)e^{-\frac{1}{2}\beta}$ and from the second we obtain $C' + \frac{\beta^2}{4}C = 0$ which has the solution $C(t) = A_0 e^{-\frac{1}{4}\beta^2 t}$ for some constant A_0. This then gives

$$A(x,t) = A_0 e^{-\frac{1}{2}\beta x - \frac{1}{4}\beta^2 t}\ . \tag{5.246}$$

The boundary conditions become

$$v(0,t)=0 \Rightarrow C_0 e^{-\frac{1}{4}\beta^2 t} v(0,t)=0 \Rightarrow v(0,t)=0, \quad (5.247)$$

$$v(1,t)=0 \Rightarrow C_0 e^{-\frac{1}{2}-\frac{1}{4}\beta^2 t} v(1,t)=0 \Rightarrow v(1,t)=0, \quad (5.248)$$

So, the boundary conditions become unchanged. Next, we consider the initial condition, so

$$u(x,0)=x-x^2 \Rightarrow v(x,0)=\left(x-x^2\right)e^{\frac{1}{2}\beta x} \quad . \quad (5.249)$$

where we have chosen $A_0=1$. So here, the initial condition actually changes. Thus, under the transformation

$$u=e^{-\frac{1}{2}\beta x-\frac{1}{4}\beta^2 t} v \quad (5.250)$$

the problem given by equation (5.241) with boundary condition equation (5.242) becomes

$$v_t=v_{xx}, \quad 0<x<1,\ t>0,$$

$$v(x,0)=0,\ v(1,t)=0,\ \mathrm{v}(x,0)=\left(x-x^2\right)e^{\frac{1}{2}\beta x}. \quad (5.251)$$

As in the previous section, the solution is given by

$$v(x,t)=\sum_{n=1}^{\infty} b_n e^{-n^2\pi^2 t} \sin n\pi x, \quad (5.252)$$

where b_n is now given by

$$b_n=\frac{2}{1}\int_0^1\left(x-x^2\right)e^{\frac{1}{2}\beta x} \sin n\pi x dx, \quad (5.253)$$

and from equation (5.250), we get

$$u(x,t)=e^{-\frac{1}{2}\beta x-\frac{1}{4}\beta^2 t}\sum_{n=1}^{\infty}b_n e^{-n^2\pi^2 t}\sin n\pi x, \tag{5.254}$$

At this point we consider two particular examples: $\beta = 6$ and$\beta = -12.$

For $\beta = 6$

$$b_n = \frac{2}{1}\int_0^1 \left(x-x^2\right)e^{3x}\sin n\pi x dx,$$

$$= -4n\pi\frac{27-n^2\pi^2+2n^2\pi^2 e^3\cos n\pi}{\left(9+n^2\pi^2\right)^3}. \tag{5.255}$$

For $\beta = -12$

$$b_n = 2\int_0^1 \left(x-x^2\right)e^{-6x}\sin n\pi x dx$$

$$= 2n\pi\frac{7n^2\pi^2+108+(5n^2\pi^2+324)e^{-6}\cos n\pi}{\left(36+n^2\pi^2\right)^3}. \tag{5.256}$$

The respective solutions for each are

$$u(x,t) \ = \ 4\pi e^{-\frac{1}{2}\beta x-\frac{1}{4}\beta^2 t}$$
$$\times \sum_{n=1}^{\infty} n\frac{27+n^2\pi^2+2n^2\pi^2 e^3\cos\ n\pi}{\left(9+n^2\pi^2\right)^3}e^{-n^2\pi^2 t}\sin n\pi x, \tag{5.257}$$

$$u(x,t) \ = \ 2\pi e^{-\frac{1}{2}\beta x-\frac{1}{4}\beta^2 t}$$
$$\times \sum_{n=1}^{\infty} n\frac{7n^2\pi^2+108+(5n^2\pi^2+324)e^{-6}\cos n\pi}{\left(36+n^2\pi^2\right)^3}. \tag{5.258}$$

For β = 6 and β = -12, graphs at various times are shown in Figure 5.8.

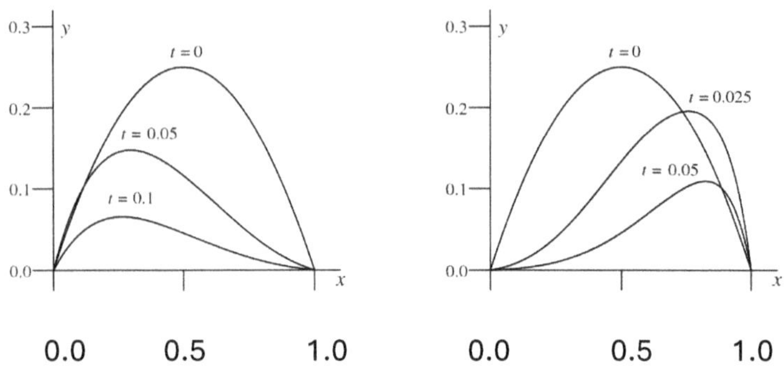

Figure 5.8: The solution (Eq.No.44) of the heat equation with convection with fixed boundary conditions with β= 6 and β =−12.

Example:

Consider $u_t = u_{xx} + \beta u_x,\ 0 < x < 2,\ t > 0$ (5.259)

with initial and boundary condition

$$u(x,\ 0) = 4x - x^3\ , u_x(0,t) = 0,\ u_x(2,t) = 0. \tag{5.260}$$

With insulated boundary conditions, this problem is identical to the one that was previously presented. To compare the changes between the two issues, we will simply transform this one into being in standard form. Since the transformation $u = e^{-\frac{1}{2}\beta x - \frac{1}{4}\beta^2 t} v$ converts the given problem in equations (5.259) and (5.260) into the heat equation, the boundary and beginning conditions will be our principal areas of concentration. When using transformation $u = e^{-\frac{1}{2}\beta x - \frac{1}{4}\beta^2 t} v$ to differentiate for the boundary conditions with respect to x, we get

$$u_x = e^{-\frac{1}{2}\beta x - \frac{1}{4}\beta^2 t} v_x - \frac{\beta}{2} e^{-\frac{1}{2}\beta x - \frac{1}{4}\beta^2 t} v, \tag{5.261}$$

and so $u_x(0,t) = 0 \Rightarrow v_x(0,t) - \frac{\beta}{2} v(0,t) = 0,$ (5.262)

$$u_x(2,t) = 0 \Rightarrow v_x(2,t) - \frac{\beta}{2} v(2,t) = 0, \tag{5.263}$$

Thus, the insulated boundary conditions become radiating boundary conditions. As for the initial condition

$$u(x,\ 0) = 4x - x^3, v(x,0) = \left(4x - x^3\right) e^{\frac{1}{2}\beta x}, \tag{5.264}$$

So again, the initial condition changes. The problem is converted into the standard form and the reader can evaluate using the previous section.

EXERCISE

1. Determine the solution of one-dimensional heat equation $u_t = c^2 u_{xx}$, subject to the boundary conditions $u(0,t) = 0, u(l,t) = 0, (t > 0)$ and the initial condition $u(x,0) = x$, l being the length of the bar.

$$\left[\text{Ans.}: u(x,t) = \frac{2l}{\pi} \sum_{n=1}^{\infty} \frac{(-1)^{n+1}}{n} \sin\left(\frac{n\pi x}{l}\right) e^{-\left(\frac{n\pi}{l}\right)^2 c^2 t}\right]$$

2. Find the temperature $u(x,t)$ in a homogeneous bar of heat conduction material of length l *cm* with its ends kept at zero temperature and initial temperature is $\frac{xd(l-x)}{l^2}$.

$$\left[\text{Ans.}: u(x,t) = \frac{8d}{\pi^3} \sum_{n=1}^{\infty} \frac{1}{(2n-1)^3} \sin\left(\frac{(2n-1)\pi x}{2l}\right) e^{-\left(\frac{(2n-1)\pi}{l}\right)^2 c^2 t}\right]$$

3. The heat flow in a bar of length $10\ cm$ of homogeneous material is governed by the partial differential equation $u_t = c^2 u_{xx}$, the ends of the bar are kept at temperature $0^\circ C$ and the initial temperature is $f(x) = x(10-x)$. Find the temperature in the bar at any instant of time.

$$\left[\text{Ans.}: u(x,t) = \frac{800}{\pi^3}\sum_{n=1}^{\infty}\frac{1}{(2n-1)^3}\sin\left(\frac{(2n-1)\pi x}{10}\right)e^{-\left(\frac{(2n-1)\pi}{10}\right)^2 c^2 t}\right]$$

4. Find the temperature $u(x,t)$ in a bar of length π which is perfectly insulated everywhere including the ends $x=\theta$ and $x=\pi$ This leads to the conditions $\frac{\partial u(\theta,t)}{\partial x} = \theta, \frac{\partial u(\pi,t)}{\partial x} = \theta.$ Further the initial conditions are as given below:

$$u(x,\theta) = \begin{cases} x, \theta < x < \frac{\pi}{2} \\ \pi x, \frac{\pi}{2} < x < \pi \end{cases}$$

Find the solution by separation of variables.

5. An isolated rod of length l has its ends *A* and *B* maintained at $0^\circ C$ and $100^\circ C$ respectively until steady state conditions prevail. If *B* is suddenly reduced to $0^\circ C$ and maintained $0^\circ C$ .Find the temperature at distance x from *A* at time t.

$$\left[\text{Ans.}: u = (-1)^{n+1}\left(\frac{200}{n\pi}\right)\sin\left(\frac{n\pi x}{l}\right)e^{-\left(\frac{n\pi}{l}\right)^2 c^2 t}\right]$$

6. A rod of length l is heated so that its ends *A* and *B* are at zero temperature. If initially its temperature is $u = \frac{cx(l-x)}{l^2}$, Find the temperature function $u(x,t)$.

$$\left[\text{Ans.}: u(x,t) = \sum_{n=odd}^{\infty} 8c\left(\frac{1}{n\pi}\right)^3 \sin\left(\frac{n\pi x}{l}\right) e^{-\left(\frac{n\pi}{l}\right)^2 a^2 t} \right]$$

7. A rod of length l has its ends *A* and *B* kept at $0^\circ C$ and $120^\circ C$ respectively until steady state conditions prevail. if the at B is reduced to $0^\circ C$ and kept so while that of A is maintained, Find the resulting temperature distribution $u(x,t)$ taking origin at A.

$$\left[\text{Ans.}: u(x,t) = \sum_{n=1}^{\infty} \left(\frac{240}{n\pi}\right)(-1)^{n+1} \sin\left(\frac{n\pi x}{l}\right) e^{-\left(\frac{n\pi}{l}\right)^2 a^2 t} \right]$$

8. An insulated rod of length 30 *cm* has its ends *A* and *B* kept at $20^\circ C$ and $80^\circ C$ respectively until steady state conditions prevail. The temperature at each end is then suddenly reduced to $0^\circ C$ and kept so. Find the resulting temperature distribution $u(x,t)$ taking origin at A.

$$\left[\text{Ans.}: u(x,t) = \sum_{n=1}^{\infty} \left(\frac{40}{n\pi}\right)\left[1 - 4(-1)^{n+1}\right] \sin\left(\frac{n\pi x}{30}\right) e^{-\left(\frac{n\pi}{30}\right)^2 a^2 t} \right]$$

9. Two ends *A* and *B* of rod of length l *cm* have the temperature at $0^\circ C$ and $100^\circ C$ respectively until steady state conditions prevail. Then the temperature at ends *A* and *B* changed to $50^\circ C$ and $150^\circ C$ respectively. Find the resulting temperature distribution $u(x,t)$ taking origin at A.

$$\left[\text{Ans.}: u(x,t) = \frac{100x}{l} + 50 - \sum_{n=odd}^{\infty} \left(\frac{200}{n\pi}\right)(-1)^{n+1} \sin\left(\frac{n\pi x}{l}\right) e^{-\left(\frac{n\pi}{l}\right)^2 a^2 t} \right]$$

10. The ends A and B of a rod $20cm$ long have the temperatures at $0^\circ C$ and $80^\circ C$ until steady state conditions prevail. If the temperature at B is reduced to $0^\circ C$ and kept so while that of

A is maintained, find the temperature distribution in the rod at any subsequent time.

$$\left[\text{Ans.}: u(x,t) = \frac{200}{\pi}\sum_{n=1}^{\infty}\left(\frac{(-1)^{n+1}}{n}\right)\sin\left(\frac{n\pi x}{l}\right)e^{-\left(\frac{n\pi}{l}\right)^2 a^2 t}\right]$$

11. A bar, $10cm$ long, with insulated sides, has its ends A and B kept at $20^{\circ}C$ and $40^{\circ}C$ respectively, until steady state conditions prevail. The temperature at the ends A is then suddenly raised to $50^{\circ}C$ and at the same instant that at B is lowered to $10^{\circ}C$. Find the subsequent temperature function $u(x,t)$.

$$\left[\text{Ans.}: u(x,t) = 50 - 4x - \frac{60}{\pi}\sum_{n=1}^{\infty}\left(\frac{1}{n}\right)\sin\left(\frac{n\pi x}{5}\right)e^{-\left(\frac{n\pi}{5}\right)^2 a^2 t}\right]$$

12. A bar, $10cm$ long, with insulated sides, has its ends A and B kept at $50^{\circ}C$ and $100^{\circ}C$ respectively, until steady state conditions prevail. The temperature at the ends A is then suddenly raised to $90^{\circ}C$ and at the same instant that at B is lowered to $60^{\circ}C$. Find the temperature at a distance x from one end at time t.

$$\left[\text{Ans.}: u(x,t) = 90 - 3x - \frac{80}{\pi}\sum_{n=1}^{\infty}\left(\frac{1}{n}\right)\sin\left(\frac{n\pi x}{5}\right)e^{-\left(\frac{n\pi}{5}\right)^2 a^2 t}\right].$$

5.7 Solution of Heat equation using Finite difference method

The heat equation is given by

$$\frac{\partial u}{\partial t} = C^2\frac{\partial^2 u}{\partial x^2} \tag{5.265}$$

where $C^2 = \frac{k}{s\rho}$ is the diffusivity of the substance $(cm^2/\sec)$.

Consider a rectangular mesh in the $x-t$ plane with spacing h along x direction and k along time t direction. Denoting a mesh point $(x,t)=(ih,ij)$ as simply $i,j,$, we have

$$\frac{\partial u}{\partial t} = \frac{u_{i,j+1}-u_{i,j}}{k}, \text{ (Refer Chapter 4)} \tag{5.266}$$

$$\text{and } \frac{\partial^2 u}{\partial x^2} = \frac{u_{i-1,j}-2u_{i,j}+u_{i+1,j}}{h^2}. \text{ (Refer Chapter 4)} \tag{5.267}$$

Substituting the values from equations (5.266) and (5.267), we get

$$u_{i,j+1}-u_{i,j} = \frac{kc^2}{h^2}\left[u_{i-1,j}-2u_{i,j}+u_{i+1,j}\right],$$

$$\text{Or } u_{i,j+1} = \alpha u_{i-1,j}+(1-2\alpha)u_{i,j}+\alpha u_{i+1,j}\,. \tag{5.268}$$

where $\alpha = \frac{kc^2}{h^2}$ is the mesh ratio parameter. This formula given by equation (5.268) determine the value of u at the $(i, j+1)^{\text{th}}$ mesh point in terms of the known function values at the points x_{i-1}, x_i and x_{i+1} at the instant t_j. It is a relation between the function values at the two time levels $j+1$ and j, is therefore, called a 2-level formula. In schematic form, the equation (5.268) is shown in Fig. 5.9 which is called the Schmidt explicit formula which is valid only for $0<\alpha\le\frac{1}{2}$.

Note: In particular when $\alpha=\frac{1}{2}$, equation (5.266) reduces to

$$u_{i,j+1} = \frac{1}{2}(u_{i-1,j}+u_{i+1,j}) \tag{5.269}$$

which shows that the value of u at x_i a time t_{j+1} is the mean of the u-values at x_{i-1}x and x_{i+1}at time t_j. This relation, known as Bendre-

Schmidt recurrence relation, gives the values of u at the internal mesh points with the help of boundary conditions.

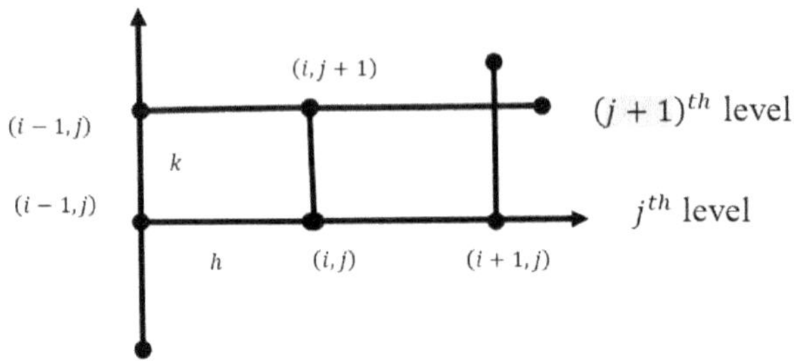

Figure 5.9: Schematic form of Schmidt explicit formula.

Example:

Solve $\frac{\partial u}{\partial t} = c^2 \frac{\partial^2 u}{\partial x^2}$ in $0 < x < 5, t \geq 0$ given that $u(x,0) = 20, u(0,\ t) = 0,\ u(5,t) = 100.$ Compute u for the time-step with $h = 1$ by Crank-Nicholson method.

Solution.

Here $c^2 = 1$ and $h = 1$.

Taking $\alpha\left(i.e. \frac{kc^2}{h}\right) = 1,$ we get $k = 1$. Also, we have

$i \rightarrow$	0	1	2	3	4	5
j↓						
0	0	20	20	20	20	100
1	0	u_1	u_2	u_3	u_4	100

Then Crank-Nicholson formula becomes

$$4u_{i,j+1} = u_{i-1,j+1} + u_{i+1,j+1} + u_{i-1,j} + u_{i+1,j}$$

$$4u_1 = 0 + 20 + 0 + u_2 \quad i.e. \quad 4u_1 - u_2 = 20 \tag{5.270}$$

$$4u_2 = 20 + 20 + u_1 + u_3 \quad i.e. \quad 4u_1 - 4u_2 + u_3 = -40 \tag{5.271}$$

$$4u_3 = 20 + 20 + u_2 + u_4 \quad i.e. \quad u_2 - 4u_3 + u_4 = -40 \tag{5.272}$$

$$4u_4 = 20 + 100 + u_3 + 100 \quad i.e. \quad 4u_3 - 4u_4 = -220 \tag{5.273}$$

Now (5.271)-4(5.270) gives $15u_2 - 4u_3 = 180$ (5.274)

Now 4(5.272)+(5.273) gives $4u_2 - 15u_3 = -380$ (5.275)

Then 15(5.274)-4(5.274) gives $209u_2 = 4220$ $i.e.$ $u_2 = 20.2$

From (5.274), we get $4u_3 = 15 \times 20.2 - 180$ $i.e.$ $u_3 = 30.75$

From (5.270), $4u_1 = 20 + 20.2$ $i.e.$ $u_1 = 10.05$

From (5.273), $4u_4 = 220 + 30.75$ $i.e.$ $u_4 = 62.69$

Thus, the required values are 10.05, 20.2, 30.75 and 62.69.

Example:

Solve the boundary value problem $u_t = u_{xx}$, under the conditions $u(0,t) = u(1,t) = 0$ and $u(x,0) = \sin \pi x, 0 \le x \le 1$ using Schmidt method $\left(\text{Take } h = 0.2 \text{ and } \alpha = \frac{1}{2}\right)$.

Solution:

Since $h = 0.2$ and $\alpha = \frac{1}{2}$,

$\therefore \alpha = \frac{k}{h^2}$ gives $k = 0.02$. Since α = 1/2, we use Bendre-Schmidt relation

$$u_{i,j+1} = \frac{1}{2}\left(u_{i-1,j} + u_{i+1,j}\right) \tag{5.276}$$

We have $u(0,0) = 0,\ u(0.2,0)\ = \sin\frac{\pi}{5} = 0.5875$.

$u(0.4,0) = \sin\frac{2\pi}{5} =\ 0.95111,\ u(0.6,0) = \sin\frac{3\pi}{5} =\ 0.9511$.

$u(0.8,0) = \sin\frac{4\pi}{5} =\ 0.5875,\ u(1,0) = \sin\pi = 0$. The value of u at the mesh points can be obtained by using the recurrence relation given by equation (5.221) as shown in table below:

	$x \rightarrow$	0	0.2	0.4	0.6	0.8	1.0
	$i \rightarrow$	0	1	2	3	4	5
t ↓	j↓						
0	0	0	0.5878	0.9511	0.9511	0.5878	0
0.02	1	0	0.4756	0.7695	0.7695	0.4756	0
0.04	2	0	0.3848	0.6225	0.6225	0.3848	0
0.06	3	0	0.3113	0.5036	0.5036	0.3113	0
0.08	4	0	0.2518	0.4074	0.4074	0.2518	0
0.1	5	0	0.2037	0.3296	0.3296	0.2037	0

Example:

Find the values of $u(x,t)$ satisfying the parabolic equation $\frac{\partial u}{\partial t} = 4\frac{\partial^2 u}{\partial x^2}$ and the boundary conditions $u(0,t) = 0 = u(8,t)$ and $u(x,0)=4x - \frac{x^2}{2}$ at the points $x = i; i = 0,1,2,......,8$ and $t = \frac{j}{8}; j == 0,1,2,......,5.$

Solution:

Here $c^2 = 4, h = 1$ and $k = \frac{1}{8}$. Then $\alpha = \frac{c^2 k}{h^2} = \frac{1}{2}$.

We have Bendre-Schmidt's recurrence relation

$$u_{i,j+1} = \frac{1}{2}\left(u_{i-1,j} + u_{i+1,j}\right). \tag{5.277}$$

Now since $u(0,t) = 0 = u(8,t)$

$\therefore u_{0,i} = 0$ and $u_{8,j} = 0$ for all values of j, i i.e., the entries in the first and last column are zero.

Since $u(x,0) = 4x - \frac{x^2}{2}$,

$\therefore u_{0,i} = 4i - \frac{i^2}{2} = 0, 3.5, 6, 7.5, 8, 7.5, 6, 3.5$ for $i = 0,1,2,3,4,5,6,7$ at $t = 0$.

These are the entries of the first row.

$i \rightarrow$ j↓	0	1	2	3	4	5	6	7	8
0	0	3.5	6	7.5	8	7.5	6	3.5	0
1	0	3	5.5	7	7.5	7	5.5	3	0
2	0	2.75	5	6.5	7	6.5	5	2.75	0
3	0	2.5	4.625	6	6.5	6	4.625	2.5	0
4	0	2.3125	4.25	5.5625	6	5.5625	4.25	2.3125	0
5	0	2.125	3.9375	5.125	5.5625	5.125	3.9375	2.125	0

Putting $j = 0$ in equation (5.277), we have $u_{i,1} = \frac{1}{2}\left(u_{i-1,0} + u_{i+1,0}\right)$

Taking $i = 1, 2, \ldots\ldots, 7$ successively, we get

$$u_{1,1} = \frac{1}{2}\left(u_{0,0} + u_{2,0}\right) = \frac{1}{2}(0+6) = 3$$

$$u_{2,1} = \frac{1}{2}\left(u_{1,0} + u_{3,0}\right) = \frac{1}{2}(3.5+7.5) = 5.5$$

$$u_{3,1} = \frac{1}{2}\left(u_{2,0} + u_{4,0}\right) = \frac{1}{2}(6+8) = 7$$

$$u_{4,1} = 7.5, u_{5,1} = 7, u_{6,1} = 5.5, u_{7,1} = 3.$$

These are the entries in the second row. Putting $j=1$ in equation (5.277), the entries of the third row are given by $u_{i,2} = \frac{1}{2}\left(u_{i-1,1} + u_{i+1,1}\right)$. Similarly putting $j = 2,3,4$ successively in equation (5.277), the entries of the fourth, fifth and sixth rows are obtained. Hence the values of $u_{i,j}$ are as given in the above table.

Example:

Solve the equation $\frac{\partial u}{\partial t} = \frac{\partial^2 u}{\partial x^2}$ subject to the conditions $u(x,0) = \sin \pi x, 0 \le x \le 1$, $u(0,t) = u(1,\ t) = 0$. Conduct computations for two levels, taking $h = \frac{1}{3}$ and $k = \frac{1}{36}$.

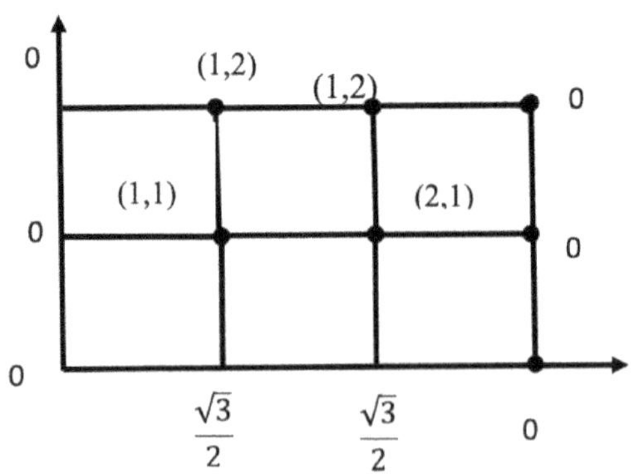

Figure 5.10: Schematic form of Schmidt explicit formula.

Solution:

Here $c^2 = 1, h = \frac{1}{3}$ and $k = \frac{1}{36}$. so that $\alpha = \frac{c^2 k}{h^2} = \frac{1}{4}$.

Also $u_{1,0} = \sin\frac{\pi}{3} = \frac{\sqrt{3}}{2}$, $u_{2,0} = \sin\frac{2\pi}{3} = \frac{\sqrt{3}}{2}$3 = √3/2

and all other boundary values are zero as shown in Figure 5.10.

Schmidt's formula given by equation (5.213) is
$u_{i,j+1} = \alpha u_{i-1,j} + (1-2\alpha)u_{i,j} + \alpha u_{i+1,j}$,

becomes $u_{i,j+1} = \frac{1}{4}\left[u_{i-1,j} + 2u_{i,j} + u_{i+1,j}\right]$

For $i = 1, 2; j = 0$:

$$u_{1,1} = \frac{1}{4}\left[u_{0,0} + 2u_{1,0} + u_{2,0}\right] = \frac{1}{4}\left[0 + 2\times\frac{\sqrt{3}}{2} + \frac{\sqrt{3}}{2}\right] = 0.65$$

$$u_{2,1} = \frac{1}{4}\left[u_{1,0} + 2u_{2,0} + u_{3,0}\right] = \frac{1}{4}\left[\frac{\sqrt{3}}{2} + 2\times\frac{\sqrt{3}}{2} + 0\right] = 0.65$$

For $i = 1, 2; j = 1$:

$$u_{1,2} = \frac{1}{4}\left[u_{0,1} + 2u_{1,1} + u_{2,1}\right] = 0.49$$

$$u_{2,2} = \frac{1}{4}\left[u_{1,1} + 2u_{2,1} + u_{3,1}\right] = 0.49.$$

EXERCISE

1 Find the solution of the parabolic equation $u_{xx} = 2u_t$ when $u(0,t) = u(4,t) = 0$ AND $u(x,0) = x(4-x)$. $h = 1$ Find the values up to $t = 5$.

2. Solve the equation $\frac{\partial^2 u}{\partial x^2} = \frac{\partial u}{\partial t}$ with the conditions $u(0,t) = 0; u(x,0) = x(1-x)$ and $u(1,t) = 0$. Assume $h = 0.1$. Tabulate u for $t = k, 2k$ and $3k$ choosing an appropriate value of k.

3. Given $\frac{\partial^2 f}{\partial x^2} - \frac{\partial f}{\partial t} = 0; f(0,t) = f(5,t) = 0, f(x,0) = x^2(25 - x^2)$; find the values of f for

$x = ih(i = 0,1,...,5)$ and $t = jk(j = 0, 1, ..., 6)$ with $h = 1$ and $k = \frac{1}{2}$,

using the explicit method.

4. Solve the heat equation $\frac{\partial u}{\partial t} = \frac{\partial^2 u}{\partial x^2}$ subject to the condition

$u(0,t) = u(1,t) = 0$ and $u(x,0) = \begin{cases} 2x \text{ for } 0 \le x \le \frac{1}{2} \\ 2(1-x) \text{ for } \frac{1}{2} \le x \le 1 \end{cases}$ .Take

$h = \frac{1}{4}$ and k according to Bendre-Schmidt equation.

Chapter 6

Wave Equation

6.1 Introduction

The wave equation is a fundamental mathematical model used to describe the behaviour of waves in various physical systems. It represents how waves propagate through a medium, whether it's a mechanical medium like a stretched string or a membrane, or electromagnetic waves traveling through space. There are several uses for the wave equation in physics, engineering, and other disciplines. For instance, it is utilized in structural engineering to examine vibrations in structures and bridges, in telecommunications to comprehend electromagnetic waves, and in seismology to investigate seismic waves. Knowing the wave equation is essential to understanding how waves move across space and interact with other objects in a variety of real-world situations.

This chapter discuss solution of the wave equation with different initial and boundary condition.

6.2 Derivation of One-Dimensional Wave Equation

Consider a tightly stretched elastic string of length l and fixed ends A and B subjected to constant tension T as shown in Figure 6.1.

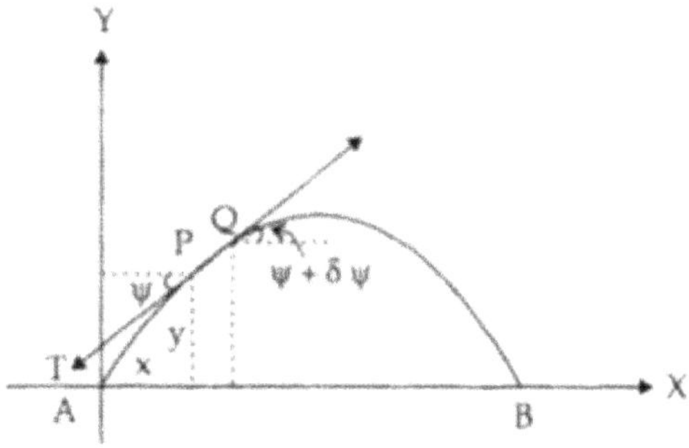

Figure 6.1: Model of a stretched string

The tension T will be large as compared to the weight of the string so that the effects of gravity are negligible. Let the string be released from rest and allowed to vibrate. We shall study the subsequent motion of the string, with no external forces acting on it, if each point of the string makes small vibrations at right angles to the equilibrium position AB, of the string entirely in one plane. Taking the end A as the origin, AB as the X-axis and AY perpendicular to it as the Y-axis, so that the motion takes place entirely in the XY - plane. Above figure shown the string in the position APB at times t. Consider the motion of the element PQ of the string between its points $P(x,y)$ and $Q(x+\delta x, y+\delta y)$ where the tangents make angles ψ and $\psi+\delta\psi$ with the x-axis. Clearly the element is moving upwards with the acceleration $\frac{\partial^2 y}{\partial t^2}$. Also, the vertical component of the force acting on this element

$$=T\sin(\psi+\delta\psi)-T\sin(\psi),$$

$$=T(\psi+\delta\psi-\psi), \ (\because \sin(\psi)=\psi, \text{as } \psi \text{is very small})$$

$$\approx T(\delta\psi).$$

The acceleration of the elements in the QY direction is $\frac{\partial^2 y}{\partial t^2}$. If the length of PQ is δs, then the mass of PQ is $m\cdot\delta s$. Hence by Newton's second law, the equation of motion becomes

$$m\cdot\delta s\frac{\partial^2 y}{\partial t^2}=T(\delta\psi)\Rightarrow\frac{\partial^2 y}{\partial t^2}=\frac{T(\delta\psi)}{m\cdot\delta s}. \tag{6.1}$$

As $Q \to P$, i.e., $\delta s \to 0$, we have $\frac{\partial^2 y}{\partial t^2} = \frac{T}{m}\frac{\delta \psi}{\delta s}$, where $\frac{\delta \psi}{\delta s}$ = curvature at P of the deflection curve .Using the formula for the radius of curvature we have $\frac{\delta \psi}{\delta s} = \frac{\frac{\partial^2 y}{\partial x^2}}{\left[1+\left(\frac{\partial y}{\partial x}\right)^2\right]^{\frac{3}{2}}}$. Here $\frac{\partial y}{\partial x}$ is very small so $\left(\frac{\partial y}{\partial x}\right)^2$ is negligible. Therefore,

$$\frac{\delta \psi}{\delta s} = \frac{\partial^2 y}{\partial x^2}. \tag{6.2}$$

Using equations (6.1) and (6.2), we have

$$\frac{\partial^2 y}{\partial t^2} = \frac{T}{m}\frac{\partial^2 y}{\partial x^2}. \tag{6.3}$$

Let $\frac{T}{m} = C^2 \text{(positive)}$, equation (6.3) represents displacement $y(x,t)$, given by

$$\frac{\partial^2 y}{\partial t^2} = C^2 \frac{\partial^2 y}{\partial x^2}. \tag{6.4}$$

This partial differential equation is known as one dimensional wave equation. Now we want to find the solution of one-dimensional wave equation subjected to the following boundary condition:

Boundary conditions	**Initial conditions**
BC1: $y(x=0,t)=0,\ \forall t>0$ (6.4a)	**IC1**: $y(x,t=0)=f(x),\ \forall x\in(0,l)$ (6.4 c)
BC2: $y(x=l,t)=0,\ \forall t>0$ (6.4b)	**IC2**: $\frac{\partial y}{\partial t}(x,0)=0,\ \forall x\in(0,l)$ (6.4 d)

6.2.1 Solution of One-Dimensional Wave Equation

Assume that a solution of equation (6.4) is of the form

$$y(x,t) = X(x)T(t). \tag{6.5}$$

where X is a function of x and T is a function of t only. Differentiate (6.5) w.r.t. x and t, we have

$$\frac{\partial y}{\partial x} = X'T \Rightarrow \frac{\partial^2 y}{\partial x^2} = X''T, \tag{6.6}$$

and

$$\frac{\partial y}{\partial t} = XT' \Rightarrow \frac{\partial^2 y}{\partial t^2} = XT'' \tag{6.7}$$

where $X' = \dfrac{dX}{dx}, X'' = \dfrac{d^2X}{dx^2}, T' = \dfrac{dT}{dx}, T'' = \dfrac{d^2T}{dx^2}$.

Using equations (6.6) and (6.7) in (6.4), we have

$$XT'' = C^2 X''T.$$

Separating the variables in the above equation, we get

$$\frac{1}{C^2}\frac{T''}{T} = \frac{X''}{X}. \tag{6.8}$$

In equation (6.8), L.H.S is a function of x only and R.H.S is a function of t only and here x and t are independent variables, so each side is constant. So rewriting equation (6.8) as

$$\frac{X''}{X} = k \Rightarrow X'' = kX \Rightarrow X'' - kX = 0, \tag{6.9}$$

and

$$\frac{1}{C^2}\frac{T''}{T} = k \Rightarrow T'' = kT \Rightarrow T'' - C^2kT = 0 \tag{6.10}$$

Equations (6.9) and (6.10) are second order homogeneous linear differential equations.

Now let us look at the boundary condition

BC1: $y(x=0,t)=0,\ \forall t>0$

BC2: $y(x=l,t)=0,\ \forall t>0$

The solution of equation (6.4) is given by equation (6.5) i.e., $y(x,t)=X(x)T(t)$. Using equation (6.5) and **BC1**, we get

$$y(0,t) = X(0)T(t) \Rightarrow X(0)T(t) = 0 \Rightarrow X(0) = 0. \tag{6.11}$$

($\because T(t) \neq 0$ as we are look for non –zero solution)

Similarly, using equation (6.5) and **BC2**, we get

$$y(l,t) = X(l)T(t) \Rightarrow X(l)T(t) = 0 \Rightarrow X(l) = 0. \tag{6.12}$$

($\because T(t) \neq 0$ as we are look for non –zero solution)

Now, we solve

$$X'' - kX = 0 \tag{6.13}$$

with boundary condition

$$X(0)=0$$

and

$$X(l)=0, \tag{6.14}$$

which is called an Eigen value problem and *k* is called Eigen value.

Case (i) when k is positive.

Let $k=p^2$ for some $p\in R$.

The equation (6.13) reduces to $X''-p^2X=0$. The auxiliary equation associate with $X''-p^2X=0$ is

$$m^2-p^2=0$$

$$\Rightarrow m=p,-p$$

Since the roots of the auxiliary equation are real and distinct, therefore

$$X(x)=Ae^{px}+Be^{-px}, \tag{6.15}$$

where $A \text{ and } B$ are arbitrary constants. Now we apply the boundary condition given by equation (6.14) to find the value of *A* and *B*. Put $x=0$ in equation (6.15), we get

$$X(0)=Ae^0+Be^0\Rightarrow X(0)=A+B. \tag{6.16}$$

Using equation (6.14) in equation (6.16), we get

$A = -B$.

Put $x = l$ in equation (6.15), we get

$$X(l) = Ae^{pl} + Be^{-pl} \Rightarrow X(l) = Ae^{pl} + Be^{-pl}. \qquad (6.17)$$

Using equation (6.14) and (6.16) in equation (6.17), we get $X(l) = Ae^{-pl}\left(e^{2pl} - 1\right) = 0$.

$$\Rightarrow A = 0 \; \left(\because e^{pl}\left(e^{2pl} - 1\right) \neq 0\right)$$

Since $A = 0$, therefore, we get $B = 0$. So, equation (6.15) reduces to $X(x) = 0$.

Substituting $X(x) = 0$ in equation (6.5), the required solution is $y(x,t) = 0$. So, **when k is positive,** there is no Eigen value, and we get a trivial solution.

Case (ii) when k is zero.

Then auxiliary equation of equations (6.9) is $m^2 = 0 \Rightarrow m = 0, 0$. Since, the roots are real and repeated, therefore solution of equation (6.9) is

$$X(x) = (A + Bx) \qquad (6.18)$$

Now, we use boundary condition $X(0) = 0$ and $X(l) = 0$ to find the value of arbitrary constant.

Put $x = 0$ in equation (6.18), we get

$$X(0) = (A + B0) \Rightarrow A = 0, \qquad (6.19)$$

Put $x = l$ in equation (6.18), we get $X(l) = A + Bl \Rightarrow Bl = 0$, (From equation (6.19))

$$\Rightarrow B = 0 \,.\, (\because l \neq 0) \tag{6.20}$$

Since $A = 0$ and $B = 0$, therefore $X(x) = 0$.Substituting $X(x) = 0$ in equation (6.19), the required solution is $y(x,t) = 0$.

So, **when k is zero,** there is no Eigen value, and we get a trivial solution.

Case (iii) when k is negative.

Let $k = -p^2$.

Then auxiliary equation of equation (6.9) is $m^2 + p^2 = 0 \Rightarrow m = \pm i\, p$
Since, the roots are complex, therefore solution of equation (6.9) is

$$X(x) = (A\cos px + B\sin px) \tag{6.21}$$

where *A* and *B* are arbitrary constants. Now, we use boundary condition $X(0) = 0$ and $X(l) = 0$ to find the value of arbitrary constant.

Put $x = 0$ in equation (6.21), we get

$$X(0) = (A\cos p0 + B\sin p0) \Rightarrow A = 0. \tag{6.22}$$

Put $x = l$ in equation (6.21), we get from equation (6.22)

$$X(l) = (A\cos pl + B\sin pl) \Rightarrow B\sin pl = 0,$$

$\Rightarrow B = 0$ or $\sin pl = 0$, (From equation (6.22))
$\Rightarrow \sin pl = 0$, $(\because B \neq 0$ as $X(x) = 0)$

$$\Rightarrow pl = \sin^{-1}(0)$$

$$\Rightarrow pl = n\pi,$$

$$\Rightarrow p = \frac{n\pi}{l},\ n = 1, 2, 3, \ldots.. \tag{6.23}$$

For $n = 1, 2, 3, \ldots..$, we get different value of p, so we denote p by p_n.Therefore $p = p_n = \frac{n\pi}{l}$. Since $p_n = \frac{n\pi}{l}$, therefore from equation (6.23)

$$X(x) = B\sin\left(\frac{n\pi}{l}\right)x. \tag{6.24}$$

For $n = 1, 2, 3, \ldots..$, we get different value of $X(x)$ using equation (6.24), so we denote $X(x)$by $X_n(x)$.

Therefore,

$$X_n(x) = B_n \sin\left(\frac{n\pi}{l}\right)x. \tag{6.25}$$

Similarly, when k is negative i.e., $k = -p^2$ then auxiliary equation associated with equation (6.10) is $m'' + C^2 p^2 = 0 \Rightarrow m = \pm iCp$.

Since the roots are complex. Therefore,

$T(t) = (C\cos Cpt + D\sin Cpt)$. From equation (6.23),

$T(t) = (C\cos Cpt + D\sin Cpt)$ reduces to

$$T(t) = \left(C\cos C\left(\frac{n\pi}{l}\right)t + D\sin C\left(\frac{n\pi}{l}\right)t\right).$$

For $n = 1, 2, 3, ...$, we get different value of $T(t)$, so we denote $T(t)$ by $T_n(t)$.

Therefore,

$$T_n(t) = \left(C_n \cos C\left(\frac{n\pi}{l}\right)t + D_n \sin C\left(\frac{n\pi}{l}\right)t \right). \tag{6.26}$$

Substituting equations (6.26) and (6.27), the solution satisfying the boundary condition is given by

$$y(x,t) = B_n \sin\left(\frac{n\pi}{l}\right)x\left(C_n \cos C\left(\frac{n\pi}{l}\right)t + D_n \sin C\left(\frac{n\pi}{l}\right)t \right). \tag{6.27}$$

For $n = 1, 2, 3, ...$, we get different value of $y(x,t)$, so we denote $y(x,t)$ by $y_n(x,t)$. (6.28)

Therefore, equation (6.27) can be written as

$$y_n(x,t) = B_n \sin\left(\frac{n\pi}{l}\right)x\left(C_n \cos C\left(\frac{n\pi}{l}\right)t + D_n \sin C\left(\frac{n\pi}{l}\right)t \right). \tag{6.29}$$

Equation (6.29) is the solution of equation (6.4) and satisfies the boundary conditions BC1 and BC2.We know that if y_1 and y_2 are two solutions of a linear, homogenous equation satisfying linear homogenous boundary conditions, then $y_1 + y_2$ is also a solution of that equation and satisfies the same boundary condition. This is called the principle of superposition. Now equation (6.4) with boundary conditions BC1 and BC2 are linear and homogenous. Therefore, by the principle of superposition, the series

$$y(x,t) = \sum_{n=1}^{\infty} y_n(x,t), \tag{6.30}$$

$$y(x,t)=\sum_{n=1}^{\infty}B_n\sin\left(\frac{n\pi}{l}\right)x\left(C_n\cos C\left(\frac{n\pi}{l}\right)t+D_n\sin C\left(\frac{n\pi}{l}\right)t\right). \quad (6.31)$$

$$y(x,t)=\sum_{n=1}^{\infty}\sin\left(\frac{n\pi}{l}\right)x\left(a_n\cos C\left(\frac{n\pi}{l}\right)t+b_n\sin C\left(\frac{n\pi}{l}\right)t\right)$$

where

$a_n = B_nC_n$ and $b_n = B_nD_n$

converges and is solution of one-dimensional wave equation given by equation (6.4) satisfying BC1 and BC2. In fact, we assume that term by term differentiation is possible and that derivative series is also convergent. Now a_n and b_n must be chosen such that y as given in equation (6.41) satisfies the initial condition **IC1** and **IC2.** Putting $t=0$ in equation (6.31), we get

$y(x,0)=\sum_{n=1}^{\infty}a_n\sin\left(\frac{n\pi}{l}\right)x$. From **IC1,** we get

$f(x)=\sum_{n=1}^{\infty}a_n\sin\left(\frac{n\pi}{l}\right)x$, which is half range Fourier sine series

and $a_n=\frac{2}{l}\int_0^l f(x)\sin\left(\frac{n\pi x}{l}\right)dx$. (6.32)

Differentiating equation (6.31) partially w.r.t t, we get

$$y_t(x,t)=\sum_{n=1}^{\infty}\sin\left(\frac{n\pi}{l}\right)x\left(-a_n\frac{n\pi C}{l}\sin C\left(\frac{n\pi}{l}\right)t+b_n\frac{n\pi C}{l}\cos C\left(\frac{n\pi}{l}\right)t\right). \quad (6.33)$$

Put $t=0$ in equation (6.33), we get

$y_t(x,0)=\sum_{n=1}^{\infty}b_n\frac{n\pi C}{l}\sin\left(\frac{n\pi}{l}\right)x$. From **IC2,** we get

$$g(x)=\sum_{n=1}^{\infty} b_n \frac{n\pi C}{l}\sin\left(\frac{n\pi}{l}\right)x,$$

which is half range Fourier sine series and

$$b_n = \frac{2}{n\pi C}\int_0^l g(x)\sin\left(\frac{n\pi x}{l}\right)dx. \tag{6.34}$$

Therefore, the required solution to the boundary value problem given by equation (6.4) with (6.4.a)- (6.4.d) is given by

$$y(x,t)=\sum_{n=1}^{\infty}\sin\left(\frac{n\pi}{l}\right)x\left(a_n\cos C\left(\frac{n\pi}{l}\right)t+b_n\sin C\left(\frac{n\pi}{l}\right)t\right),$$

where

$a_n = \frac{2}{l}\int_0^l f(x)\sin\left(\frac{n\pi x}{l}\right)dx$, and $b_n = \frac{2}{n\pi C}\int_0^l g(x)\sin\left(\frac{n\pi x}{l}\right)dx$.

Example:

Solve the boundary value problem $u_{tt}=a^2u_{xx}\ \ 0<x<l;t>0$ with $u(0,t)=0;\ \ u(l,t)=0$ and $u(x,0)=0;u_t(x,0)=u_0x(1-x)$.

Solution:

It is given that $u_{tt}=a^2u_{xx}\ \ 0<x<l;t>0$, (6.35)

with boundary conditions $u(0,t)=0;$ (6.35a)

$$u(l,t)=0 \tag{6.35b}$$

IC1: $u(x,0)=0;$ (6.35c)

IC2: $u_t(x,0)=u_0x(1-x)$. (6.35 d)

Comparing the above problem with boundary value problem given by (6.4), (6.4 a)- (6.4 d), the solution of (6.35) satisfy the above boundary and initial conditions is given by

$$u(x,t)=\sum_{n=1}^{\infty}\sin\left(\frac{n\pi}{l}\right)x\left(a_n\cos C\left(\frac{n\pi}{l}\right)t+b_n\sin C\left(\frac{n\pi}{l}\right)t\right), \qquad (6.36)$$

where $a_n=\frac{2}{l}\int_0^l f(x)\sin\left(\frac{n\pi x}{l}\right)dx$, and

$$b_n=\frac{2}{n\pi C}\int_0^l g(x)\sin\left(\frac{n\pi x}{l}\right)dx.$$

From equation (6.35c) i.e. **IC1**, $f(x)=0$ and therefore

$$a_n=0. \qquad (6.37)$$

Now to evaluate b_n, we use equation (6.35d) i.e., **IC2**.

$$b_n=\frac{2}{n\pi C}\int_0^l g(x)\sin\left(\frac{n\pi x}{l}\right)dx=\frac{2}{n\pi a}\int_0^l u_0x(1-x)\sin\left(\frac{n\pi x}{l}\right)dx.$$

Using integration by parts, we get

$$b_n=\frac{2u_0}{n\pi a}\int_0^l x(1-x)\sin\left(\frac{n\pi x}{l}\right)dx,$$

$$=\frac{2u_0}{n\pi a}\left[-(x-x^2)\frac{\cos\left(\frac{n\pi x}{l}\right)}{\left(\frac{n\pi}{l}\right)}-(1-2x)\left(-\frac{\sin\left(\frac{n\pi x}{l}\right)}{\left(\frac{n\pi}{l}\right)^2}\right)+(-2)\left(\frac{\cos\left(\frac{n\pi x}{l}\right)}{\left(\frac{n\pi}{l}\right)^3}\right)\right]_0^l,$$

$$=\frac{2u_0}{n\pi a}\left[-\left(l-l^2\right)\frac{\cos(n\pi)}{\left(\frac{n\pi}{l}\right)}+(-2)\left(\frac{\cos(n\pi)}{\left(\frac{n\pi}{l}\right)^3}\right)\right]-\left[-2\frac{1}{\left(\frac{n\pi}{l}\right)^3}\right]$$

Example:

A string of length *l* has its ends *x=0* and *x=l* fixed. It is released in the position $y=\frac{4\lambda x(l-x)}{l^2}$. Find an expression for the displacement of the string at any subsequent time.

Solution:

The displacement $y(x,t)$ is the solution of the wave equation

$$\frac{\partial^2 y}{\partial t^2}=C^2\frac{\partial^2 y}{\partial x^2}. \tag{6.38}$$

The boundary and initial conditions are:

Boundary conditions

BC1:

$$y(x=0,t)=0,\ \forall t>0$$

(6.38a)

BC2:

$$y(x=l,t)=0,\ \forall t>0$$

(6.38b)

Initial conditions

IC1:

$$y(x,t=0)=\frac{4\lambda x(l-x)}{l^2}=f(x),\ \forall x\in(0,l)$$

(6.38c)

IC2: $\frac{\partial y}{\partial t}(x,0)=0,\ \forall x\in(0,l).$

(6.38d)

We know that the solution of equation (6.38) with boundary and initial condition equations (6.38a) -(6.38d) is given by

$$y(x,t)=\sum_{n=1}^{\infty}\sin\left(\frac{n\pi}{l}\right)x\left(a_n\cos C\left(\frac{n\pi}{l}\right)t+b_n\sin C\left(\frac{n\pi}{l}\right)t\right), \tag{6.39}$$

where

$$a_n = \frac{2}{l}\int_0^l f(x)\sin\left(\frac{n\pi x}{l}\right)dx,$$

$$a_n = \frac{2}{l}\int_0^l \frac{4\lambda x(l-x)}{l^2}\sin\left(\frac{n\pi x}{l}\right)dx,$$

$$a_n = \frac{8\lambda}{l^3}\left[-\left(lx-x^2\right)\frac{\cos\left(\frac{n\pi x}{l}\right)}{\left(\frac{n\pi}{l}\right)}-(l-2x)\left(-\frac{\sin\left(\frac{n\pi x}{l}\right)}{\left(\frac{n\pi}{l}\right)^2}\right)+(-2)\left(\frac{\cos\left(\frac{n\pi x}{l}\right)}{\left(\frac{n\pi}{l}\right)^3}\right)\right]_0^l,$$

$$a_n = \frac{8\lambda}{l^3}\left[-\frac{2l^3}{n^3\pi^3}(-1)^n+\frac{2l^3}{n^3\pi^3}\right],$$

$$a_n = \frac{16\lambda}{n^3\pi^3}\left[-(-1)^n+1\right].$$

and $b_n = \frac{2}{n\pi C}\int_0^l g(x)\sin\left(\frac{n\pi x}{l}\right)dx = 0$. ($g(x)=0$)

Putting the value of a_n and b_n in equation (6.39), we get

$$y(x,t) = \sum_{n=1}^{\infty}\sin\left(\frac{n\pi}{l}\right)x\left(\frac{16\lambda}{n^3\pi^3}\left[-(-1)^n+1\right]\cos C\left(\frac{n\pi}{l}\right)t\right),$$

$$y(x,t) = \sum_{n=1}^{\infty}\frac{16\lambda}{n^3\pi^3}\left[-(-1)^n+1\right]\sin\left(\frac{n\pi x}{l}\right)\cos\left(\frac{Cn\pi t}{l}\right),$$

Replacing n by $2n-1$, we get

$$y(x,t) = \frac{32\lambda}{\pi^3}\sum_{n=1}^{\infty}\frac{1}{(2n-1)^3}\sin\left(\frac{(2n-1)\pi x}{l}\right)\cos\left(\frac{C(2n-1)\pi t}{l}\right).$$

Example:

A taut string of length $'l'$ is fastened at both ends as shown in Figure 6.2. The midpoint of the string is taken to a height $'h'$ and then released from rest in that position. Find the displacement of the string at any subsequent time.

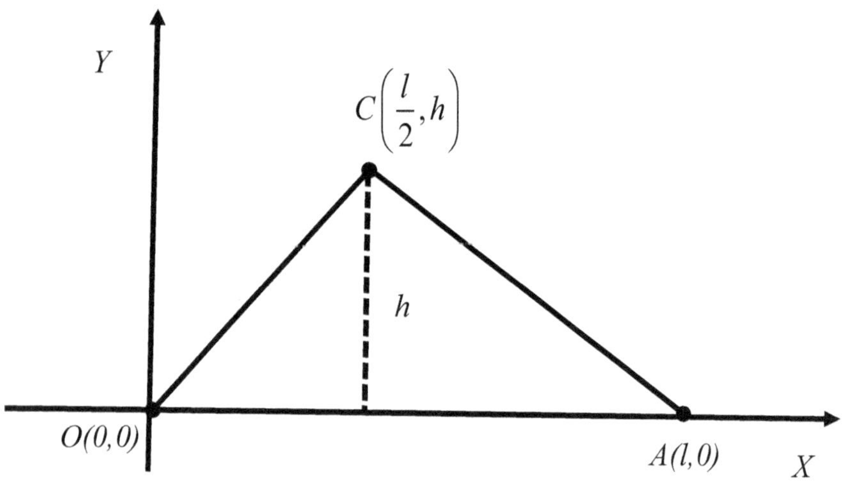

Figure 6.2: Initial position of the string

Solution:

The displacement function $y(x,t)$ of the string is the solution of wave equation

$$\frac{\partial^2 y}{\partial t^2} = C^2\frac{\partial^2 y}{\partial x^2}, \tag{6.40}$$

subject to the boundary condition

$$y(0,t) = y(l,t) = 0 \tag{6.41}$$

for all $t \geq 0$. For initial condition, it is given that midpoint of the string is taken to a height $'h'$ and then released from rest in that position as shown in Figure 6.1. The initial position of the string is shown in the Figure 6.1. Taking $OA = l,$ length of the string with *B* as its point of bisection. From Figure 6.1, initial position of the string at $t = 0$ is made up of two straight line segment *OC* and *CA* and the string is released from rest. The initial displacement of string is given by

$$y = f(x) = \begin{cases} Equation\, of\ line\ OC, & 0 \leq x \leq \dfrac{l}{2} \\ Equation\, of\ line\ CA, & \dfrac{l}{2} \leq x \leq l \end{cases}.$$

Equation of the line *OC* joining points $O(0,0)$ and $C\left(\dfrac{l}{2}, h\right)$ is given

$$\frac{y - y_1}{y_2 - y_1} = \frac{x - x_1}{x_2 - x_1} \Rightarrow \frac{y - 0}{h - 0} = \frac{x - 0}{\dfrac{l}{2} - 0} \Rightarrow y = \frac{2h}{l}, \ 0 \leq x \leq \frac{l}{2}.$$

Equation of the line *AC* joining points $C\left(\dfrac{l}{2}, h\right)$ and $A(l,0)$ is given

$$\frac{y - y_1}{y_2 - y_1} = \frac{x - x_1}{x_2 - x_1} \Rightarrow \frac{y - h}{0 - h} = \frac{x - \dfrac{l}{2}}{l - \dfrac{l}{2}} \Rightarrow y = 2h - \frac{2hx}{l}, \ \frac{l}{2} \leq x \leq l.$$

Therefore, initial displacement of string (initial condition) is given by

$$y = f(x) = \begin{cases} \dfrac{2h}{l}x, & 0 \leq x \leq \dfrac{l}{2} \\ 2h - \dfrac{2hx}{l}, & \dfrac{l}{2} \leq x \leq l \end{cases}, \tag{6.42}$$

and initial velocity is given by

$$\left(\frac{\partial y}{\partial t}\right)_{t=0} = 0. \tag{6.43}$$

The above problem can be stated as: $\frac{\partial^2 y}{\partial t^2} = C^2 \frac{\partial^2 y}{\partial x^2}$

Subject to initial condition **IC1**: $y(x,0) = f(x)$, where

$$f(x) = \begin{cases} \frac{2h}{l}x, & 0 \le x \le \frac{l}{2} \\ 2h - \frac{2hx}{l}, & \frac{l}{2} \le x \le l \end{cases},$$

IC2: $\left(\frac{\partial y}{\partial t}\right)_{t=0} = 0 = g(x)$,

and boundary condition **BC1**: $y(x=0,t) = 0,\ \forall t > 0$,

BC2: $y(x=l,t) = 0,\ \forall t > 0$.

Comparing the above problem with boundary value problem given by (6.4), (6.4 a)- (6.4 d), the solution of (6.43) satisfy the above boundary and initial conditions is given by

$$y(x,t) = \sum_{n=1}^{\infty} \sin\left(\frac{n\pi}{l}\right)x\left(a_n \cos C\left(\frac{n\pi}{l}\right)t + b_n \sin C\left(\frac{n\pi}{l}\right)t\right), \tag{6.44}$$

where $a_n = \frac{2}{l}\int_0^l f(x)\sin\left(\frac{n\pi x}{l}\right)dx$, and

$$b_n = \frac{2}{n\pi C}\int_0^l g(x)\sin\left(\frac{n\pi x}{l}\right)dx.$$

From equation (6.44) i.e. **IC2**, $g(x)=0$ and therefore

$$b_n = \frac{2}{n\pi C}\int_0^l g(x)\sin\left(\frac{n\pi x}{l}\right)dx = 0. \qquad (6.45)$$

Now to evaluate a_n, we use **IC1**. According to **IC1**, *f(x)* is given by

$$f(x)=\begin{cases} \dfrac{2h}{l}x, & 0 \le x \le \dfrac{l}{2} \\ 2h-\dfrac{2hx}{l}, & \dfrac{l}{2} \le x \le l \end{cases}.$$

$$\therefore\ a_n = \frac{2}{l}\int_0^l f(x)\sin\left(\frac{n\pi x}{l}\right)dx = \frac{2}{l}\int_0^{\frac{l}{2}} f(x)\sin\left(\frac{n\pi x}{l}\right)dx + \frac{2}{l}\int_{\frac{l}{2}}^l f(x)\sin\left(\frac{n\pi x}{l}\right)dx,$$

$$a_n = \frac{2}{l}\int_0^{\frac{l}{2}} \frac{2h}{l}x\sin\left(\frac{n\pi x}{l}\right)dx + \frac{2}{l}\int_{\frac{l}{2}}^l \left(2h-\frac{2hx}{l}\right)\sin\left(\frac{n\pi x}{l}\right)dx,$$

$$a_n = \frac{4h}{l^2}\int_0^{\frac{l}{2}} x\sin\left(\frac{n\pi x}{l}\right)dx + \frac{2}{l}\int_{\frac{l}{2}}^l \left(2h\sin\left(\frac{n\pi x}{l}\right) - \frac{2hx}{l}\sin\left(\frac{n\pi x}{l}\right)\right)dx,$$

$$a_n = \frac{4h}{l^2}\left[-\frac{xl}{n\pi}\cos\left(\frac{n\pi x}{l}\right)+\frac{l^2}{n^2\pi^2}\sin\left(\frac{n\pi x}{l}\right)\right]_0^{\frac{l}{2}} +$$

$$\frac{2}{l}\left[-\frac{2h}{n\pi}\cos\left(\frac{n\pi x}{l}\right)\right]_{\frac{l}{2}}^l$$

$$-\frac{4h}{l^2}\left[-\frac{xl}{n\pi}\cos\left(\frac{n\pi x}{l}\right)+\frac{l^2}{n^2\pi^2}\sin\left(\frac{n\pi x}{l}\right)\right]_{\frac{l}{2}}^l,$$

$$a_n = \frac{4h}{l^2}\left[-\frac{l^2}{2n\pi}\cos\frac{n\pi}{2} + \frac{l^2}{n^2\pi^2}\sin\left(\frac{n\pi}{2}\right)\right]$$

$$+\frac{4h}{l^2}\left[\frac{l^2}{2n\pi}\cos\frac{n\pi}{2} + \frac{l^2}{n^2\pi^2}\sin\left(\frac{n\pi}{2}\right)\right],$$

$$a_n = \frac{8h}{n^2\pi^2}\sin\left(\frac{n\pi}{2}\right),$$

$$a_n = \begin{cases} \dfrac{8(-1)^{m+1}h}{(2m-1)^2\pi^2} & , if\ n = 2m-1(odd)\ and\ m = 1,2,3,.... \\ 0 & , if\ n = 2m(even)\ and\ m = 1,2,3,... \end{cases}.$$

(For $n = 2m-1$, $\sin\left(\frac{n\pi}{2}\right) = \sin\left\{(2m-1)\frac{\pi}{2}\right\} = \sin\left(m\pi - \frac{\pi}{2}\right)$

$\sin m\pi\ \cos\frac{\pi}{2} - \cos\ m\pi\sin\left(\frac{\pi}{2}\right) = 0-(-1)^n = (-1)^{n+1}$)

Substituting the value a_n and b_n in equation (6.44), the required displacement function is

$$y(x,t) = \frac{8h}{\pi^2}\sum_{m=1}^{\infty}\frac{(-1)^{m+1}}{(2m-1)^2}\sin\frac{(2m-1)\pi x}{l}\cos\frac{(2m-1)\pi Ct}{l}.$$

Example:

A uniform string fixed at the ends $x=0$ and $x=l$ are fixed. One end is taken at the origin and at a distance $'b'$ from this end the string is displaced a distance $'h'$ transversely and is released from rest in that position. Find the displacement of any of the string at any subsequent time.

Solution:

The dis placement function $y(x,t)$ of the string is the solution of wave equation

$$\frac{\partial^2 y}{\partial t^2} = C^2 \frac{\partial^2 y}{\partial x^2}, \qquad (6.46)$$

where $C^2 = \frac{T}{M} = \frac{\text{Tension}}{\text{Mass}}$.

The boundary and initial conditions are:

Boundary conditions	**Initial conditions**
BC1:	**IC1:**
$y(x=0,t)=0,\ \forall t>0$	$y(x,t=0)=f(x),\ \forall x\in(0,l)$
(6.46a)	(6.46c)
BC2:	**IC2:** $\frac{\partial y}{\partial t}(x,0)=0,\ \forall x\in(0,l)$
$y(x=l,t)=0,\ \forall t>0$	(6.46d)
(6.46b)	

Where *f(x)* is the initial displacement in the string and can be obtained as follows: Taking $OA=l$,length of the string with B as its points of bisection.

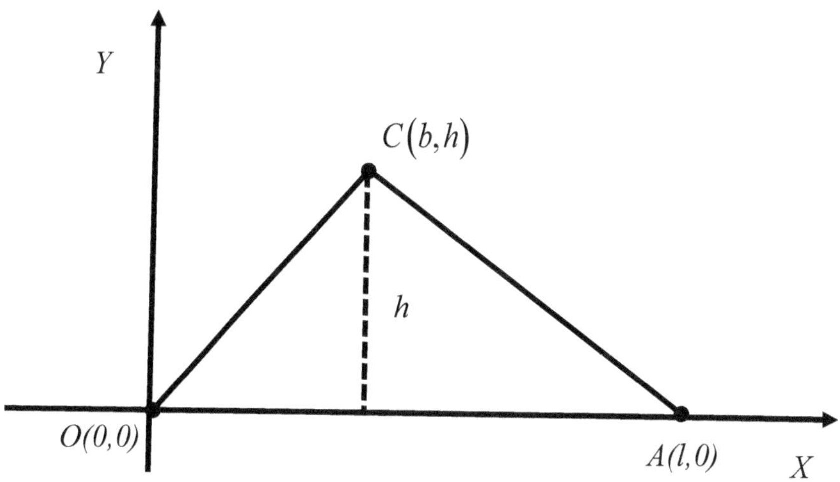

Figure 6.3: Initial displacement of the string

The initial position of the string given in the figure.

To find $y=f\left(x\right)=\begin{cases} Equation\ of\ OC, & 0<x<b \\ Equation\ of\ CA, & b<x<l \end{cases}$

Equation of the line *OC* joining points $(0,0)$ & (b,h) is given by

$$\frac{y-y_1}{y_2-y_1}=\frac{x-x_1}{x_2-x_1}\Rightarrow\frac{y-0}{h-0}=\frac{x-0}{b-0}\Rightarrow y=\frac{h}{b}x,0\le x\le b$$

Equation of the line *CA* joining (b,h) & $(l,0)$is given by

$$\frac{y-y_1}{y_2-y_1}=\frac{x-x_1}{x_2-x_1}\Rightarrow\frac{y-h}{0-h}=\frac{x-b}{l-b}\Rightarrow y-h=-h\left(\frac{x-b}{l-b}\right),b\le x\le l$$

$$\Rightarrow y=\frac{h\left(x-l\right)}{b-l},b\le x\le l$$

$\therefore$ The initial displacement of the string is

$$y=f(x)=\begin{cases} \frac{h}{b}x,0\le x\le b \\ \frac{h(x-l)}{b-l},b\le x\le l \end{cases}. \qquad (6.47)$$

The solution of wave equation given by equation (6.46) with boundary and initial condition equation (6.46 a)- (6.46 d) is given by

$$y(x,t)=\sum_{n=1}^{\infty}\sin\left(\frac{n\pi}{l}\right)x\left(c_n\cos C\left(\frac{n\pi}{l}\right)t+d_n\sin C\left(\frac{n\pi}{l}\right)t\right), \qquad (6.47)$$

where $c_n=\frac{2}{l}\int_0^l f\left(x\right)\sin\left(\frac{n\pi x}{l}\right)dx$, and $d_n = \frac{2}{n\pi C}\int_0^l g(x)\sin\left(\frac{n\pi x}{l}\right)dx = 0$ as g(x)=0

Now, from equation (6.47)

$$c_n = \frac{2}{l}\left[\int_0^b f(x)\sin\left(\frac{n\pi x}{l}\right)dx + \int_b^l f(x)\sin\left(\frac{n\pi x}{l}\right)dx\right] \tag{6.48}$$

Now we evaluate first integral of equation (6.48).

$$\int_0^b \frac{h}{b}x\sin\left(\frac{n\pi x}{l}\right)dx = \frac{h}{b}\left[x\left(\frac{-\cos\left(\frac{n\pi x}{l}\right)}{\frac{n\pi}{l}}\right) - \left(\frac{-\sin\left(\frac{n\pi b}{l}\right)}{\left(\frac{n\pi}{l}\right)^2}\right)\right]_0^b,$$

$$\int_0^b \frac{h}{b}x\sin\left(\frac{n\pi x}{l}\right)dx = \frac{h}{b}\left[b\left(\frac{-\cos\left(\frac{n\pi b}{l}\right)}{\frac{n\pi}{l}}\right) - \left(\frac{-\sin\left(\frac{n\pi b}{l}\right)}{\left(\frac{n\pi}{l}\right)^2}\right) - (0+0)\right]$$

,

$$\therefore \int_0^b \frac{h}{b}x\sin\left(\frac{n\pi x}{l}\right)dx = \frac{hl}{n\pi}\left(-\cos\left(\frac{n\pi b}{l}\right)\right) + \frac{hl^2}{b}\left(\frac{\sin\left(\frac{n\pi b}{l}\right)}{(n\pi)^2}\right). \tag{6.49}$$

Now the second integral of equation (6.48)

$$\int_b^l \frac{h(x-l)}{b-l}\sin\left(\frac{n\pi x}{l}\right)dx = \left[\left(\frac{h(x-l)}{b-l}\right)\left(\frac{-\cos\left(\frac{n\pi x}{l}\right)}{\frac{n\pi}{l}}\right) - \left(\frac{h}{b-l}\right)\left(\frac{-\sin\left(\frac{n\pi x}{l}\right)}{\left(\frac{n\pi}{l}\right)^2}\right)\right]_b^l,$$

$$\int_b^l \frac{h(x-l)}{(b-l)}\sin\left(\frac{n\pi x}{l}\right)dx = \left[\begin{array}{l}\left(\frac{h(l-l)}{b-l}\right)\left(\frac{-\cos\left(\frac{n\pi l}{l}\right)}{\frac{n\pi}{l}}\right)-\left(\frac{h}{b-l}\right)\left(\frac{-\sin\left(\frac{n\pi l}{l}\right)}{\left(\frac{n\pi}{l}\right)^2}\right)- \\ \left(\left(\frac{h(b-l)}{b-l}\right)\left(\frac{-\cos\left(\frac{n\pi b}{l}\right)}{\frac{n\pi}{l}}\right)-\left(\frac{hl^2}{b-l}\right)\left(\frac{-\sin\left(\frac{n\pi b}{l}\right)}{\left(\frac{n\pi}{l}\right)^2}\right)\right)\end{array}\right],$$

$$\int_b^l \frac{h(x-l)}{(b-l)}\sin\left(\frac{n\pi x}{l}\right)dx = \left[(0-0)-\left(h\left(\frac{-\cos\left(\frac{n\pi b}{l}\right)}{\frac{n\pi}{l}}\right)-\left(\frac{h}{b-l}\right)\left(\frac{-\sin\left(\frac{n\pi b}{l}\right)}{\left(\frac{n\pi}{l}\right)^2}\right)\right)\right]$$

,

$$\therefore \int_b^l \frac{h(x-l)}{(b-l)}\sin\left(\frac{n\pi x}{l}\right)dx = \left[hl\left(\frac{\cos\left(\frac{n\pi b}{l}\right)}{n\pi}\right)-\left(\frac{hl^2}{b-l}\right)\frac{\sin\left(\frac{n\pi b}{l}\right)}{(n\pi)^2}\right] \tag{6.50}$$

Putting value from equations (6.49) and (6.50) in (6.48), we get

$$c_n = \left[\frac{hl^2}{b}\left(\frac{\sin\left(\frac{n\pi b}{l}\right)}{(n\pi)^2}\right)-\left(\frac{hl^2}{b-l}\right)\frac{\sin\left(\frac{n\pi b}{l}\right)}{(n\pi)^2}\right],$$

$$\therefore c_n = \frac{2}{l}\left[hl^2\left(\frac{\sin\left(\frac{n\pi b}{l}\right)}{(n\pi)^2}\right)\left(\frac{1}{b}-\frac{1}{b-l}\right)\right],$$

$$\therefore c_n = \frac{2}{l}\left[hl^2 \left(\frac{\sin\left(\frac{n\pi b}{l}\right)}{(n\pi)^2} \right)\left(\frac{-l}{b(b-l)} \right) \right],$$

$$\therefore c_n = \frac{-2hl^2 \sin\left(\frac{n\pi b}{l}\right)}{b(b-l)(n\pi)^2}$$

Putting the value of c_n and d_n in equation (6.47), we get the general solution

$$y(x,t) = \sum_{n=1}^{\infty} \frac{-2hl^2 \sin\left(\frac{n\pi b}{l}\right)}{b(b-l)(n\pi)^2} \sin\left(\frac{n\pi x}{l}\right)\cos\left(\frac{n\pi Ct}{l}\right) \quad ,$$

$$\therefore y(x,t) = \frac{-2hl^2}{b(b-l)(\pi)^2} \sum_{n=1}^{\infty} \frac{\sin\left(\frac{n\pi b}{l}\right)}{(n)^2} \sin\left(\frac{n\pi x}{l}\right)\cos\left(\frac{n\pi Ct}{l}\right),$$

which is the required displacement solution $y(x,t)$.

Example:

The points of trisection of a string are pulled aside through a distance $'h'$ on opposite sides of the position of equilibrium and the string is released from rest. Derive an expression for the string at any subsequent time and show that the middle point of the string always remains at rest.

Solution:

The dis placement function $y(x,t)$ of the string is the solution of wave equation

$$\frac{\partial^2 y}{\partial t^2} = C^2 \frac{\partial^2 y}{\partial x^2}, \tag{6.51}$$

where $C^2 = \frac{T}{M} = \frac{\text{Tension}}{\text{Mass}}$. The boundary and initial conditions are:

Boundary conditions

BC1:

$$y(x=0,t)=0,\ \forall t>0$$

(6.51a)

BC2:

$$y(x=l,t)=0,\ \forall t>0$$

(6.51b)

Initial conditions

IC1:

$$y(x,t=0)=f(x),\ \forall x\in(0,l)$$

(6.51c)

IC2:

$$\frac{\partial y}{\partial t}(x,0)=0,\ \forall x\in(0,l)$$

(6.51d)

where *f(x)* is the initial displacement in the string and can be obtained as follows:

Taking $OA=l$,length of the string with *B* and *C* as its points of trisection as shown in the Figure 6.4, which represents the initial displacement of the string.

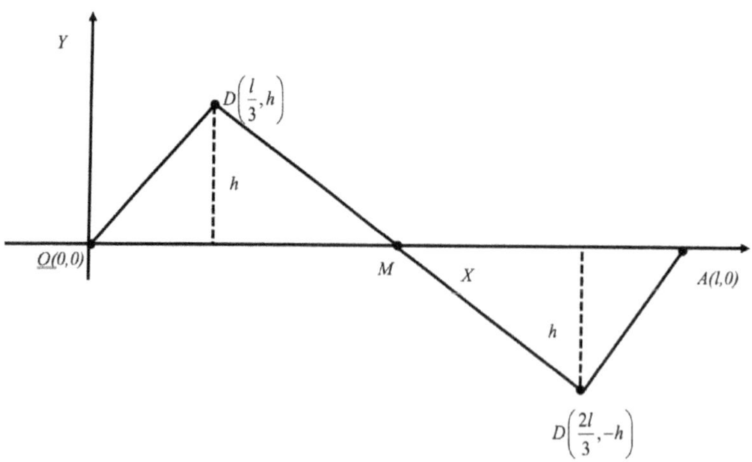

Figure 6.4: Initial displacement of the string.

Now we find the initial displacement using Figure 6.4 and is given by

$$y = f(x) = \begin{cases} \text{Equation of } OD; 0 < x < \dfrac{l}{3} \\ \text{Equation of } DE; \dfrac{l}{3} < x < \dfrac{2l}{3} \\ \text{Equation of } EA; \dfrac{2l}{3} < x < l \end{cases}$$

Equation of the line OD joining $(0,0)$ & $\left(\frac{l}{3}, h\right)$ is given by

$$\therefore \frac{y-y_1}{y_2-y_1} = \frac{x-x_1}{x_2-x_1} \Rightarrow \frac{y-0}{h-0} = \frac{x-0}{\frac{l}{3}-0} \Rightarrow y = \frac{3hx}{l}, 0 \le x \le \frac{l}{3}$$

Equation of the line DE joining $\left(\frac{l}{3}, h\right)$ & $\left(\frac{2l}{3}, -h\right)$ is given by

$$\therefore \frac{y-y_1}{y_2-y_1} = \frac{x-x_1}{x_2-x_1} \Rightarrow \frac{y-h}{-h-h} = \frac{x-\frac{l}{3}}{\frac{2l}{3}-\frac{l}{3}} \Rightarrow y = h - 2h\left(\frac{3x-l}{l}\right), \frac{l}{3} \le x \le \frac{2l}{3}$$

$$\therefore y = \frac{hl - 6hx + 2hl}{l} = \frac{3h(l-2x)}{l}, \frac{l}{3} \le x \le \frac{2l}{3}$$

Equation of the line EA joining $\left(\frac{2l}{3}, -h\right)$ & $(l,0)$ is given by

$$\therefore \frac{y-y_1}{y_2-y_1} = \frac{x-x_1}{x_2-x_1} \Rightarrow \frac{y+h}{0+h} = \frac{x-\frac{2l}{3}}{l-\frac{2l}{3}} \Rightarrow y + h = h\left(\frac{3x-2l}{l}\right), \frac{2l}{3} \le x \le l$$

$$\therefore y = -h + h\left(\frac{3x-2l}{l}\right) \Rightarrow y = \frac{3hx - 3hl}{l} = \frac{3h}{l}(x-l), \frac{2l}{3} \le x \le l$$

Therefore, the initial displacement of the string is

$$\therefore \quad y = f(x) = \begin{cases} \dfrac{3hx}{l}, 0 \le x \le \dfrac{l}{3} \\ \dfrac{3h(l-2x)}{l}, \dfrac{l}{3} \le x \le \dfrac{2l}{3}. \\ \dfrac{3h}{l}(x-l), \dfrac{2l}{3} \le x \le l \end{cases} \tag{6.52}$$

$\therefore$

The solution of wave equation given by equation (6.46) with boundary and initial condition equation (6.46 a)- (6.46 d) is given by

$$y(x,t) = \sum_{n=1}^{\infty} \sin\left(\frac{n\pi}{l}\right)x\left(c_n \cos C\left(\frac{n\pi}{l}\right)t + d_n \sin C\left(\frac{n\pi}{l}\right)t\right), \tag{6.53}$$

where $c_n = \dfrac{2}{l}\int_0^l f(x)\sin\left(\dfrac{n\pi x}{l}\right)dx$,

and $d_n = \dfrac{2}{n\pi C}\int_0^l g(x)\sin\left(\dfrac{n\pi x}{l}\right)dx = 0$ *as g(x)=0.*

Now, from equation (6.47)

$$c_n = \frac{2}{l}\left[\int_0^{\frac{l}{3}} f(x)\sin\left(\frac{n\pi x}{l}\right)dx + \int_{\frac{l}{3}}^{\frac{2l}{3}} f(x)\sin\left(\frac{n\pi x}{l}\right)dx + \int_{\frac{2l}{3}}^{l} f(x)\sin\left(\frac{n\pi x}{l}\right)dx\right] \tag{6.54}$$

Now we evaluate first integral of equation (6.54).

$$\int_0^{\frac{l}{3}} f(x)\,\sin\left(\frac{n\pi x}{l}\right)dx = \int_0^{\frac{l}{3}} \frac{3hx}{l}\,\sin\left(\frac{n\pi x}{l}\right)dx$$

$$\therefore \int_0^{\frac{l}{3}} \frac{3hx}{l}\sin\left(\frac{n\pi x}{l}\right)dx = \left[\frac{3hx}{l}\left(\frac{-\cos\left(\frac{n\pi x}{l}\right)}{\frac{n\pi}{l}}\right) - \frac{3h}{l}\left(\frac{-\sin\left(\frac{n\pi x}{l}\right)}{\left(\frac{n\pi}{l}\right)^2}\right)\right]_0^{\frac{l}{3}}$$

$$\therefore \int_0^{\frac{l}{3}} \frac{3hx}{l}\sin\left(\frac{n\pi x}{l}\right)dx = \left[\left(\frac{3hl}{n\pi 3}\left(-\cos\left(\frac{n\pi l}{3l}\right)\right) - 3hl\left(\frac{-\sin\left(\frac{n\pi l}{3l}\right)}{(n\pi)^2}\right)\right) - (0-0)\right]$$

$$\therefore \int_0^{\frac{l}{3}} \frac{3hx}{l}\sin\left(\frac{n\pi x}{l}\right)dx = \frac{-hl}{n\pi}\cos\left(\frac{n\pi}{3}\right) + \frac{3hl}{(n\pi)^2}\sin\left(\frac{n\pi}{3}\right) \qquad (6.55)$$

Now, we evaluate the second integral of equation (6.54)

$$\int_{\frac{l}{3}}^{\frac{2l}{3}} f(x)\sin\left(\frac{n\pi x}{l}\right)dx = \int_{\frac{l}{3}}^{\frac{2l}{3}} \frac{3h(l-2x)}{l}\sin\left(\frac{n\pi x}{l}\right)dx$$

$$\int_{\frac{l}{3}}^{\frac{2l}{3}} \frac{3h(l-2x)}{l}\sin\left(\frac{n\pi x}{l}\right)dx = \left[\frac{3h(l-2x)}{l}\left(\frac{-\cos\left(\frac{n\pi x}{l}\right)}{\left(\frac{n\pi}{l}\right)}\right) - \left(\frac{-6h}{l}\right)\left(\left(\frac{-\sin\left(\frac{n\pi x}{l}\right)}{\left(\frac{n\pi}{l}\right)^2}\right)\right)\right]_{\frac{l}{3}}^{\frac{2l}{3}}$$

$$\int_{\frac{l}{3}}^{\frac{2l}{3}} \frac{3h(l-2x)}{l}\sin\left(\frac{n\pi x}{l}\right)dx = \left[\left(\left(\frac{hl\cos\left(\frac{2n\pi}{3}\right)}{n\pi}\right) - 6hl\left(\frac{\sin\left(\frac{2n\pi}{3}\right)}{(n\pi)^2}\right)\right) - \left(\left(\frac{-hl\cos\left(\frac{n\pi}{3}\right)}{n\pi}\right) - 6hl\left(\frac{\sin\left(\frac{n\pi}{3}\right)}{(n\pi)^2}\right)\right)\right]$$

$$\int_{\frac{l}{3}}^{\frac{2l}{3}} \frac{3h(l-2x)}{l}\sin\left(\frac{n\pi x}{l}\right)dx = \frac{hl}{n\pi}\cos\left(\frac{2n\pi}{3}\right) - \frac{6hl}{(n\pi)^2}\sin\left(\frac{2n\pi}{3}\right) + \frac{hl}{n\pi}\cos\left(\frac{n\pi}{3}\right) + \frac{6hl}{(n\pi)^2}\sin\left(\frac{n\pi}{3}\right) \qquad (6.56)$$

Now, we evaluate the third integral of equation (6.54).

$$\int_{2\frac{l}{3}}^{l} f(x)\sin\left(\frac{n\pi x}{l}\right)dx = \int_{\frac{l}{3}}^{\frac{2l}{3}} \frac{3h(x-l)}{l}\sin\left(\frac{n\pi x}{l}\right)dx$$

$$\int_{\frac{2l}{3}}^{l} \frac{3h}{l}(x-l)\sin\left(\frac{n\pi x}{l}\right) = \left[\frac{3h(x-l)}{l}\frac{\left(-\cos\left(\frac{n\pi x}{l}\right)\right)}{\left(\frac{n\pi}{l}\right)} - \frac{3h}{l}\frac{\left(-\sin\left(\frac{n\pi x}{l}\right)\right)}{\left(\frac{n\pi}{l}\right)^2}\right]_{\frac{2l}{3}}^{l}$$

$$\int_{\frac{2l}{3}}^{l} \frac{3h}{l}(x-l)\sin\left(\frac{n\pi x}{l}\right) = \left[\frac{3h(x-l)}{n\pi}\left(-\cos\left(\frac{n\pi x}{l}\right)\right) + 3hl\frac{\left(\sin\left(\frac{n\pi x}{l}\right)\right)}{(n\pi)^2}\right]_{\frac{2l}{3}}^{l}$$

$$\int_{\frac{2l}{3}}^{l} \frac{3h}{l}(x-l)\sin\left(\frac{n\pi x}{l}\right) = \left[\left(\frac{3h(l-l)}{n\pi}\left(-\cos\left(\frac{n\pi l}{l}\right)\right) + 3hl\frac{\left(\sin\left(\frac{n\pi l}{l}\right)\right)}{(n\pi)^2}\right) - \left(\frac{3h\left(\frac{2l}{3}-l\right)}{n\pi}\left(-\cos\left(\frac{2n\pi l}{3l}\right)\right) + 3hl\frac{\left(\sin\left(\frac{2n\pi l}{3l}\right)\right)}{(n\pi)^2}\right)\right]$$

$$\int_{\frac{2l}{3}}^{l} \frac{3h}{l}(x-l)\sin\left(\frac{n\pi x}{l}\right) = \left[(0+0) - \left(\frac{h}{n\pi}\left(\cos\left(\frac{2n\pi}{3}\right)\right) + 3hl\frac{\left(\sin\left(\frac{2n\pi}{3}\right)\right)}{(n\pi)^2}\right)\right]$$

$$\int_{\frac{2l}{3}}^{l} \frac{3h}{l}(x-l)\sin\left(\frac{n\pi x}{l}\right) = \frac{-hl}{n\pi}\left(\cos\left(\frac{2n\pi}{3}\right)\right) - 3hl\frac{\left(\sin\left(\frac{2n\pi}{3}\right)\right)}{(n\pi)^2} \qquad (6.57)$$

Putting the values from equations (6.55), (6.56) and (6.57), we get

$$c_n=\frac{2}{l}\begin{bmatrix} -\frac{h}{n\pi}\cos\left(\frac{n\pi}{3}\right)+\frac{3hl}{(n\pi)^2}\sin\left(\frac{n\pi}{3}\right)+\frac{hl}{n\pi}\cos\left(\frac{2n\pi}{3}\right)-\frac{6hl}{(n\pi)^2}\sin\left(\frac{2n\pi}{3}\right) \\ +\frac{hl}{n\pi}\cos\left(\frac{n\pi}{3}\right)+\frac{6hl}{(n\pi)^2}\sin\left(\frac{n\pi}{3}\right)-\frac{hl}{n\pi}\cos\left(\frac{2n\pi}{3}\right)-\frac{3hl}{(n\pi)^2}\sin\left(\frac{2n\pi}{3}\right) \end{bmatrix}$$

$$\therefore c_n=\frac{2}{l}\left[\frac{9hl}{(n\pi)^2}\sin\left(\frac{n\pi}{3}\right)-\frac{9hl}{(n\pi)^2}\sin\left(\frac{2n\pi}{3}\right)\right]$$

$$\therefore c_n=\frac{18h}{(n\pi)^2}\left[\sin\left(\frac{n\pi}{3}\right)-\sin\left(\frac{2n\pi}{3}\right)\right]$$

$$\therefore c_n=\frac{18h}{(n\pi)^2}\left[\sin\left(\frac{n\pi}{3}\right)-\sin\left(n\pi-\frac{n\pi}{3}\right)\right]$$

$$\therefore c_n=\frac{18h}{(n\pi)^2}\left[\sin\left(\frac{n\pi}{3}\right)-\left[\sin n\pi\cos\frac{n\pi}{3}-\cos n\pi\sin\frac{n\pi}{3}\right]\right]$$

But we know that $\sin n\pi=0, \cos n\pi=(-1)^n$

$$\therefore c_n=\frac{18h}{(n\pi)^2}\left[\sin\left(\frac{n\pi}{3}\right)-\left[0-(-1)^n\sin\frac{n\pi}{3}\right]\right]$$

$$\therefore c_n=\frac{18h}{(n\pi)^2}\left[\sin\left(\frac{n\pi}{3}\right)+(-1)^n\sin\frac{n\pi}{3}\right]$$

$$\therefore c_n = \frac{18h}{(n\pi)^2}\left[\sin\left(\frac{n\pi}{3}\right)\left(1+(-1)^n\right)\right]$$

$$\therefore c_n = \begin{cases} \dfrac{36h}{(n\pi)^2}\sin\left(\dfrac{n\pi}{3}\right); \text{if } n \text{ is even} \\ 0; \text{if } n \text{ is odd} \end{cases}$$

The most general solution from equation (6.51) we have

$$y(x,t) = \sum_{n=\text{even}}^{\infty} \frac{36h}{(n\pi)^2}\sin\left(\frac{n\pi}{3}\right)\sin\left(\frac{n\pi x}{l}\right)\cos\left(\frac{n\pi Ct}{l}\right),$$

$$y(x,t) = \frac{36h}{(\pi)^2}\sum_{n=\text{even}}^{\infty} \frac{1}{(n)^2}\sin\left(\frac{n\pi}{3}\right)\sin\left(\frac{n\pi x}{l}\right)\cos\left(\frac{n\pi Ct}{l}\right)$$

$$y(x,t) = \frac{36h}{(\pi)^2}\sum_{n=1}^{\infty} \frac{1}{(2n)^2}\sin\left(\frac{2n\pi}{3}\right)\sin\left(\frac{2n\pi x}{l}\right)\cos\left(\frac{2n\pi Ct}{l}\right)$$

$$y(x,t) = \frac{9h}{\pi^2}\sum_{n=1}^{\infty} \frac{1}{n^2}\sin\left(\frac{2n\pi}{3}\right)\sin\left(\frac{2n\pi x}{l}\right)\cos\left(\frac{2n\pi Ct}{l}\right)$$

is the required displacement solution $y(x,t)$. By putting $x = \frac{l}{2}$ in the above equation, the displacement at the midpoint is

$$y\left(x = \frac{l}{2}, t\right) = \frac{9h}{\pi^2}\sum_{n=1}^{\infty} \frac{1}{n^2}\sin\left(\frac{2n\pi}{3}\right)\sin\left(\frac{2n\pi l}{2l}\right)\cos\left(\frac{2n\pi Ct}{l}\right)$$

$$\therefore y\left(x = \frac{l}{2}, t\right) = \frac{9h}{\pi^2}\sum_{n=1}^{\infty} \frac{1}{n^2}\sin\left(\frac{2n\pi}{3}\right)\sin(n\pi)\cos\left(\frac{2n\pi Ct}{l}\right)$$

$$\therefore y\left(x = \frac{l}{2}, t\right) = 0 \qquad (\because \sin n\pi = 0)$$

The middle point of the string always remains at rest.

Example:

A string of length l is stretched and fastened to two fixed points. Find the solution of the wave equation $\frac{\partial^2 y}{\partial t^2} = C^2 \frac{\partial^2 y}{\partial x^2}$ when initial displacement is $y(x,0) = b\sin\left(\frac{\pi x}{l}\right) = f(x)$ where symbols have usual meaning.

Solution:

Consider an elastic string tightly stretched between two points O and A. (as shown in Figure 6.5).

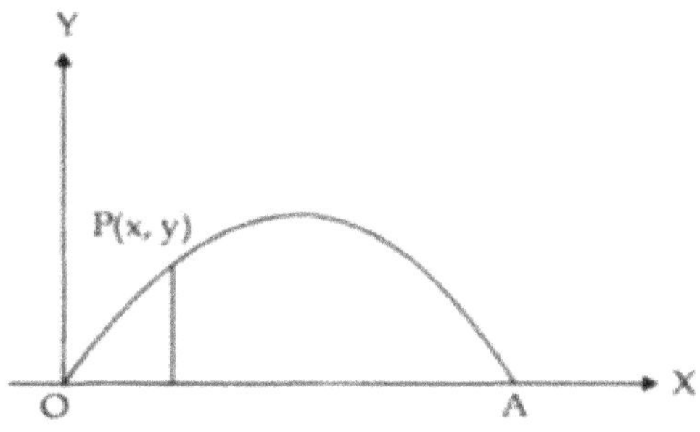

Figure 6.5: Position of the string

Let O be the origin and OA as *X*-axis on giving a small transverse displacement i.e. the displacement perpendicular to its length. Let y be the displacement at the point $P(x, y)$ at any time, the wave equation

$$\frac{\partial^2 y}{\partial t^2} = C^2 \frac{\partial^2 y}{\partial x^2}. \tag{6.58}$$

As the end points of the string are fixed for all time, the boundary condition is given by

$$y(0,t)=0, \tag{6.59}$$

$$y(l,t)=0, \tag{6.60}$$

Initial displacement is given by

$$\textbf{IC1}: y(x,0)=b\sin\left(\frac{\pi x}{l}\right)=f(x). \tag{6.61}$$

Since the initial transverse velocity of any point of the string is zero, so

$$\textbf{IC 2}: \left(\frac{\partial y}{\partial t}\right)_{t=0}=0=g(x). \tag{6.62}$$

Comparing the above problem with boundary value problem given by (6.4), (6.4 a)- (6.4 d), the solution of (6.58) satisfy the above boundary and initial conditions is given by

$$y(x,t)=\sum_{n=1}^{\infty}\sin\left(\frac{n\pi}{l}\right)x\left(a_n\cos C\left(\frac{n\pi}{l}\right)t+b_n\sin C\left(\frac{n\pi}{l}\right)t\right), \tag{6.63}$$

where $a_n=\frac{2}{l}\int_0^l f(x)\sin\left(\frac{n\pi x}{l}\right)dx$, and

$$b_n=\frac{2}{n\pi C}\int_0^l g(x)\sin\left(\frac{n\pi x}{l}\right)dx.$$

From equation (6.62) i.e. **IC2**, $g(x)=0$ and therefore

$$b_n=\frac{2}{n\pi C}\int_0^l g(x)\sin\left(\frac{n\pi x}{l}\right)dx=0. \tag{6.64}$$

Using equation (6.64), equation (6.63) reduces to

$$y(x,t)=\sum_{n=1}^{\infty} a_n \sin\left(\frac{n\pi}{l}\right)x \cos C\left(\frac{n\pi}{l}\right)t. \qquad (6.65)$$

Putting $t=0$ in equation (6.65) and expanding the summation, we get

$$y(x,0)=a_1 \sin\left(\frac{\pi x}{l}\right)+a_2 \sin\left(\frac{2\pi x}{l}\right)+a_3 \sin\left(\frac{3\pi x}{l}\right)+\ldots\ldots\ldots\ldots \qquad (6.66)$$

Substituting the value of from equation (6.61) in equation (6.65), we get

$$b\sin\left(\frac{\pi x}{l}\right)=a_1 \sin\left(\frac{\pi x}{l}\right)+a_2 \sin\left(\frac{2\pi x}{l}\right)+a_3 \sin\left(\frac{3\pi x}{l}\right)+\ldots\ldots\ldots\ldots$$

Comparing the coefficients of the like on both sides, we have

$a_1=b$ for $n=1$,

and $a_n=0$ for $n\neq 1$. Using the $a_1=b$ for $n=1$, and $a_n=0$ for $n\neq 1$, the solution of the wave equation given by equation (6.58) with initial and boundary condition (6.59)-(6.62) is

$$y(x,t)=b\sin\left(\frac{\pi x}{l}\right)\cos\left(\frac{C\pi t}{l}\right).$$

Example:

If a string of length l is initially at rest in its equilibrium position and each of its points is given the velocity $y_0 \sin^3\left(\frac{\pi x}{l}\right)$ determine the displacement $y(x,t)$.

Solution:

The vibration of the string is given by $\frac{\partial^2 y}{\partial t^2} = C^2 \frac{\partial^2 y}{\partial x^2}$. (6.67)

Boundary conditions

BC1:

$$y(x=0,t) = 0, \ \forall t \geq 0$$

(6.67a)

BC2:

$$y(x=l,t) = 0, \ \forall t \geq 0$$

(6.67b)

Initial conditions

IC1:

$$y(x,t=0) = 0 = f(x), \ \forall x \in (0,l)$$

(6.67c)

IC2:

$$\frac{\partial y}{\partial t}(x,0) = y_0 \sin^3\left(\frac{\pi x}{l}\right) = g(x) \ , \ \forall x \in (0,l)$$

(6.67d)

Comparing the above problem with boundary value problem given by (6.4), (6.4 a)- (6.4 d), the solution of (6.67) satisfying the above boundary and initial conditions is given by

$$y(x,t) = \sum_{n=1}^{\infty} \sin\left(\frac{n\pi}{l}\right) x \left(a_n \cos C\left(\frac{n\pi}{l}\right) t + b_n \sin C\left(\frac{n\pi}{l}\right) t \right), \quad (6.68)$$

Where,

$a_n = \frac{2}{l} \int_0^l f(x) \sin\left(\frac{n\pi x}{l}\right) dx = 0$ as $f(x) = 0$.

Therefore, equation (6.68) reduces to

$$y(x,t) = \sum_{n=1}^{\infty} b_n \sin\left(\frac{n\pi}{l}\right) x \sin C\left(\frac{n\pi}{l}\right) t . \quad (6.69)$$

Differentiating equation (6.69) w.r.t t, we get

$$\frac{\partial y(x,t)}{\partial t} = \sum_{n=1}^{\infty} b_n \sin\left(\frac{n\pi}{l}\right) x \frac{n\pi C}{l} \cos C\left(\frac{n\pi}{l}\right) t . \quad (6.70)$$

Using **IC2:** $\frac{\partial y}{\partial t}(x,0) = y_0 \sin^3\left(\frac{\pi x}{l}\right) = g(x)$.

$$\frac{\partial y}{\partial t}(x,0) = y_0 \sin^3\left(\frac{\pi x}{l}\right) = \frac{y_0}{4}\left(3\sin\frac{\pi x}{l} - \sin\frac{3\pi x}{l}\right). \tag{6.71}$$

Putting $t = 0$ in equation (6.70) and expanding the summation, we get

$$\frac{\partial y(x,0)}{\partial t} = \frac{\pi C}{l}\left\{b_1 \sin\left(\frac{\pi x}{l}\right) + 2b_2 \sin\left(\frac{2\pi x}{l}\right) + 3b_3 \sin\left(\frac{3\pi x}{l}\right) + \dots\dots\dots\right\}. \tag{6.72}$$

Substituting the value of from equation (6.71) in equation (6.71), we get

$$\frac{y_0}{4}\left(3\sin\frac{\pi x}{l} - \sin\frac{3\pi x}{l}\right) = \frac{\pi C}{l}\left\{b_1 \sin\left(\frac{\pi x}{l}\right) + 2b_2 \sin\left(\frac{2\pi x}{l}\right) + 3b_3 \sin\left(\frac{3\pi x}{l}\right) + \dots\dots\dots\right\}$$

$$\frac{3y_0}{4}\sin\frac{\pi x}{l} - \frac{y_0}{4}\sin\frac{3\pi x}{l} = \frac{\pi C}{l}\left\{b_1 \sin\left(\frac{\pi x}{l}\right) + 2b_2 \sin\left(\frac{2\pi x}{l}\right) + 3b_3 \sin\left(\frac{3\pi x}{l}\right) + \dots\dots\dots\right\}$$

Comparing the coefficients of the like on both sides, we have

$$\frac{3y_0}{4} = \frac{b_1 \pi C}{l} \Rightarrow b_1 = \frac{3y_0 l}{4\pi C},$$

$$-\frac{y_0}{4} = \frac{3b_3 \pi C}{l} \Rightarrow b_3 = \frac{-y_0 l}{12\pi C},$$

and $b_2 = b_4 = b_5 = 0$.

Substituting the value of $b_1 = \frac{3y_0 l}{4\pi C}$, $b_2 = 0$, $b_3 = \frac{-y_0 l}{12\pi C}$, $b_4 = b_5 = 0$ in equation (6.69), we get

$$y(x,t)=\frac{3y_0 l}{4\pi C}\sin\left(\frac{\pi x}{l}\right)\sin\left(\frac{C\pi t}{l}\right)-\frac{1y_0}{12\pi C}\sin\left(\frac{3\pi x}{l}\right)\sin\left(\frac{3C\pi t}{l}\right)$$

the solution of the wave equation given by equation (6.67) with initial and boundary condition (6.67a) -(6.67d) is the required solution.

Example:

Solve the boundary value problem $\frac{\partial^2 y}{\partial t^2}=4\frac{\partial^2 y}{\partial x^2}$, given that $y(0,t)=0$, $y(5,t)=0$

$y(x,0)=0$ and $\left(\frac{\partial y}{\partial t}\right)_{t=0}=5\sin\pi x$.

Solution:

The vibration of the string is given by $\frac{\partial^2 y}{\partial t^2}=4\frac{\partial^2 y}{\partial x^2}$. (6.72)

Boundary conditions	**Initial conditions**
BC1:	**IC1:**
$y(x=0,t)=0,\ \forall t>0$	$y(x,t=0)=0=f(x),\ \forall x\in(0,5)$
(6.72a)	(6.72c)
BC2:	**IC2:**
$y(x=5,t)=0,\ \forall t>0$	$\left(\frac{\partial y}{\partial t}\right)_{t=0}=5\sin\pi x=g(x)\ ,\ \forall x\in(0,5)$
(6.72b)	(6.72d)

Comparing the above problem with boundary value problem given by (6.4), (6.4 a)- (6.4 d), the solution of (6.58) satisfy the above boundary and initial conditions is given by

$$y(x,t)=\sum_{n=1}^{\infty}\sin\left(\frac{n\pi}{5}\right)x\left(a_n\cos C\left(\frac{n\pi}{5}\right)t+b_n\sin C\left(\frac{n\pi}{5}\right)t\right), \quad (6.73)$$

where $a_n = \frac{2}{l}\int_0^l f(x)\sin\left(\frac{n\pi x}{l}\right)dx = 0$, as $f(x) = 0$ (6.74)

Using equation (6.74), equation (6.73) reduces to

$$y(x,t) = \sum_{n=1}^{\infty} a_n \sin\left(\frac{n\pi}{5}\right)x \sin\left(\frac{2n\pi}{5}\right)t . (\because C = 1) \tag{6.75}$$

Differentiating equation (6.75) partially w.r.t t, we get

$$\frac{\partial y(x,t)}{\partial t} = \sum_{n=1}^{\infty} a_n \sin\left(\frac{n\pi}{5}\right)x \frac{2n\pi}{5}\cos\left(\frac{2n\pi}{5}\right)t \tag{6.76}$$

Putting $t = 0$ in equation (6.76) and expanding the summation, we get

$$\frac{\partial y(x,0)}{\partial t} = a_1\frac{2\pi}{5}\sin\left(\frac{\pi x}{5}\right) + a_2\frac{4\pi}{5}\sin\left(\frac{2\pi x}{5}\right) + a_3\frac{6\pi}{5}\sin\left(\frac{3\pi x}{5}\right) + a_4\frac{8\pi}{5}\sin\left(\frac{4\pi x}{5}\right) + a_5\frac{10\pi}{5}\sin\left(\frac{5\pi x}{5}\right) + ...$$

Substituting the value of from equation (6.61) in equation (6.65), we get

$$5\sin \pi x = a_1\frac{2\pi}{5}\sin\left(\frac{\pi x}{5}\right) + a_2\frac{4\pi}{5}\sin\left(\frac{2\pi x}{5}\right) + a_3\frac{6\pi}{5}\sin\left(\frac{3\pi x}{5}\right) + a_4\frac{8\pi}{5}\sin\left(\frac{4\pi x}{5}\right) + a_5\frac{10\pi}{5}\sin(\pi x) + ...$$

Comparing the coefficients of the like on both sides, we have $a_5 = \frac{5}{2\pi}$, $a_i = 0$ for $i \neq 5$. Using the $a_5 = \frac{5}{2\pi}$, $a_i = 0$ for $i \neq 5$, the solution of the wave equation given by equation (6.72) with initial and boundary condition (6.72a)-(6.72d) is

$y(x,t) = \frac{5}{2\pi}\sin \pi x \sin 2\pi t$, which is the required solution.

Example:

The vibration of an elastic string is governed by the partial differential equation value problem $\frac{\partial^2 y}{\partial t^2}=\frac{\partial^2 y}{\partial x^2}$ The length of the string is π and the ends are fixed. The initial velocity is zero and the initial deflection is $y(x,0)=2(\sin x+\sin 3x)$ find the deflection $y(x,t)$ of the vibrating string for $t>0.$

Solution:

Let $y(x,t)$ be deflection of the vibrating string of above IBVP represented by following PDE with initial and boundary condition as follows:

$$\frac{\partial^2 y}{\partial t^2}=\frac{\partial^2 y}{\partial x^2} \tag{6.77}$$

with $y(0,t)=y(\pi,t)=0$ for all $t\geq 0$, and for $0\leq x\leq \pi$

Initial deflection **IC1**: $y(x,0)=2(\sin\, x+\sin 3x),$ (6.78)

Initial velocity**IC2**: $y_t\,(x,0)=0\,.$ (6.79)

Comparing the above problem with boundary value problem given by (6.4), (6.4 a)- (6.4 d), the solution of (6.77) satisfy the above boundary and initial conditions is given by

$$y(x,t)=\sum_{n=1}^{\infty}\sin\left(\frac{n\pi}{\pi}\right)x\left(a_n\cos C\left(\frac{n\pi}{\pi}\right)t+b_n\sin C\left(\frac{n\pi}{\pi}\right)t\right), \tag{6.80}$$

where $a_n=\frac{2}{\pi}\int_0^{\pi} f(x)\sin\left(\frac{n\pi x}{\pi}\right)dx,$

and

$$b_n = \frac{2}{n\pi C}\int_0^{\pi} g(x)\sin\left(\frac{n\pi x}{\pi}\right)dx.$$

From equation (6.70) i.e. **IC2,** $g(x)=0$ and therefore

$$b_n = \frac{2}{n\pi C}\int_0^{\pi} g(x)\sin\left(\frac{n\pi x}{\pi}\right)dx = 0. \qquad (6.81)$$

Using equation (6.81), equation (6.80) reduces to

$$y(x,t) = \sum_{n=1}^{\infty} a_n \sin(nx)\cos(nt). \quad (\because C = 1) \qquad (6.82)$$

Putting $t = 0$ in equation (6.82) and expanding the summation, we get

$$y(x,0) = a_1 \sin x + a_2 \sin 2x + a_3 \sin 3x + a_4 \sin 4x. + \ldots\ldots\ldots\ldots \qquad (6.83)$$

Substituting the value of from equation (6.78) in equation (6.83), we get

$$2(\sin x + \sin 3x) = a_1 \sin x + a_2 \sin 2x + a_3 \sin 3x + a_4 \sin 4x. + \ldots\ldots\ldots\ldots (6.84)$$

Comparing the coefficients of the like on both sides, we have
$a_1 = 2, a_3 = 2$

and $a_n = 0$ for $n \neq 1,3$. Putting value $a_1 = 2, a_3 = 2$ and $a_n = 0$ for $n \neq 1,3$ in (6.88), the solution of the wave equation given by equation (6.77) is

$$y(x,t) = 2\sin x\cos t + 2\sin 3x\cos 3t.$$

Example:

If the string of length l is initially at rest in equilibrium position and each of its points is given by the velocity $u_0 \sin\left(\frac{3\pi x}{l}\right)\cos\left(\frac{2\pi x}{l}\right)$ where $0 \le x \le l$ at $t = 0$, determine the displacement $u(x,t)$.

Solution:

The vibration of the string is given by $\frac{\partial^2 y}{\partial t^2} = C^2 \frac{\partial^2 y}{\partial x^2}$. (6.85)

Boundary conditions	**Initial conditions**
BC1: $u(x=0,t)=0,\ \forall t>0$ (6.85a)	**IC1:** $u(x,t=0)=0=f(x),\ \forall x\in(0,5)$ (6.85c)
BC2: $u(x=5,t)=0,\ \forall t>0$ (6.85b)	**IC2:** $\left(\frac{\partial u}{\partial t}\right)_{t=0} = u_0 \sin\left(\frac{3\pi x}{l}\right)\cos\left(\frac{2\pi x}{l}\right) = g(x)\ ,\ \forall x\in(0,l)$ (6.85d)

Comparing the above problem with boundary value problem given by (6.4), (6.4 a)- (6.4 d), the solution of (6.85) satisfying the above boundary and initial conditions is given by

$$u(x,t) = \sum_{n=1}^{\infty} \sin\left(\frac{n\pi}{l}\right)x\left(a_n \cos C\left(\frac{n\pi}{l}\right)t + b_n \sin C\left(\frac{n\pi}{l}\right)t\right), \quad (6.86)$$

where $a_n = \frac{2}{l}\int_0^l f(x)\sin\left(\frac{n\pi x}{l}\right)dx = 0$, as $f(x)=0$ (6.87)

Using equation (6.87), equation (6.86) reduces to

$$y(x,t) = \sum_{n=1}^{\infty} b_n \sin\left(\frac{n\pi x}{l}\right)\sin\left(\frac{Cn\pi t}{l}\right). \quad (6.88)$$

Differentiating equation (6.86) partially w.r.t t, we get

$$\frac{\partial y(x,t)}{\partial t}=\sum_{n=1}^{\infty} b_n \sin\left(\frac{n\pi x}{l}\right)\frac{Cn\pi}{l}\cos\left(\frac{Cn\pi t}{l}\right). \tag{6.89}$$

Putting $t=0$ in equation (6.76) and expanding the summation, we get

$$\frac{\partial y(x,0)}{\partial t}=b_1\frac{C\pi}{l}\sin\left(\frac{\pi x}{l}\right)+b_2\frac{2C\pi}{l}\sin\left(\frac{2\pi x}{l}\right)+b_3\frac{3C\pi}{l}\sin\left(\frac{3\pi x}{l}\right)$$
$$+b_4\frac{4C\pi}{l}\sin\left(\frac{4\pi x}{l}\right)+b_5\frac{5C\pi}{l}\sin\left(\frac{5\pi x}{l}\right)+\ldots.$$

Substituting the value from equation (6.85d) in equation (6.65), we get

$$u_0\sin\left(\frac{3\pi x}{l}\right)\cos\left(\frac{2\pi x}{l}\right)=b_1\frac{C\pi}{l}\sin\left(\frac{\pi x}{l}\right)+b_2\frac{2C\pi}{l}\sin\left(\frac{2\pi x}{l}\right)+b_3\frac{3C\pi}{l}\sin\left(\frac{3\pi x}{l}\right)$$
$$+b_4\frac{4C\pi}{l}\sin\left(\frac{4\pi x}{l}\right)+b_5\frac{5C\pi}{l}\sin\left(\frac{5\pi x}{l}\right)+\ldots.,$$

$$\frac{u_0}{2}\left\{\sin\left(\frac{\pi x}{l}\right)+\sin\left(\frac{5\pi x}{l}\right)\right\}=b_1\frac{C\pi}{l}\sin\left(\frac{\pi x}{l}\right)+b_2\frac{2C\pi}{l}\sin\left(\frac{2\pi x}{l}\right)+b_3\frac{3C\pi}{l}\sin\left(\frac{3\pi x}{l}\right)$$
$$+b_4\frac{4C\pi}{l}\sin\left(\frac{4\pi x}{l}\right)+b_5\frac{5C\pi}{l}\sin\left(\frac{5\pi x}{l}\right)+\ldots.,$$

Comparing the coefficients of the like on both sides, we have

$$\frac{u_0}{2}=b_1\frac{\pi C}{l}\Rightarrow b_1=\frac{lu_0}{2\pi C},$$

$$\frac{u_0}{2}=b_5\frac{5\pi C}{l}\Rightarrow b_5=\frac{lu_0}{10\pi C},$$

and $b_2=b_3=b_4=b_6=b_7=\ldots..=0$.

Using $b_1 = \frac{lu_0}{2\pi C}$, $b_5 = \frac{lu_0}{10\pi C}$, $b_2 = b_3 = b_4 = b_6 = b_7 = ... = 0$, $a_n = 0$ in Eq. (6.86) ,the solution of the wave equation given by Eq. (6.85) with initial and boundary condition (6.85a)-(6.85d) is $u(x,t) = \frac{lu_0}{2\pi C}\sin\frac{\pi x}{l}\cos\frac{\pi Ct}{l} + \frac{lu_0}{10\pi C}\sin\frac{5\pi x}{l}\cos\frac{5\pi Ct}{l}$ the required solution.

EXERCISE

1. The vibrations of an elastic string is governed by the P.D.E $\frac{\partial^2 u}{\partial t^2} = \frac{\partial^2 u}{\partial x^2}$. The length of the string is π and ends are fixed. The initial velocity is zero and the initial deflection is $u(x,0) = 2(\sin x + \sin 3x)$. Find the deflection $u(x,t)$ of the vibrating string for $t > 0$.

   $[\text{Ans.}: u(x,t) = 4\cos x \cos 2t \sin 2x]$

2. Solve:

   $$\frac{\partial^2 u}{\partial t^2}(x,t) = C^2 \frac{\partial^2 u}{\partial x^2}(x,t); \forall\ 0 < x < 1 \text{ and } t > 0$$

   $$u(0,t) = u(1,t) = 0; \forall\ t > 0$$

   $$u(x,0) = x(1-x); \forall\ 0 < x < 1$$

   $$u_t(x,0) = 0; \forall\ 0 < x < 1.$$

   $$\left[\text{Ans.}: u(x,y) = \sum_{\substack{k=1 \\ k \text{ is odd}}}^{\infty} \frac{8}{k^3\pi^3}\cos(\pi kCt)\sin(\pi kx)\right]$$

3. A tightly stretched string of length l fastened at both ends, is disturbed from the position of equilibrium imparting to each of its points an initial velocity of magnitude $f(x)$. Show that

the solution of the problem is $u(x,t)=\frac{2}{l}\sum_{n=1}^{\infty}\int_{0}^{l}\left(f(x)\sin\left(\frac{n\pi x}{l}\right)dx\right)\sin\left(\frac{n\pi x}{l}\right)\cos\left(\frac{n\pi Ct}{l}\right)$.

4. Find the displacement of a string stretched between the fixed points $(0,0)$ and $(1,0)$ and released from rest from position $a\sin\pi x+b\cos 2\pi x$.
$[\text{Ans.}: u(x,t)=a\sin\pi x\cos\pi Ct+b\sin 2\pi x\cos 2\pi Ct]$

5. Find the solution of the equation $\frac{\partial^2 u}{\partial t^2}=C^2\frac{\partial^2 u}{\partial x^2}$. subject to the boundary conditions
$y(0,t)=0, y(l,t)=0, y(x,0)=\phi(x), \frac{\partial y}{\partial t}(x,0)=\psi(x)$.
$$\left[\text{Ans.}: y=\phi(x)\cos\left(\frac{n\pi Ct}{l}\right)+\frac{l\psi(x)}{n\pi C}\frac{\sin\left(\frac{n\pi Ct}{l}\right)}{\sin\left(\frac{n\pi x}{l}\right)}\right]$$

6. $\frac{\partial^2 u}{\partial t^2}(x,t)=C^2\frac{\partial^2 u}{\partial x^2}(x,t); \forall\ 0<x<1 \text{ and } t>0$,

$u(0,t)=u(1,t)=0; \forall\ t>0$,

$u(x,0)=\sin(5\pi x)+2\sin(7\pi x); \forall\ 0<x<1$,

$u_t(x,0)=0; \forall\ 0<x<1$.
$[\text{Ans.}: u(x,y)=\sin(5\pi x)\cos(5\pi Ct)+2\sin(7\pi x)\cos(7\pi Ct)]$

7. Solve the wave equation $\frac{\partial^2 u}{\partial t^2}=C^2\frac{\partial^2 u}{\partial x^2}$. subjected to the conditions

(i) $u(0,t)=u(l,t)=0; \forall\ t\geq 0$

(ii) $u(x,0)=\begin{cases}\frac{2x}{l};0\le x\le\frac{l}{2}\\ \frac{2(l-x)}{l};\frac{l}{2}\le x\le l\end{cases}$

(iii) $\left(\frac{\partial u}{\partial t}\right)_{t=0}=0;0\le x\le l$

$$\left[\text{Ans.}: u(x,t)=\frac{8}{\pi^2}\sum_{n=1}^{\infty}\frac{1}{n^2}\sin\left(\frac{n\pi}{2}\right)\sin\left(\frac{n\pi x}{l}\right)\cos\left(\frac{n\pi Ct}{l}\right)\right]$$

8. The vibration of an elastic string of length l are governed by the one dimensional wave equation $\frac{\partial^2 u}{\partial t^2}=C^2\frac{\partial^2 u}{\partial x^2}$. The string is fixed at the ends. $u(0,t)=u(l,t)=0;\forall\ t\ge 0$. The initial deflection is $u(x,0)=\begin{cases}\frac{2x}{l};0\le x\le\frac{l}{2}\\ \frac{2(l-x)}{l};\frac{l}{2}\le x\le l\end{cases}$

   And the initial velocity is zero. Find the deflection of the string at any instant of time.

$$\left[\text{Ans.}: u(x,t)=\frac{4l}{\pi^2}\sum_{n=1}^{\infty}\frac{1}{n^2}\sin\left(\frac{n\pi}{2}\right)\sin\left(\frac{n\pi x}{l}\right)\cos\left(\frac{n\pi Ct}{l}\right)\right]$$

9. A taut string of length $2l$ is fastened at both ends. The mid-point of the string is taken to height b and then released from the rest in that position. Find the displacement of the string.

$$\left[\text{Ans.}: u(x,t)=\frac{8b}{\pi^2}\sum_{m=1}^{\infty}\frac{1}{(2m-1)^2}\sin\left(\frac{(2m-1)\pi}{2}\right)\sin\left(\frac{(2m-1)\pi x}{2l}\right)\cos\left(\frac{(2\right.\right.$$

10. A tightly stretched string with fixed end points $x=0$ and $x=l$ is initially at rest in its equilibrium position. If it is set vibrating by giving to each of its points a velocity $\lambda x(l-x)$, Find the displacement of the string at any distance x from one end at any time t.

$$\left[\text{Ans.}: u(x,t)=\frac{8\lambda l^3}{C\pi^4}\sum_{m=1}^{\infty}\frac{1}{(2m-1)^4}\sin\left(\frac{(2m-1)\pi x}{l}\right)\sin\left(\frac{(2m-1)\pi Ct}{l}\right)\right].$$

11. A uniform elastic strings of length $60\ cm$ is subjected to a constant tension $2\ kg$. If the ends are fixed and the initial displacement $y=60x-x^2, 0<x<60$ while the initial velocity is zero. Find displacement function $y(x,t)$.

$$\left[\text{Ans.}: y(x,t)=\frac{28800}{\pi^3}\sum_{n=odd}^{\infty}\frac{1}{n^3}\sin\left(\frac{n\pi x}{l}\right)\cos\left(\frac{n\pi Ct}{l}\right)\right].$$

12. A string of length l is tightly stretched and fixed at its ends at the points $(0,0)$ and $(2l,0)$ of the $xy-$ plane. It is made to vibrate transversely in the $xy-$ plane by giving to each of its points transverse velocity v in the $xy-$ plane, where v is given by

$$v=\begin{cases} kx; 0\le x\le l \\ k(2l-x); l\le x\le 2l \end{cases}$$

Find the expression for the transverse displacement of the string at any time t.

$$\left[\text{Ans.}: y(x,t)=\frac{16kl^2}{\pi^3 C}\sum_{n=1}^{\infty}\frac{1}{n^3}\sin\left(\frac{n\pi}{2}\right)\sin\left(\frac{n\pi x}{2l}\right)\sin\left(\frac{n\pi Ct}{2l}\right)\right].$$

13. A string is stretched between two fixed points at a distance of $4l$ apart and the point of the string are given velocity v where v is given by

$$v=\begin{cases} \dfrac{cx}{l}; 0\le x\le 2l \\ \dfrac{c(4l-x)}{l}; 2l\le x\le 4l \end{cases}$$

Find the displacement of the string at any time.

$$\left[\text{Ans.}: y(x,t)=\frac{64cl}{\pi^3 a}\sum_{n=1}^{\infty}\frac{1}{n^3}\sin\left(\frac{n\pi}{2}\right)\sin\left(\frac{n\pi x}{4l}\right)\sin\left(\frac{n\pi at}{4l}\right)\right].$$

14. A string is stretched between two fixed points at a distance of l apart and the point of the string are given velocity v where v is given by

$$v=\begin{cases}\frac{cx}{l};0\le x\le\frac{l}{2}\\ \frac{c(l-x)}{l};\frac{l}{2}\le x\le l\end{cases}$$

x being the distance from an end point. Find the displacement of the string at any time.

$$\left[\text{Ans.}: y(x,t)=\frac{4cl}{\pi^3 a}\sum_{n=1}^{\infty}\frac{1}{n^3}\sin\left(\frac{n\pi}{2}\right)\sin\left(\frac{n\pi x}{l}\right)\sin\left(\frac{n\pi at}{l}\right)\right].$$

15. A string is stretched between two fixed points at a distance of $60\ cm$ apart and the point of the string are given velocity v where v is given by

$$v=\begin{cases}\frac{\lambda x}{30};0<x\le 30\\ \frac{\lambda(60-x)}{60};30<x\le 60\end{cases}$$

x being the distance from an end point. Find the displacement of the string at any time.

$$\left[\text{Ans.}: y(x,t)=\frac{240\lambda}{\pi^3 a}\sum_{n=1}^{\infty}\frac{1}{n^3}\sin\left(\frac{n\pi}{2}\right)\sin\left(\frac{n\pi x}{60}\right)\sin\left(\frac{n\pi at}{60}\right)\right].$$

16. A string is stretched fastened at two points $x=0$ and $x=l$ .Motion is stated by displacing the string into form of the curve $y(x,0)=lx-x^2$ and also by imparting a constant velocity v_c to every point of the string in this position at $t=0$. Find the subsequent displacement of the string as a function of x and t.

$$\left[\text{Ans.}: u(x,t)=\frac{4l^3}{\pi^3}\sum_{n=odd}^{\infty}\frac{1}{n^3}\sin\left(\frac{n\pi x}{l}\right)\sin\left(\frac{n\pi at}{l}\right)+\frac{4v_c l}{a\pi^3}\sum_{n=odd}^{\infty}\frac{1}{n^2}\sin\left(\frac{n\pi x}{l}\right)\sin\left(\frac{n\pi at}{l}\right)\right].$$

6.3 D' Alembert Solution of One-dimensional Wave Equation

The one-dimensional wave problem has a general solution known as D'Alembert's formula in mathematics, and more especially in partial differential equations (PDEs):

$\frac{\partial^2 y}{\partial t^2}=C^2\frac{\partial^2 y}{\partial x^2}$ and assume initial condition $y(x,0)=f(x)$ and $\left(\frac{\partial y}{\partial t}\right)_{t=0}=g(x)$ for $-\infty<x<\infty,\ t>0$.It is named after the mathematician Jean le Rond D'Alembert, who discovered it in 1747 while working on a vibrating string problem.

6.3.1 Vibration of an Infinite String

Transform the equation $\frac{\partial^2 y}{\partial t^2}=C^2\frac{\partial^2 y}{\partial x^2}$ to its normal form using the transformation $u=x+ct, v=x-ct$ and hence solve it. Show that the solution may be put in the form $y=\frac{1}{2}(f(x+ct)+f(x-ct))$. Assume initial condition $y(x,0)=f(x)$ and $\left(\frac{\partial y}{\partial t}\right)_{t=0}=g(x)$.

Proof: Consider one dimensional wave equation

$$\frac{\partial^2 y}{\partial t^2}=C^2\frac{\partial^2 y}{\partial x^2}. \tag{6.90}$$

Let $u=x+ct, v=x-ct$ be a transformation of x and t into u and v. Then

$$\frac{\partial y}{\partial x}=\frac{\partial y}{\partial u}.\frac{\partial u}{\partial x}+\frac{\partial y}{\partial v}.\frac{\partial v}{\partial x}. \tag{6.91}$$

Using $u=x+ct, v=x-ct$, we get

$$\frac{\partial u}{\partial x}=1,\frac{\partial u}{\partial t}=c,\frac{\partial v}{\partial x}=1,\frac{\partial v}{\partial t}=-c.$$

Therefore, equation (6.91) reduces to

$$\frac{\partial y}{\partial x}=\frac{\partial y}{\partial u}.1+\frac{\partial y}{\partial v}.1 \Rightarrow \frac{\partial y}{\partial x}=\frac{\partial y}{\partial u}+\frac{\partial y}{\partial v}. \tag{6.92}$$

Again, differentiating equation (6.92) partially w.r.t x, we get

$$\frac{\partial^2 y}{\partial x^2}=\frac{\partial}{\partial x}\left(\frac{\partial y}{\partial u}+\frac{\partial y}{\partial v}\right),$$

$$\frac{\partial^2 y}{\partial x^2}=\frac{\partial^2 y}{\partial u^2}.\frac{\partial u}{\partial x}+\frac{\partial^2 y}{\partial u \partial v}.\frac{\partial u}{\partial x}+\frac{\partial^2 y}{\partial u \partial v}.\frac{\partial v}{\partial x}+\frac{\partial^2 y}{\partial v^2}.\frac{\partial v}{\partial x}. \tag{6.93}$$

Using $u=x+ct, v=x-ct$,

we get

$$\frac{\partial u}{\partial x}=1, \frac{\partial u}{\partial t}=c, \frac{\partial v}{\partial x}=1, \frac{\partial v}{\partial t}=-c.$$

Therefore, equation (6.93)

$$\frac{\partial^2 y}{\partial x^2}=\frac{\partial^2 y}{\partial u^2}.1+\frac{\partial^2 y}{\partial u \partial v}.1+\frac{\partial^2 y}{\partial u \partial v}.1+\frac{\partial^2 y}{\partial v^2}\cdot 1,$$

$$\frac{\partial^2 y}{\partial x^2}=\frac{\partial^2 y}{\partial u^2}+\frac{\partial^2 y}{\partial u \partial v}+\frac{\partial^2 y}{\partial u \partial v}+\frac{\partial^2 y}{\partial v^2},$$

$$\frac{\partial^2 y}{\partial x^2}=\frac{\partial^2 y}{\partial u^2}+2\frac{\partial^2 y}{\partial u \partial v}+\frac{\partial^2 y}{\partial v^2}. \tag{6.94}$$

Again $\frac{\partial y}{\partial t}=\frac{\partial y}{\partial u}.\frac{\partial u}{\partial t}+\frac{\partial y}{\partial v}.\frac{\partial v}{\partial t}$. (6.95)

Using $u = x + ct, v = x - ct$,

we get $\frac{\partial u}{\partial x} = 1, \frac{\partial u}{\partial t} = c, \frac{\partial v}{\partial x} = 1, \frac{\partial v}{\partial t} = -c$.

Therefore, equation (6.95) reduces to

$$\frac{\partial y}{\partial x} = \frac{\partial y}{\partial u}.c + \frac{\partial y}{\partial v}.(-c) \Rightarrow \frac{\partial y}{\partial x} = c\frac{\partial y}{\partial u} - c\frac{\partial y}{\partial v}. \qquad (6.96)$$

Again, differentiating equation (6.96) partially w.r.t t, we get

$$\frac{\partial^2 y}{\partial t^2} = \frac{\partial}{\partial y}\left(c\frac{\partial y}{\partial u} - c\frac{\partial y}{\partial v}\right),$$

$$\frac{\partial^2 y}{\partial t^2} = c\frac{\partial^2 y}{\partial u^2}.\frac{\partial u}{\partial t} + c\frac{\partial^2 y}{\partial u \partial v}.\frac{\partial v}{\partial t} - c\frac{\partial^2 y}{\partial u \partial v}.\frac{\partial u}{\partial t} - c\frac{\partial^2 y}{\partial v^2}.\frac{\partial v}{\partial t}. \qquad (6.97)$$

Using $u = x + ct, v = x - ct$,

we get

$$\frac{\partial u}{\partial x} = 1, \frac{\partial u}{\partial t} = c, \frac{\partial v}{\partial x} = 1, \frac{\partial v}{\partial t} = -c.$$

Therefore, equation (6.97) reduces to

$$\frac{\partial^2 y}{\partial t^2} = c^2\frac{\partial^2 y}{\partial u^2} - c^2\frac{\partial^2 y}{\partial u \partial v} - c^2\frac{\partial^2 y}{\partial u \partial v} + c^2\frac{\partial^2 y}{\partial v^2},$$

$$\frac{\partial^2 y}{\partial t^2} = c^2\frac{\partial^2 y}{\partial u^2} + -2c^2\frac{\partial^2 y}{\partial u \partial v} + c^2\frac{\partial^2 y}{\partial v^2}. \qquad (6.98)$$

Substituting the values from equations (6.94) and equation (6.98) in equation (6.90), we get

$$c^2\left(\frac{\partial^2 y}{\partial u^2}+2\frac{\partial^2 y}{\partial u\partial v}+\frac{\partial^2 y}{\partial v^2}\right)=c^2\frac{\partial^2 y}{\partial u^2}+-2c^2\frac{\partial^2 y}{\partial u\partial v}+c^2\frac{\partial^2 y}{\partial v^2},$$

$$4c^2\frac{\partial^2 y}{\partial u\partial v}=0,$$

$$\frac{\partial^2 y}{\partial u\partial v}=0. \tag{6.99}$$

Integrating (6.80) w.r.t. v , we get

$$\frac{\partial^2 y}{\partial u}=\phi(v), \tag{6.100}$$

where $\phi(v)$ is a constant in respect to v .Integrating (6.100) w.r.t. u , we get

$$y=\int\phi(u)du+\phi_2(v),$$

$$y=\phi_1(u)+\phi_2(v),$$

$$y(x,t)=\phi_1(x+ct)+\phi_2(x-ct). \tag{6.101}$$

To find the solution of $y(x,t)$, we need to evaluate the arbitrary function $\phi_1(x+ct)$ and $\phi_2(x+ct)$The solution (6.101) is Alembert's solution of wave equation.

Now we apply initial condition $y=f(x)$ at $t=0.$ to equation (6.101) changes to

$$f(x)=\phi_1(x)+\phi_2(x). \tag{6.102}$$

Differentiating equation (6.101), we get

$$\frac{\partial y}{\partial t} = c\phi_1^{'}\left(x+ct\right)+\phi_2{'}\left(x-ct\right). \tag{6.103}$$

Now we apply initial condition $\left(\frac{\partial y}{\partial t}\right)_{t=0} = g\left(x\right)$, equation (6.103) reduces to

$$g\left(x\right) = c\phi_1^{'}\left(x\right) - c\phi_2^{'}\left(x\right). \tag{6.104}$$

Integrating equation (6.104) both sides, we get

$$\int_{x_0}^{x} g\left(x\right)\,dx = c\phi_1\left(x\right) - c\phi_2\left(x\right). \tag{6.105}$$

Multiplying equation (6.102) by c and adding it to equation (6.105), we get

$$cf\left(x\right) + \int_{x_0}^{x} g\left(x\right)\,dx = 2c\phi_1\left(x\right),$$

$$\Rightarrow \phi_1\left(x\right) = \frac{1}{2c}\left[cf\left(x\right) + \int_{x_0}^{x} g\left(x\right)dx\right]. \tag{6.106}$$

Multiplying equation (6.102) by c and subtracting it to equation (6.105), we get

$$cf\left(x\right) - \int_{x_0}^{x} g\left(x\right)\,dx = 2c\phi_2\left(x\right),$$

$$\Rightarrow \phi_2\left(x\right) = \frac{1}{2c}\left[cf\left(x\right) - \int_{x_0}^{x} g\left(x\right)dx\right]. \tag{6.107}$$

From equations (6.106) and (6.107), equation (6.101) reduces to

$$y(x,t)=\frac{1}{2c}\left[cf(x+ct)+\int_{x_0}^{x+ct}g(x)dx\right]+\frac{1}{2c}\left[cf(x-ct)-\int_{x_0}^{x+ct}g(x)dx\right],$$

$$y(x,t)=\frac{1}{2c}\left[cf(x+ct)+cf(x-ct)+\int_{x-ct}^{x_0}g(x)dx+\int_{x_0}^{x+ct}g(x)dx\right],$$

$$y(x,t)=\frac{1}{2c}\left[cf(x+ct)+cf(x-ct)+\int_{x-ct}^{x+ct}g(x)dx\right]. \quad (6.108)$$

equation (6.108) is D' Alembert formula for the solution of the Cauchy problem for the wave equation on the entire line. It is an explicit formula for the solution of the Cauchy problem, in terms of the given initial position and velocity functions.

Example:

Solve the boundary value problem $u_{tt}=4u_{xx}$ for $-\infty<x<\infty,\ t>0$, $u(x,0)=x,\ u_t(x,0)=0$ for $-\infty<x<\infty$.

Solution:

By D' Alembert solution is given by

$$u(x,t)=\frac{1}{2}\left[f(x-ct)+f(x+ct)\right]+\frac{1}{2c}\int_{x-ct}^{x+ct}g(\xi)d\xi.$$

Here $f(x)=x;\ g(x)=0$, therefore

$$f(x-ct)=x-ct;\ f(x+ct)=x+ct;\ g(x)=0.$$

$\therefore\ u(x,t)=\frac{1}{2}[x-ct+x+ct]+0$ (From equation (6.108)

$u(x,t) = x$.

Example:

Solve the boundary value problem $u_{tt} = 4u_{xx}$ for $-\infty < x < \infty,\ t > 0$, $u(x,0) = e^{-|x|}$, $u_t(x,0) = \cos 4x$ for $-\infty < x < \infty$.

Solution:

By D' Alembert solution is given by

$$u(x,t) = \frac{1}{2}\left[f(x-ct) + f(x+ct)\right] + \frac{1}{2c}\int_{x-ct}^{x+ct} g(\xi)\,d\xi .$$

Here $f(x) = e^{-|x|}$; $g(x) = \cos 4x$, therefore

$$f(x-ct) = x-ct;\ f(x+ct) = x+ct;\ g(\xi) = \cos 4\xi .$$

From equation (6.108)

$$\therefore u(x,t) = \frac{1}{2}\left(e^{-|x-2t|} + e^{-|x+2t|}\right) + \frac{1}{4}\int_{x-2t}^{x+2t} \cos 4\xi \, d\xi ,$$ (From equation

$$u(x,t) = \frac{1}{2}\left(e^{-|x-2t|} + e^{-|x+2t|}\right) + \frac{1}{16}\left[\sin 4\xi\right]_{x-2t}^{x+2t},$$

$$u(x,t) = \frac{1}{2}\left(e^{-|x-2t|} + e^{-|x+2t|}\right) + \frac{1}{16}\left[\sin 4(x+2t) - \sin 4(x-2t)\right],$$

$$u(x,t) = \frac{1}{2}\left(e^{-|x-2t|} + e^{-|x+2t|}\right) + \frac{1}{8}\cos 4x \sin 8t .$$

Example:

Solve the boundary value problem $u_{tt} - 9u_{xx} = 0$ for $-\infty < x < \infty,\ t > 0$, $u(x,0) = x^3$, $u_t(x,0) = x\cos x$ for $-\infty < x < \infty$.

Solution:
By D' Alembert solution is given by

$$u(x,t) = \frac{1}{2}\left[f(x-ct) + f(x+ct)\right] + \frac{1}{2c}\int_{x-ct}^{x+ct} g(\xi)\,d\xi.$$

Here $f(x) = x;\ g(x) = x\cos x$ therefore

$$f(x-ct) = (x-ct)^3;\ f(x+ct) = (x+ct)^3;\ g(\xi) = \xi\cos 4\xi.$$

From equation (6.108)

$$\therefore\ u(x,t) = \frac{1}{2}\left((x-3t)^3 + (x+3t)^3\right) + \frac{1}{6}\int_{x-3t}^{x+3t} \xi\cos\xi\ d\xi$$

$$u(x,t) = \frac{1}{2}\left([x-3t+x+3t]\left[(x-3t)^2 + (x-3t)(x+3t) + (x+3t)^2\right]\right) + \frac{1}{6}\int_{x-3t}^{x+3t} \xi\cos\xi\ d\xi$$

,

$$u(x,t) = \frac{1}{2}\left([2x]\left[(x-3t)^2 + (x-3t)(x+3t) + (x+3t)^2\right]\right) + \frac{1}{6}\left[\xi\sin\xi - \cos\xi\right]_{x-3t}^{x+3t},$$

$$u(x,t) = \frac{1}{2}\left(54t^2x + 2x^3\right) + \frac{1}{6}\left[\{(x+3t)\sin(x+3t) - \cos(x+3t)\} - \{(x-3t)\sin(x-3t) - \cos(x-3t)\}\right]$$

$$u(x,t) = \frac{1}{2}\left(54t^2x + 2x^3\right) + \frac{1}{6}\left[x\cos x\sin 3t + 3t\ \sin x\cos 3t - \sin x\sin 3t\right].$$

Exercise

Solve the following boundary value problem using D' Alembert solution

1 $u_{tt} = u_{xx}$ for $-\infty < x < \infty$, $t > 0$,

$u(x,0) = x^2$, $u_t(x,0) = -x$ for $-\infty < x < \infty$.

[Ans $u(x,t) = x^2 + t^2 - xt$]

2 $u_{tt} = 7u_{xx}$ for $-\infty < x < \infty$, $t > 0$

$u(x,0) = \cos \pi x$, $u_t(x,0) = 1 - x^2$ for $-\infty < x < \infty$.

[Ans
$u(x,t) = \frac{1}{2}\left[\cos\left(\pi(x-t)\right) + \cos\left(\pi(x+t)\right)\right] + t - x^2 t - \frac{49}{3}t^3$]

3 $u_{tt} = 14u_{xx}$ for $-\infty < x < \infty$, $t > 0$

$u(x,0) = e^x$, $u_t(x,0) = x$ for $-\infty < x < \infty$.

[Ans $u(x,t) = \frac{1}{2}\left[e^{x-14t} + e^{x-+4t}\right] + xt$]

4 $u_{tt} = 4u_{xx}$ for $-\infty < x < \infty$, $t > 0$

$u(x,0) = xe^x$, $u_t(x,0) = \cos hx$ for $-\infty < x < \infty$.

[Ans
$u(x,t) = \frac{1}{2}\left[(x-4t)e^{x-4t} + (x+4t)e^{x+4t}\right] + \frac{1}{8}\left[\sinh(x+4t) - \sinh(x-4t)\right]$]

5 $u_{tt} = 15u_{xx}$ for $-\infty < x < \infty,\ t > 0$

$u(x,0) = |x|,\ u_t(x,0) = e^{2x}$ for $-\infty < x < \infty$.

[Ans $u(x,t) = \frac{1}{2}\left[|x-15t| + |x+15t|\right] + \frac{1}{60}\left[e^{2(x+15t)} - e^{2(x-15t)}\right]$]

6.4 Vibrating Membrane: Two-Dimensional Wave Equation

Consider a lightly stretched uniform membrane (such as the membrane of a drum) with tension T per unit length is the same in all directions at every point. Consider the forces on an element $\delta x\ \delta y$ of the membranes. Due to its displacement u perpendicular to the xy plane, the forces $T\delta x$ and $T\delta y$ act on the edges along the tangent to the membrane.

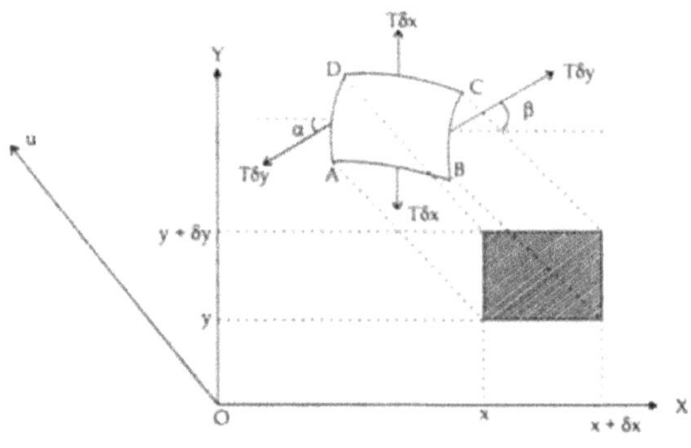

Figure 6.6: Vibrating membrane

The forces $T\delta y$ (tangential to the membrane) on its opposite edges of length δy act at angles α and β to the horizontal. So, their vertical component $= (T\delta y)\sin\beta - (T\delta y)\sin\alpha$.

But angles α and β are very small i.e. $\sin\alpha \approx \tan\alpha$ etc.

vertical component $=(T\delta y)(\tan\beta-\tan\alpha)$,

$$=T\delta y\left\{\left(\frac{\partial u}{\partial x}\right)_{x+\delta x}-\left(\frac{\partial u}{\partial x}\right)_{x}\right\},$$

$$=T\delta y\delta x\frac{\left\{\left(\frac{\partial u}{\partial x}\right)_{x+\delta x}-\left(\frac{\partial u}{\partial x}\right)_{x}\right\}}{\delta x},$$

$=T\delta y\delta x\dfrac{\partial^2 u}{\partial x^2}$, up to a first order of approximation.

Similarly, the forces $T\delta x$ (the vertical component of the force) acting on the edges of length δx have the vertical component $=T\delta x\delta y\dfrac{\partial^2 u}{\partial y^2}$.If m be the mass per unit area of the membrane, then the equation of motion of the element ABCD is

$$m\delta x\delta y\frac{\partial^2 u}{\partial t^2}=T\left(\frac{\partial^2 u}{\partial x^2}+\frac{\partial^2 u}{\partial y^2}\right)\delta x\delta y,$$

or $$\frac{\partial^2 u}{\partial t^2}=\frac{T}{m}\left(\frac{\partial^2 u}{\partial x^2}+\frac{\partial^2 u}{\partial y^2}\right),$$

or $$\frac{\partial^2 u}{\partial t^2}=c^2\left(\frac{\partial^2 u}{\partial x^2}+\frac{\partial^2 u}{\partial y^2}\right), \qquad (6.109)$$

where $c^2=\dfrac{T}{m}$,is wave equation in two dimensions.

6.4.1 Solution of Equation of the Vibrating Membrane (Rectangular Membrane)

Let u be the solution of equation (6.109) and we assume that

$$u = X(x) \cdot Y(y) \cdot T(t). \tag{6.110}$$

On differentiating (6.110), we have

$$\frac{\partial^2 u}{\partial x^2} = Y \cdot T \frac{d^2 X}{dx^2},$$

$$\frac{\partial^2 u}{\partial y^2} = X \cdot T \frac{d^2 Y}{dy^2},$$

$$\frac{\partial^2 u}{\partial t^2} = XY \frac{d^2 T}{dt^2}.$$

Substituting these values in (6.110), we get

$$XY \frac{d^2 T}{dt^2} = C^2 \left[\frac{d^2 X}{dx^2} YT + XT \frac{d^2 Y}{dy^2} \right].$$

Dividing by $XYTc^2$, we get

$$\frac{1}{c^2} \frac{d^2 T}{dt^2} = \left[\frac{1}{X} \frac{d^2 X}{dx^2} + \frac{1}{Y} \frac{d^2 Y}{dy^2} \right] \tag{6.111}$$

Each member is constant

$$\frac{d^2 X}{dx^2} + k^2 X = 0, \frac{d^2 Y}{dy^2} + l^2 Y = 0$$

Putting these values in equation (6.109), we get

$$\frac{d^2T}{dt^2}+\left(k^2+l^2\right)c^2T=0$$

$\frac{d^2X}{dx^2}+k^2X=0$	$\frac{d^2Y}{dy^2}+l^2Y=0$	$\frac{d^2T}{dt^2}+\left(k^2+l^2\right)c^2T=0$
$\left(D^2+k^2\right)X=0$	$\left(D^2+l^2\right)Y=0$	$\left[D^2+\left(k^2+l^2\right)c^2\right]T=0$
$m^2+k^2=0$	$m^2+l^2=0$	$m^2+\left[k^2+l^2\right]c^2=0$
$m=\pm ik$	$m=\pm il$	$m=ic\sqrt{k^2+l^2}$
$X=c_1\cos kx+c_2\sin kx$	$Y=c_3\cos ly+c_4\sin ly$	$T=c_5\cos ct\sqrt{k^2+l^2}$ $+c_6\sin ct\sqrt{k^2+l^2}$

Putting the values of X,Y,T in (6.110), we get

$$u(x,y,t)=(c_1\cos kx+c_2\sin kx)(c_3\cos ly+c_4\sin ly)\left[c_5\cos\sqrt{(k^2+l^2)}ct+c_6\sin\sqrt{(k^2+l^2)}ct\right] \quad (6.112)$$

Suppose if the membrane is rectangular and stretched between the lines $x=0, x=a,\ y=0,\ y=b$, then the boundary conditions are

(i) $u=0$ when $x=0$

(ii) $u=0$ when $x=a$

(iii) $u=0$ when $y=0$

(iv) $u=0$ when $y=b$ for all t.

On putting $x=0,\ u=0$ in (6.112), we get

$$0 = c_1 \left(c_3 \cos ly + c_4 \sin ly\right)\left[c_5 \cos\sqrt{\left(k^2+l^2\right)}ct + c_6 \sin\sqrt{\left(k^2+l^2\right)}ct\right],$$

On putting $c_1 = 0$, equation (6.112) becomes

$$u = c_2 \sin kx\left(c_3 \cos ly + c_4 \sin ly\right)\left[c_5 \cos x\sqrt{\left(k^2+l^2\right)}ct + c_6 \sin\sqrt{\left(k^2+l^2\right)}ct\right] \quad (6.113)$$

On putting $u = 0$ and $x = a$ in (6.113), we get

$$0 = c_2 \sin ka\left(c_3 \cos ly + c_4 \sin ly\right)\left[c_5 \cos\sqrt{\left(k^2+l^2\right)}ct + c_6 \sin\sqrt{\left(k^2+l^2\right)}ct\right]$$

$$\Rightarrow \qquad \sin ka = 0$$

$$\Rightarrow \qquad \sin ka = \sin m\pi$$

$$\Rightarrow \qquad k = \frac{m\pi}{a}$$

On putting the value of k in (6.113), we get

$$u = c_2 \sin\frac{m\pi x}{a}\left(c_3 \cos ly + c_4 \sin ly\right)\left[c_5 \cos x\sqrt{\left(\frac{m^2\pi^2}{a^2}+l^2\right)}ct + c_6 \sin\sqrt{\left(\frac{m^2\pi^2}{a^2}+l^2\right)}ct\right] \quad (6.114)$$

Similarly, applying the conditions (iii) and (iv) in (6.114), we get

$c_3 = 0$ or $l = \dfrac{n\pi}{b}$, where n is an integer.

equation (6.114) becomes,

$$u = c_2 c_4 \sin\frac{m\pi x}{a}\sin\frac{n\pi y}{b}\left[c_5 \cos x\sqrt{\left(\frac{m^2\pi^2}{a^2}+\frac{n^2\pi^2}{b^2}\right)}ct + c_6 \sin\sqrt{\left(\frac{m^2\pi^2}{a^2}+\frac{n^2\pi^2}{b^2}\right)}ct\right] \quad (6.115)$$

Let $p = c\pi\sqrt{\frac{m^2}{a^2} + \frac{n^2}{b^2}}$

Now, $u = c_2 c_4 \sin\frac{m\pi x}{a} \sin\frac{n\pi y}{b}\left(c_5 \cos pt + c_6 \sin pt\right)$

These are the solution of wave equation and are called Eigen functions. Choosing the constants c_2 and c_4 so that $c_2 c_4 = 1$. We can write the general solution of (6.109) as

$$u = \sum_{m=1}^{\infty}\sum_{n=1}^{\infty} \sin\frac{m\pi x}{a} \sin\frac{n\pi y}{b}\left(A_{mn} \cos pt + B_{mn} \sin pt\right) \tag{6.116}$$

Differentiating (6.116) with respect to 't', we get

$$\frac{\partial u}{\partial t} = \sum_{m=1}^{\infty}\sum_{n=1}^{\infty} \sin\frac{m\pi x}{a} \sin\frac{n\pi y}{b}\left(-pA_{mn} \sin pt + pB_{mn} \cos pt\right) \tag{6.117}$$

On putting $\frac{\partial u}{\partial t} = 0$ when $t = 0$ in (6.117), we get

$$0 = \sum_{m=1}^{\infty}\sum_{n=1}^{\infty} \sin\frac{m\pi x}{a} \sin\frac{n\pi y}{b}\left(pB_{mn} \cos pt\right) \Rightarrow B_{mn} = 0 .$$

Also using the condition: $u = f(x, y)$ when $t = 0$, we get

$$f(x, y) = \sum_{m=1}^{\infty}\sum_{n=1}^{\infty} A_{mn} \sin\frac{m\pi x}{a} \sin\frac{n\pi y}{b} .$$

This is a double Fourier series. Multiplying both sides by $\sin\frac{m\pi x}{a} \sin\frac{n\pi y}{b}$ and integrating from $x = 0$ to $x = a$ and $y = 0$ to $y = b$, every term on the right except one all become zero. Thus, we get $\int_0^a\int_0^b f(x,y)\,A_{mn}\sin\frac{m\pi x}{a}\sin\frac{n\pi y}{b}\,dydx=\frac{ab}{4}A_{mn}$ .It is the generalized Euler's formula and gives the coefficients in the solution. On putting the value of A_{mn}, we get the solution of (6.109),

6.5 Solution of Wave equation using Finite Difference Method

The most basic illustration of a hyperbolic partial differential equation is the wave equation i.e. $\frac{\partial^2 u}{\partial t^2}=c^2\frac{\partial^2 u}{\partial x^2}$. The displacement function $u(x,t)$, which is defined for values of x from 0 to l and for t from 0 to ∞, satisfying the initial and boundary conditions, is its solution. However, we have two initial conditions and two boundary conditions in the case of hyperbolic equations. These equations originate from convective problems in gas dynamics, vibrations, and wave mechanics.

6.5.1 Solution of Wave Equation

Consider $\frac{\partial^2 u}{\partial t^2}=c^2\frac{\partial^2 u}{\partial x^2}$, (6.118)

subject to the initial conditions:

$$u=f(x),\frac{\partial u}{\partial t}=g(x),0\le x\le 1 \text{ at } t=0, \tag{6.119}$$

and the boundary conditions: $u(0,t)=\phi(t),u(1,t)=\psi(t)$. (6.120)

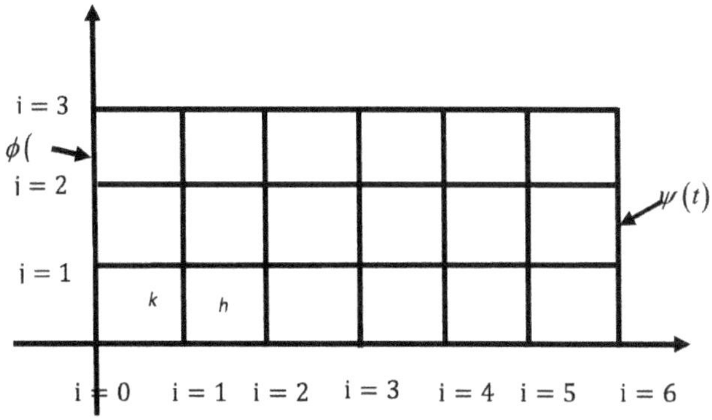

Figure 6.7: Rectangular mesh.

Consider a rectangular mesh in the $x-t$ plane spacing h along x-direction and k along time $t-$direction as shown in Figure 6.7. Among the dividing lines, mesh points, nodal points, and grid points represent the points of intersection. Denoting a mesh point $(x,t)=(ih, jk)$ as simply i, j we have

$$\frac{\partial^2 u}{\partial x^2}=\frac{u_{i-1,j}-2u_{i,j}+u_{i+1,j}}{h^2} \text{ and } \frac{\partial^2 u}{\partial t^2}=\frac{u_{i,j-1}-2u_{i,j}+u_{i,j+1}}{k^2}.$$

Replacing the derivatives in equation (6.118) by their above approximations, we obtain

$$u_{i,j-1}-2u_{i,j}+u_{i,j+1}=\frac{k^2c^2}{h^2}\left(u_{i-1,j}-2u_{i,j}+u_{i+1,j}\right).$$

or $u_{i,j+1}=2u_{i,j}\left(1-\alpha^2c^2\right)+\alpha^2c^2\left(u_{i-1,j}+u_{i+1,j}\right)-u_{i,j-1}$. (6.121)

where $\alpha=\frac{k}{h}$. Now replacing the derivative in equation (6.119) by its

central difference approximation, we get $\frac{u_{i,j+1}-u_{i,j-1}}{2k}=\frac{\partial u}{\partial t}=g(x)$.

or

$u_{i,j+1} = u_{i,j-1} + 2k\, g(x)$ at $t = 0$

i.e. $u_{i,1} = u_{i,-1} + 2k\, g(x)$ for $j = 0$. (6.122)

The initial condition given by $u = f(x)$ at $t = 0$ changes to

$$u_{i,0} = f(x) \tag{6.123}$$

Combining (6.122) and (6.123), we have

$$u_{i,1} = f(x) + 2k\, g(x). \tag{6.124}$$

Also the boundary conditions given by equation (6.120) gives

$$u_{0,j} = \phi(t) \text{ and } u_{1,j} = \psi(t). \tag{6.125}$$

Hence, equation (6.121) gives the values of $u_{i,j+1}$ at the $(j+1)^{th}$ level when the nodal values at $(j-1)^{th}$ and j^{th} levels are known from Equations (6.123) and (6.1247) as shown in Figure.6.1. Thus equation (6.121) gives an implicit scheme for the solution of the wave equation.

A special case: The coefficient of $u_{i,j}$ in (6.121) will vanish if $\alpha = \frac{1}{c}$ or $k = \frac{h}{c}$. Then equation (6.121) reduces to the simple form

$$u_{i,j+1} = u_{i-1,j} + u_{i+1,j} - u_{i,j-1}. \tag{6.126}$$

This result provides an explicit scheme for the solution of the wave equation.

Notes:

1. For $\alpha = \frac{1}{c}$, the solution of (6.121.) is stable and coincides with the solution of (6.118).

2. For $\alpha < \frac{1}{c}$, the solution is stable but inaccurate.

3. For $\alpha > \frac{1}{c}$, the solution is unstable.

4. The formula (6.121) converges for $\alpha \leq 1$ i.e. for $k \leq h$.

Example:

Evaluate the pivotal values of the equation $u_{tt} = 16u_{xx}$, taking $h = 1$ upto $t = 1.25$. The boundary conditions are $u(0,t) = u(5,t) = 0$, $u_t(x,0) = 0$ and $u(x,0) = x^2(5 - x)$.

Solution:

Here $c^2 = 16$.

For the given equation the difference equation is $u_{i,j+1} = 2u_{i,j}(1 - 16\alpha^2) + 16\alpha^2(u_{i-1,j} + u_{i+1,j}) - u_{i,j-1}$, where

$$\alpha = \frac{k}{h}. \tag{6.127}$$

Taking $h = 1$ and choosing k so that the coefficient of $u_{i,j}$ vanishes, we have $16\alpha^2 = 1$,, i.e., $k = \frac{h}{4} = \frac{1}{4}$.

$\therefore$ equation (6.127) reduces to

$$u_{i,j+1} = u_{i-1,j} + u_{i+1,j} - u_{i,j-1} \tag{6.128}$$

which gives a convergent solution (since $\frac{k}{h} < 1$). Its solution coincides with the solution of the given differential equation. Now since $u(0,t) = u(5,t) = 0$.

$\therefore\ u_{0,j} = 0$ and $u_{5,j} = 0$ for all values of j i.e. the entries in the first and last columns are zero.

Since $u(x,0) = x^2(5-x)$, therefore $u_{i,0} = i^2(5-i) = 4, 12, 18, 16$ for $i = 1, 2, 3, 4$ at $t = 0$.

These are the entries of the first row. Finally, the initial condition

$u_i(x,0) = 0$ changes to $\frac{u_{i,j+1} - u_{i,j-1}}{2k} = 0$ when $j = 0$,

which further implies $u_{i,1} = u_{i,-1}$. (6.129)

Putting $j = 0$ in equation (6.128), we get $u_{i,1} = u_{i-1,0} + u_{i+1,0} - u_{i,-1}$,

$= u_{i-1,0} + u_{i+1,0} - u_{i,1}$, (From equation (6.129))

$$u_{i,1} = \frac{1}{2}\left(u_{i-1,0} + u_{i+1,0}\right) \tag{6.130}$$

Taking $i = 1, 2, 3, 4$ successively, we obtain

$$u_{1,1} = \frac{1}{2}\left(u_{0,0} + u_{2,0}\right) = \frac{1}{2}(0 + 12) = 6.$$

$$u_{2,1} = \frac{1}{2}\left(u_{1,0} + u_{3,0}\right) = \frac{1}{2}(4 + 18) = 11.$$

$$u_{3,1} = \frac{1}{2}(u_{2,0} + u_{3,0}) = \frac{1}{2}(12+16) = 14.$$

$$u_{4,1} = \frac{1}{2}(u_{3,0} + u_{5,0}) = \frac{1}{2}(18+0) = 9.$$

The above entries are of the second row. Putting $j=1$ in equation (6.128), we get $u_{i,2} = u_{i-1,1} + u_{i+1,1} - u_{i,0}$. Taking $i = 1,2,3,4$ successively, we obtain

$$u_{1,2} = u_{0,,1} + u_{2,1} - u_{1,0} = 0+11-4 = 7.$$

$$u_{2,2} = u_{1,,1} + u_{3,1} - u_{2,0} = 6+14-12 = 8.$$

$$u_{3,2} = u_{2,,1} + u_{4,1} - u_{3,0} = 11+9-18 = 2.$$

$$u_{4,2} = u_{3,,1} + u_{5,1} - u_{4,0} = 14+0-16 = -2$$

The above entries are of the third row. Similarly putting $j = 2,3,4$ successively in equation (6.128), the entries of the fourth, fifth and sixth rows are obtained. Hence the values of $u_{i,j}$ are as shown in the table below:

$i \rightarrow$ j↓	0	1	2	3	4	5
0	0	4	12	18	16	0
1	0	6	11	14	9	0
2	0	7	8	2	−2	0
3	0	2	−2	−8	−7	0
4	0	−9	−14	−11	−6	0
5	0	−16	−18	−12	−4	0

Here we provide the MATHLAB code to solve the given problem:

```
% Parameters

L = 5;% Length of the domain

T = 1.25;% Final time

h = 0.1;% Spatial step size

k = 0.01;% Temporal step size

x = 0:h:L;% Spatial grid

t = 0:k:T;% Temporal grid

nx = length(x); % Number of spatial grid points

nt = length(t); % Number of temporal grid points


% Initialize solution matrix

u = zeros(nx, nt);


% Initial condition

u(:,1) = x.^2 .* (5 - x);


% Boundary conditions

u(1,:) = 0; % u(0,t) = 0

u(end,:) = 0; % u(5,t) = 0
```

```
% Compute solution using finite difference method

for j = 2:nt

for i = 2:nx-1

u(i,j+1) = 2*u(i,j) - u(i,j-1) + 0.16*(u(i+1,j) - 2*u(i,j) + u(i-1,j));

end

end
```

Example:

The transverse displacement u of a point at a distance x from one end and at any time t of a vibrating string satisfies the equation $\frac{\partial^2 u}{\partial t^2} = 4\frac{\partial^2 u}{\partial x^2}$, with boundary conditions $u = 0$ at $x = 0, t > 0$ and $u = 0$ at $x = 4$ $t > 0$ and initial conditions $u = x(4 - x)$ and $\frac{\partial u}{\partial t} = 0$ at $t = 0$, $0 \le x \le 4$. Solve this equation numerically for one half period of vibration, taking $h = 1$ and $k = \frac{1}{2}$.

Solution. Comparing the given PDE $\frac{\partial^2 u}{\partial t^2} = 4\frac{\partial^2 u}{\partial x^2}$ with standard form, $c = 2$. Now $\frac{h}{k} = 2 = c$.

The difference equation for the given equation is

$$u_{i,j+1} = u_{i-1,j} + u_{i+1,j} - u_{i,j-1} \tag{6.131}$$

which gives a convergent solution as $k < h$ .Now

$u(0,t) = u(4,t) = 0$,

$\therefore\ u_{0,j} = 0$ and $u_{4,j} = 0$ for all values of *j*

i.e. the entries in the first and last columns are zero. Since $u(x,0) = x(4-x)$, therefore $u_{i,0} = i(4-i) = 3, 4, 2$ for $i = 1, 2, 3$ at $t = 0$.

These are the entries of the first row. Also $u_t(x,0) = 0$ becomes $\frac{u_{i,j+1} - u_{i,j-1}}{2k} = 0$ when $j = 0$, giving

$$u_{i,1} = u_{i,-1}. \tag{6.132}$$

Putting $j = 0$ in equation (6.131), we get $u_{i,1} = u_{i-1,0} + u_{i+1,0} - u_{i,-1}$,

$u_{i,1} = u_{i-1,0} + u_{i+1,0} - u_{i,1}$ (From equation (6.132)

or $u_{i,1} = \frac{1}{2}\left(u_{i-1,0} + u_{i+1,0}\right)$ (6.133)

Taking $i = 1, 2, 3$ successively, we obtain

$$u_{1,1} = \frac{1}{2}\left(u_{0,0} + u_{2,0}\right) = 2$$

$$u_{2,1} = \frac{1}{2}\left(u_{1,0} + u_{3,0}\right) = 3$$

$$u_{3,1} = \frac{1}{2}\left(u_{2,0} + u_{4,0}\right) = 2$$

These are the entries of the 2nd row. Putting $j = 1$ in equation (6.131), $u_{i,2} = u_{i-1,1} + u_{i+1,1} - u_{i,0}$.

Taking $i = 1,2,3$ successively, we obtain

$$u_{1,2} = u_{0,1} + u_{2,1} - u_{1,0} = 0 + 3 - 3 = 0$$

$$u_{2,2} = u_{1,1} + u_{3,1} - u_{2,0} = 2 + 2 - 4 = 0$$

$$u_{3,2} = u_{2,1} + u_{4,1} - u_{3,0} = 3 + 0 - 3 = 0$$

These are the entries of the 3rd row and so on.

Now the equation of the vibrating string of length l is $u_{tt} = c^2 u_{xx}$. Its period of vibration is given by $= \frac{2l}{c} = \frac{2 \times 4}{2} = 4\,\text{sec}$.This shows that we have to compute $u_{(x,t)}$ up to $t = 0$.

i.e., similarly we obtain the values of $u_{i,2}$ (4th row) and $u_{i,3}$ (5th row). Hence the values of $u_{i,j}$ are as shown in the table below:

$i \rightarrow$	0	1	2	3	4
j↓					
0	0	3	4	3	0
1	0	2	3	2	0
2	0	0	0	0	0
3	0	-2	-3	-2	0
4	0	-3	-4	-3	0

Exercise

1. Solve the boundary value problem $u_{tt} = u_{xx}$ with the conditions $u(0,t) = u(1,t) = 0,\ u(x,0) = 0$ and $u(x,0) = \frac{1}{2}x(1-x)$ and $u_t\,(x,0) = 0$ taking $h = k = 0.1$ for $0 \le t \le 0.4$. Compare your solution with the exact solution at $x = 0.5$ and $t = 0.3$.

2. The transverse displacement u of a point at a distance x from one end and at any time t of a vibrating string satisfies the equation $\frac{\partial^2 u}{\partial t^2} = 25\frac{\partial^2 u}{\partial x^2}$, with the boundary conditions $u(0,t) = u(5,t) = 0$ and the initial conditions

$u(x,0) = \begin{cases} 20x & \text{for } 0 \le x \le 1 \\ 5(5-x) & \text{for } 1 \le x \le 5 \end{cases}$ and $u_t(x,0) = 0$. Solve this equation numerically for one half period of vibration, taking $h = k = 0.2$.

3. Solve $y_{tt} = y_{xx}$ up to $t = 0.5$ with a spacing of 0.1 subject to $y(0,t) = y(1,t) = 0$, $y_t(x,0) = 0$ and $y(x,0) = 10 + x(1-x)$.

4. The function u satisfies the equation $\frac{\partial^2 u}{\partial t^2} = \frac{\partial^2 u}{\partial x^2}$ and the conditions:

$$u(x,0) = \frac{1}{8}\sin \pi x, u_t(x,0) = 0 \text{ for } 0 \le x \le 1, u(0,t) = u(1,t) = 0 \text{ for } t \ge 0.$$

Use the explicit scheme to calculate u for $x = 0(0.1)1$ and $t = 0(0.1)0.5$.

Chapter 7

Laplace Equation

7.1 Introduction

Laplace's equation is a second-order partial differential equation usually beneficial in physics because its solutions (known as harmonic functions) occur in problems of electrical, magnetic, and gravitational potentials, of steady-state temperatures, and of hydrodynamics. The equation was discovered by the French mathematician and astronomer Pierre-Simon Laplace (1749–1827).

The two-dimensional heat equation is given by

$$\frac{\partial u}{\partial t} = C^2 \nabla^2 u = C^2 \left(\frac{\partial^2 u}{\partial x^2} + \frac{\partial^2 u}{\partial y^2} \right). \qquad (7.1)$$

Consider that the heat flow is time independent that is steady state condition, then $\frac{\partial u}{\partial t} = 0$, and Eq. (7.1) reduces to Laplace's equation

$$\nabla^2 u = \frac{\partial^2 u}{\partial x^2} + \frac{\partial^2 u}{\partial y^2} = 0. \qquad (7.2)$$

A heat problem then considered of the above Eq. (7.2) to be considered in some region R of the xy plane and a given boundary condition on the boundary curve C of R. This is called a boundary value problem and can be classified as

1. **Dirichlet Boundary value problem**: If u is defined on boundary curve C.
2. **Neumann Boundary value problem**: If the normal derivative $\frac{\partial u}{\partial n}$ is prescribed on boundary curve C.

3. **Mixed Boundary Value problem**: If u is defined on some boundary curve C and normal derivative $\frac{\partial u}{\partial n}$ is prescribed remaining on boundary curve C. The solution of Eq. (7.2) is called harmonic function.

Laplace's equation states that the sum of the second-order partial derivatives of *R*, the unknown function, with respect to the Cartesian coordinates, equals zero . Because it may be used to describe a wide range of physical processes, the Laplace equation is vital to both physics and engineering. The following are some important domains in which the Laplace equation is indispensable:

Heat Conduction:
The temperature distribution in a steady-state heat conduction scenario is governed by the Laplace equation. The temperature field T in such cases satisfies $\nabla^2 T = \frac{\partial^2 T}{\partial x^2} + \frac{\partial^2 T}{\partial y^2} = 0$.

Fluid Dynamics:
In certain situations, such as irrational and incompressible flow, the velocity potential of a fluid satisfies the Laplace equation $\nabla^2 u = \frac{\partial^2 u}{\partial x^2} + \frac{\partial^2 u}{\partial y^2} = 0.$

Potential Theory:
The Laplace equation is central to potential theory, where it describes the behavior of scalar and vector potentials in various physical systems. Solutions to the Laplace equation are known as harmonic functions.

Image Processing:
The Laplace operator is used in image processing for edge detection. It highlights regions of rapid intensity change in an image.

Quantum Mechanics:
In quantum mechanics, the time-independent Schrödinger equation for a particle with no potential energy (free particle) is a form of the Laplace equation.

Geophysics:

The Laplace equation is employed in geophysics to model the gravitational potential field of the Earth.

Control Theory:

In control theory and signal processing, Laplace transforms are used to analyze dynamic systems and their responses.

The Laplace equation's importance in explaining equilibrium and steady-state behaviour in physical systems is highlighted by its wide range of applications. Its solutions direct the design and study of numerous scientific and engineering challenges as well as offer insights into the underlying physics. This chapter deals with the analytical approach to solve the Laplace equation and Poission equation.

Corrollary: In the steady state, u is independent of t, so that $\frac{\partial u}{\partial t}=0$ and the above equation reduces to, $\frac{\partial^2 u}{\partial x^2}+\frac{\partial^2 u}{\partial y^2}$=0 and this is called Laplace's equation in two dimensions.

7. 2 General Solution of Laplace's Equation in Two Dimensions

Laplace's equation in two dimensions is given by $\frac{\partial^2 u}{\partial x^2}+\frac{\partial^2 u}{\partial y^2}=0$.

(7.3)

Let $u(x,y)=X(x)Y(y)$ or $u=XY$ (7.4)

be a solution of (7.3) where X is a function of *x* alone and Y is a function of *y* alone.

Differentiating twice Eq. (7.4) partially w.r.t x and y, we get

$$\frac{\partial^2 u}{\partial x^2}=X''Y,\ \frac{\partial^2 u}{\partial y^2}=XY''. \tag{7.5}$$

Substituting Eq. (7.5) in (7.3), we get

$$\therefore X''Y+XY''=0\Rightarrow\frac{X''}{X}=-\frac{Y''}{Y}. \tag{7.6}$$

Now in Eq. (7.6) variables are separable since x and y are independent variables, this equation can hold only when both sides reduce to a constant say k.

$$\therefore X''Y + XY'' = 0 \Rightarrow \frac{X''}{X} = -\frac{Y''}{Y} = k, \tag{7.7}$$

$$\therefore \frac{X''}{X} = k \Rightarrow X'' - kX = 0, \tag{7.8}$$

and $$-\frac{Y''}{Y} = k \Rightarrow Y'' + kY = 0. \tag{7.9}$$

Eqs. (7.8) and (7.9) is second order ordinary differential equation. Solving Eqs. (7.8) and (7.9), the following three cases arises:

Case I: when k is positive say $k = p^2$ then Eqs. (7.8) and (7.9) reduces to

$$\therefore X'' - p^2 X = 0 \tag{7.10}$$

and $$Y'' + p^2 Y = 0 \tag{7.11}$$

Auxiliary equation of the Eqs. (7.10) is $m^2 - p^2 = 0 \Rightarrow m = \pm p$ (7.12)

The roots of the auxiliary equation are real and distinct, therefore the general solution Eq. (7.10) is given by $X(x) = C_1 e^{px} + C_2 e^{-px}$.

Auxiliary equation of the Eqs. (7.11) is $m^2 + p^2 = 0 \Rightarrow m = \pm ip$ (7.13)

The roots of the auxiliary equation are complex, therefore the general solution Eq. (7.11) is given by $Y(y) = C_3 \cos py + C_4 \sin py$. (7.14)

Therefore, the general solution of Eq. (7.3) is given by $u(x,y) = X(x)Y(y)$,

i.e $u(x,y) = \left(C_1 e^{px} + C_2 e^{-px}\right)\left(C_3 \cos py + C_4 \sin py\right)$ where C_i are arbitrary constant for $i = 1,2,3,4$.

Case II: When k is negative say $-p^2$, then Eqs. (7.8) and (7.9) reduces to

$X'' + p^2 X = 0,$ (7.15)

and $Y'' - p^2 Y = 0.$ (7.16)

Auxiliary equation of the Eq. (7.15) is $m^2 + p^2 = 0 \Rightarrow m = \pm ip.$ (7.17)

Since the roots of the auxiliary equation are complex, therefore the general solution Eq. (7.15) is given by

$$X = c_5 \cos px + c_6 \sin px.$$

Auxiliary equation of the Eq. (7.16), we get $m^2 - p^2 = 0 \Rightarrow m = \pm p.$ (7.18)

Since the roots are real and distinct, the general solution of the auxiliary equation is given by

$$Y = c_7 e^{py} + c_8 e^{-py}.$$

Therefore, the general solution of Eq. (7.3) is given by $u(x,y) = X(x)Y(y)$

i.e., $u(x,y) = (c_5 \cos px + c_6 \sin px)(c_7 e^{py} + c_8 e^{-py})$ where c_i are arbitrary constants for $i = 1,2,3,4.$

Case III: When $k = 0$, then then Eqs. (7.8) and (7.9) reduce to $X'' = 0$ (7.19)

and $Y'' = 0.$ (7.20)

Auxiliary equation of the Eq. (7.19) is $m_1^2 = 0 \Rightarrow m = 0,0.$ Since, the roots of the auxiliary equation are real and repeated, therefore the general solution Eq. (7.19) is given by

$$X = c_9 x + c_{10}.$$

Auxiliary equation of the Eq. (7.20) is $m_2^2 = 0 \Rightarrow m_2 = 0, 0$. Since, the roots of the auxiliary equation has real and repeated root, therefore the general solution Eq. (7.20) is given by

$$Y = c_9 y + c_{10}.$$

Therefore, the general solution of Eq. (7.3) is given by $u(x, y) = X(x)Y(y)$

i.e., $u(x, y) = (c_9 x + c_{10})(c_{11} y + c_{12})$ where c_i are arbitrary constants for $i = 1, 2, 3, 4$.

Thus, the various possible solution of Eq. (7.3) are

$$u(x, y) = (C_1 e^{px} + C_2 e^{-px})(C_3 \cos py + C_4 \sin py) \quad \text{for } k > 0.$$
$$u(x, y) = (c_5 \cos px + c_6 \sin px)(c_7 e^{py} + c_8 e^{-py}) \quad \text{for } k = 0.$$
$$u(x, y) = (c_9 x + c_{10})(c_{11} y + c_{12}) \quad \text{for } k < 0.$$

Of these we take that solution, which is consistent with the given boundary conditions, i.e., physical nature of the problem.

7.2.1 Dirichlet Boundary Value Problem in Rectangle

The Dirichlet boundary value problem (DBVP) in a rectangle is given as follows:

Laplace Equation: $\frac{\partial^2 u}{\partial x^2} + \frac{\partial^2 u}{\partial y^2} = 0$, $0 \le x \le a$, $0 \le x \le b$. (7.21)

Boundary conditions: $u(0, y) = 0$, $0 \le y \le b$, (7.21a)

$$u(a, y) = 0, \; 0 \le y \le b, \tag{7.21b}$$

$$u(x, b) = 0, \; 0 \le x \le a, \tag{7.21c}$$

$$u(x, 0) = f(x), 0 \le x \le a. \tag{7.21d}$$

Now we evaluate the solution of the Eq. (7.21) with boundary conditions Eq. (7.21a)-(7.21d).

Assume that $u(x,y)=X(x)Y(y)$ or $u=XY$ (7.22)

be a solution of (7.21) where X is a function of *x* alone and Y is a function of *y* alone.

Differentiating twice Eq. (7.22) partially w.r.t *x* and *y*, we get

$$\frac{\partial^2 u}{\partial x^2}=X''Y,\ \frac{\partial^2 u}{\partial y^2}=XY''. \tag{7.23}$$

Substituting Eqs. (7.22) and (7.23) in (7.21), we get

$$\therefore X''Y+XY''=0\Rightarrow\frac{X''}{X}=-\frac{Y''}{Y}. \tag{7.24}$$

Now in Eq. (7.24), variables are separable since x and y are independent variables, this equation can hold only when both sides reduce to a constant say k.

$$\therefore X''Y+XY''=0\Rightarrow\frac{X''}{X}=-\frac{Y''}{Y}=k, \tag{7.25}$$

$$\therefore \frac{X''}{X}=k\Rightarrow X''-kX=0, \tag{7.26}$$

and $-\frac{Y''}{Y}=k\Rightarrow Y''+kY=0.$ (7.27)

Eqs. (7.26) and (7.27) is second order ordinary differential equation. Now let us look at the boundary condition

BC1: $u(0,y)=0,$ **BC2**: $u(a,y)=0.$

The solution of Eq. (7.21) is given by Eq. (7.22) i.e., $u(x,t)=X(x)T(t)$. Using Eq. (7.22) and **BC1**, we get

$$u(0,y)=X(0)Y(y) \Rightarrow X(0)Y(y)=0 \Rightarrow X(0)=0. \tag{7.28}$$

($\because Y(y) \neq 0$ as we are look for non –zero solution)

Similarly, using Eq. (7.22) and **BC2**, we get

$$u(a,t)=X(a)T(t) \Rightarrow X(a)T(t)=0 \Rightarrow X(a)=0. \tag{7.29}$$

($\because Y(y) \neq 0$ as we are look for non –zero solution)

Now we solve Eq. $X'' - kX = 0$, (7.30)

with boundary conditions $X(0)=0$ and $X(a)=0$. (7.31)

which is called an Eigen value problem and *k* is called Eigen value.

Case (i) when k is positive. Let $k = p^2$,for some $p \in R - \{0\}$.

The Eq. (7.30) reduces to $X'' - p^2 X = 0$. The auxiliary equation associate with $X'' - p^2 X = 0$ is $m^2 - p^2 = 0$

$\Rightarrow m = p, -p$

Since the roots of the auxiliary equation are real and distinct, therefore

$$X(x) = Ae^{px} + Be^{-px}, \tag{7.32}$$

where *A* and *B* are arbitrary constant. Now we apply the boundary condition given by Eq. (7.31) to find the value of *A* and *B*. Put $x = 0$ in Eq. (7.32), we get $X(0) = Ae^0 + Be^0 \Rightarrow X(0) = A + B$. (7.33)

Using Eq. (7.31) in Eq. (7.33), we get $A = -B$. (7.34)

Put $x = a$ in Eq. (7.32), we get

$$X(a) = Ae^{pa} + Be^{-pa} \Rightarrow X(a) = Ae^{pa} + Be^{-pa}. \tag{7.35}$$

Using Eqs. (7.34) and (7.31) in Eq. (7.32), we get

$$X(a) = Ae^{-pa}\left(e^{2pa} - 1\right) = 0. \tag{7.36}$$

$$\Rightarrow A = 0 \ \left(\because e^{pa}\left(e^{2pa} - 1\right) \neq 0\right).$$

Since $A = 0$, therefore from Eq. (7.34) , we get $B = 0$. So, Eq. (7.32) reduces to $X(x) = 0$.

Substituting $X(x) = 0$ in Eq. (7.22), the required solution is $u(x,t) = 0$. So, when k is positive, there is no Eigen value and we get a trivial solution.

Case (ii) when k is zero.

Then auxiliary equation of Eqs. (7.30) is $m^2 = 0 \Rightarrow m = 0,0$. Since, the roots are real and repeated, therefore solution of Eq. (7.30) is

$$X(x) = (A + Bx) \tag{7.37}$$

Now, we use boundary condition $X(0) = 0$ and $X(l) = 0$ to find the value of arbitrary constant.

Put $x = 0$ in Eq. (7.37), we get $X(0) = (A + B0) \Rightarrow A = 0$,(From Eq. (7.31)) (7.38)

Put $x = a$ in Eq. (7.37), we get $X(a) = A + Ba \Rightarrow Ba = 0$, (From Eq. (7.31))

$\Rightarrow B = 0 .(\because l \neq 0)$ (7.39)

Since $A = 0$ and $B = 0$, therefore $X(x) = 0$.Substituting $X(x) = 0$ in Eq. (7.37) , the required solution is $u(x,t) = 0$. So, when k is zero, there is no Eigen value and we get a trivial solution.

Case (iii) when k is negative. Let $k = -p^2$, where $p \in R - \{0\}$.

Then auxiliary equation of Eqs. (7.30) is $m^2 + p^2 = 0 \Rightarrow m = \pm ip$. Since, the roots are complex, therefore solution of Eq. (7.30) is

$$X(x) = (A\cos px + B\sin px) \tag{7.40}$$

where *A* and *B* are arbitrary constant. Now, we use boundary condition $X(0) = 0$ and $X(l) = 0$ to find the value of arbitrary constant. Put $x = 0$ in Eq. (7.40), we get

$$X(0) = (A\cos p0 + B\sin p0) \Rightarrow A = 0. \tag{7.41}$$

Put $x = a$ in Eq. (7.40), we get

$$X(a) = (A\cos pa + B\sin pa) \Rightarrow B\sin pa = 0,$$

(From Eq. (7.31))

$\Rightarrow B = 0$ or $\sin pa = 0$,

$\Rightarrow \sin pa = 0, (\because B \neq 0$ as $X(x) = 0)$

$$\Rightarrow pa = \sin^{-1}(0)$$

$$\Rightarrow pa = n\pi \quad ,$$

$$\Rightarrow p = \frac{n\pi}{a},\ n = 1, 2, 3, \ldots.. \tag{7.42}$$

For $n = 1, 2, 3, ...$, we get different values of p, so we denote p by p_n. Therefore $p = p_n = \frac{n\pi}{a}$. Since $p_n = \frac{n\pi}{a}$, therefore from Eq. (7.40)

$$X(x) = B\sin\left(\frac{n\pi}{a}\right)x. \quad (7.43)$$

For $n = 1, 2, 3, ...$, we get different value of $X(x)$ using Eq. (7.43), so we denote $X(x)$ by $X_n(x)$. Therefore $X_n(x) = B_n \sin\left(\frac{n\pi}{a}\right)x$.

(7.44)

Similarly, when k is negative i.e., $k = -p^2$ then Eq. (7.27) changes to $Y'' - \frac{n^2\pi^2}{a^2}Y = 0$.

The auxiliary equation associate with $Y'' - \frac{n^2\pi^2}{a^2}Y = 0$ is

$$m^2 - \frac{n^2\pi^2}{a^2} = 0$$

$$\Rightarrow m = \frac{n\pi}{a}, -\frac{n\pi}{a}.$$

Since the roots of the auxiliary equation are real and distinct, therefore

$$Y(y) = Ce^{\frac{n\pi}{a}y} + De^{-\frac{n\pi}{a}y}, \quad (7.45)$$

where *A* and *B* are arbitrary constant. For $n = 1, 2, 3, ...$, we get different value of $Y(y)$ using Eq. (7.45), so we denote $Y(y)$ by $Y_n(y)$. Therefore $Y_n(y) = C_n e^{\frac{n\pi}{a}y} + D_n e^{-\frac{n\pi}{a}y}$. (7.46)

Using boundary condition given by Eq. (7.21c) in Eq. (7.22), we get

$$u(x,b) = X(b)Y(b) = 0 \Rightarrow X(x) = 0 \text{ or } Y(b) = 0$$
$$\Rightarrow Y(b) = 0. \tag{7.47}$$

($\because$ $X(x) \neq 0$, as $X(x) = 0$ gives $u(x,y) = 0$, which is a trivial solution and we are looking for non-trivial solution)

From Eq. (7.47), we have $Y_n(b) = 0.$ (7.48)

Substituting $y = b$ in Eq. (7.46), we get $Y_n(b) = C_n e^{\frac{n\pi b}{a}} + D_n e^{-\frac{n\pi b}{a}}.$ (7.49)

From Eqs. (7.48) and (7.49), we get $0 = C_n e^{\frac{n\pi b}{a}} + D_n e^{-\frac{n\pi b}{a}}$,

$$\Rightarrow 0 = e^{\frac{n\pi b}{a}} \left(C_n e^{\frac{2n\pi b}{a}} + D_n \right),$$

$$\Rightarrow 0 = \left(C_n e^{\frac{2n\pi b}{a}} + D_n \right) \quad (e^{\frac{2n\pi b}{a}} \neq 0)$$

$$\Rightarrow D_n = -C_n e^{\frac{2n\pi b}{a}}$$

Substituting the value of $D_n = -C_n e^{\frac{2n\pi b}{a}}$ in Eq. (7.4y), we get

$$Y_n(y) = C_n e^{\frac{n\pi}{a}y} - C_n e^{\frac{2n\pi b}{a}} e^{-\frac{n\pi}{a}y},$$

$$\Rightarrow Y_n(b) = C_n \left(e^{\frac{n\pi}{a}y} - e^{\frac{2n\pi b}{a} - \frac{n\pi}{a}y} \right),$$

$$\Rightarrow Y_n(b) = C_n \left(e^{\frac{n\pi}{a}y} - e^{\frac{2n\pi b}{a} - \frac{n\pi}{a}y} \right),$$

$$\Rightarrow Y_n(b) = C_n \left(e^{\frac{n\pi}{a}y} - e^{\frac{n\pi(2b-y)}{a}} \right).$$

Thus, by the principle of superposition, we have

$$u(x,y)=\sum_{n=1}^{\infty} X_n(x)Y_n(x),$$

$$u(x,y)=\sum_{n=1}^{\infty} B_n \sin\left(\frac{n\pi}{a}\right)x\,C_n\left(e^{\frac{n\pi}{a}y}-e^{\frac{n\pi(2b-y)}{a}}\right),$$

$$u(x,y)=\sum_{n=1}^{\infty} A_n \sin\left(\frac{n\pi}{a}\right)x\left(e^{\frac{n\pi}{a}y}-e^{\frac{n\pi(2b-y)}{a}}\right), \tag{7.50}$$

where $A_n = B_nC_n$. Eq. (7.50) is the solution satisfying the boundary conditions Eq. (7.21a) -(7.21c). Now to find A_n, we use the boundary condition (7.21 d) i.e. $u(x,0)=f(x)$. Substituting $y=0$ in Eq. (7.50), we get $u(x,0)=\sum_{n=1}^{\infty} A_n \sin\left(\frac{n\pi}{a}\right)x\left(1-e^{\frac{2bn\pi}{a}}\right)$.

From (7.21d), the above equation changes to $f(x)=\sum_{n=1}^{\infty} A_n \sin\left(\frac{n\pi}{a}\right)x\left(1-e^{\frac{2bn\pi}{a}}\right)$,

which represent the half range Fourier sine series of $f(x)$ over the interval $(0\ a)$.Hence , we get

$$A_n\left(1-e^{\frac{2bn\pi}{a}}\right)=\frac{2}{a}\int_0^a f(x)\sin\left(\frac{n\pi x}{a}\right)dx,$$

$$A_n=\frac{2}{a\left(1-e^{\frac{2bn\pi}{a}}\right)}\int_0^a f(x)\sin\left(\frac{n\pi x}{a}\right)dx.$$

Therefore, the required solution of the Laplace equation given by Eq. (7.21) with the boundary conditions (7.21a) -(7.21d) is

$$u(x,y)=\sum_{n=1}^{\infty} A_n \sin\left(\frac{n\pi}{a}\right)x\left(e^{\frac{n\pi}{a}y}-e^{\frac{n\pi(2b-y)}{a}}\right),$$

(7.51)

where $$A_n=\frac{2}{a\left(1-e^{\frac{2bn\pi}{a}}\right)}\int_0^a f(x)\sin\left(\frac{n\pi x}{a}\right)dx. \quad (7.52)$$

Example: Solve the Laplace equation $\frac{\partial^2 u}{\partial x^2}+\frac{\partial^2 u}{\partial y^2}=0$, $0<x,y<\pi$, subjected to the conditions

$$u(0,y)=u(\pi,y)=u(x,\pi)=0 \text{ and } u(x,0)=\sin^3 x.$$

Solution: Comparing the given PDE $\frac{\partial^2 u}{\partial x^2}+\frac{\partial^2 u}{\partial y^2}=0$ and boundary conditions $u(0,y)=u(\pi,y)=u(x,\pi)=0$ and $u(x,0)=\sin^3 x$ with Eq.(5.21), Eq.(5.21a-5.21d), we get $f(x)=\sin^3 x, a=\pi$ and $b=\pi$. The solution of Laplace equation is given by Eq. (7.51) – (7.52) as $$u(x,y)=\sum_{n=1}^{\infty} A_n \sin(nx)\left(e^{ny}-e^{\frac{n\pi(2\pi-y)}{\pi}}\right). \quad (7.53)$$

Substituting $y=0$ in Eq. (7.53), we get

$$u(x,0)=\sum_{n=1}^{\infty} A_n \sin(nx)\left(1-e^{2n\pi}\right). \quad (7.54)$$

Using the boundary condition $u(x,0)=\sin^3 x$ in Eq. (7.54), we get

$$\sin^3 x=\sum_{n=1}^{\infty} A_n \sin(nx)\left(1-e^{2n\pi}\right),$$

$$\frac{3\sin x}{4}-\frac{\sin 3x}{4}=A_1\sin x\left(1-e^{2\pi}\right)+A_2\sin 2x\left(1-e^{4\pi}\right)+A_3\sin 3x\left(1-e^{6\pi}\right)+A_4\sin 4x\left(1-e^{8\pi}\right)+\ldots$$

(7.55)

Comparing on both the side, we get

$A_1 = \dfrac{3}{4\left(1-e^{2\pi}\right)}$, $A_3 = -\dfrac{1}{4\left(1-e^{6\pi}\right)}$, $A_2 = 0$ and $A_n = 0$, for $n \geq 4$.

Substituting the above values in Eq. (7.53), we get

$$u(x,y) = \frac{3\sin x}{4\left(1-e^{2\pi}\right)} e^{y}\left(1-e^{2\pi}\right) - \frac{1}{4\left(1-e^{2\pi}\right)} \sin 3x\, e^{3y}\left(1-e^{6\pi}\right),$$

which is the required solution.

Example: A rectangular plate is bounded by the line $x=0, x=a, y=0$ and $y=b$ and the temperatures are $u(0,y)=0,\ u(a,y)=0, u(x,b)=0, u(x,0)=5\sin\dfrac{4\pi x}{a}+3\sin\dfrac{3\pi x}{a}$

Find the temperature.

Solution: Comparing the given PDE $\dfrac{\partial^2 u}{\partial x^2}+\dfrac{\partial^2 u}{\partial y^2}=0$ and boundary conditions $u(0,y)=u(a,y)=u(x,b)=0$ and $u(x,a)=5\sin\dfrac{4\pi x}{a}+3\sin\dfrac{3\pi x}{a}$ with Eq.(5.21), Eq.(5.21a-5.21d), we get $f(x)=5\sin\dfrac{4\pi x}{a}+3\sin\dfrac{3\pi x}{a}$, $a=a$ and $b=b$. The solution of Laplace equation is given by Eq. (7.51) – (7.52) as

$$u(x,y)=\sum_{n=1}^{\infty} A_n \sin\left(\frac{n\pi}{a}\right)x\left(e^{\frac{n\pi}{a}y}-e^{\frac{n\pi(2b-y)}{a}}\right). \tag{7.56}$$

Substituting $y=0$ in Eq. (7.56), we get

$$u(x,0)=\sum_{n=1}^{\infty} A_n \sin\left(\frac{n\pi}{a}\right)x\left(1-e^{\frac{2bn\pi}{a}}\right). \tag{7.57}$$

Using the boundary condition $u(x,0)=5\sin\dfrac{4\pi x}{a}+3\sin\dfrac{3\pi x}{a}$ in Eq. (7.57), we get

$$5\sin\frac{4\pi x}{a}+3\sin\frac{3\pi x}{a}=\sum_{n=1}^{\infty} A_n \sin\left(\frac{n\pi}{a}\right)x\left(1-e^{\frac{2bn\pi}{a}}\right)$$

$$5\sin\frac{4\pi x}{a}+3\sin\frac{3\pi x}{a}=A_1\sin\left(\frac{\pi}{a}\right)x\left(1-e^{\frac{2b\pi}{a}}\right)+A_2\sin\left(\frac{2\pi}{a}\right)x\left(1-e^{\frac{4b\pi}{a}}\right)+A_3\sin\left(\frac{3\pi}{a}\right)x\left(1-e^{\frac{6b\pi}{a}}\right)$$
$$+A_4\sin\left(\frac{4\pi}{a}\right)x\left(1-e^{\frac{8b\pi}{a}}\right)+\ldots$$

Comparing on both the side, we get

$A_1=0,\ A_2=0\ A_3=\dfrac{3}{\left(1-e^{\frac{6b\pi}{a}}\right)},\qquad A_4=\dfrac{5}{\left(1-e^{\frac{8b\pi}{a}}\right)}$ and $\quad A_n=0,$ for $n\geq 5$.

Substituting the above values in Eq. (7.56), we get

$$u(x,y)=\frac{3}{\left(1-e^{\frac{6\pi b}{a}}\right)}\sin\left(\frac{3\pi x}{a}\right)e^{-\frac{3\pi y}{a}}\left(e^{\frac{6\pi y}{a}}-e^{\frac{6\pi b}{a}}\right)+\frac{5}{1-e^{\frac{8\pi b}{a}}}\sin\left(\frac{4\pi x}{a}\right)e^{-\frac{4\pi y}{a}}\left(e^{\frac{8\pi y}{a}}-e^{\frac{8\pi b}{a}}\right)$$

,
which is the required solution.

Example: Find the steady state temperature distribution at points in a rectangular plate with insulated faces, the edges of the plate begin the lines $x=0, x=a, y=0$ and $y=b$ when three of the edges are kept at temperature zero and fourth at a fixed temperature $\alpha^{\circ}C$.

Solution: The given problem is : PDE $\dfrac{\partial^2 u}{\partial x^2}+\dfrac{\partial^2 u}{\partial y^2}=0$, subjected to boundary conditions $u(0,y)=u(a,y)=u(x,b)=0$ and $u(x,0)=0$. Comparing the given PDE and boundary conditions with Eq.(5.21), Eq.(5.21a-5.21d), wc get $f(x)=\alpha$, $a=a$ and $b=b$. The solution of Laplace equation is given by Eq. (7.51) – (7.52) as

$$u(x,y)=\sum_{n=1}^{\infty}A_n\sin\left(\frac{n\pi}{a}\right)x\left(e^{\frac{n\pi}{a}y}-e^{\frac{n\pi(2b-y)}{a}}\right), \tag{7.58}$$

where $A_n = \dfrac{2}{a\left(1-e^{\frac{2bn\pi}{a}}\right)}\displaystyle\int_0^a f(x)\sin\left(\frac{n\pi x}{a}\right)dx.$ (7.59)

Now $A_n = \dfrac{2}{a\left(1-e^{\frac{2bn\pi}{a}}\right)}\displaystyle\int_0^a f(x)\sin\left(\frac{n\pi x}{a}\right)dx,$

$$A_n = \frac{2}{a\left(1-e^{\frac{2bn\pi}{a}}\right)}\int_0^a \alpha\sin\left(\frac{n\pi x}{a}\right)dx,$$

$$A_n = \frac{2\alpha}{a\left(1-e^{\frac{2bn\pi}{a}}\right)}\left[-\frac{\cos\left(\frac{n\pi x}{a}\right)}{\frac{n\pi}{a}}\right]_0^a,$$

$$A_n = \frac{2\alpha a}{n\pi a\left(1-e^{\frac{2bn\pi}{a}}\right)}\left[-\cos(n\pi)+\cos(0)\right],$$

$$A_n = \frac{2\alpha a}{n\pi a\left(1-e^{\frac{2bn\pi}{a}}\right)}\left[-(-1)^n+1\right],$$

$$A_n = \frac{2\alpha}{n\pi\left(1-e^{\frac{2bn\pi}{a}}\right)}\left[-(-1)^n+1\right]. \tag{7.60}$$

Substituting values from Eq. (7.60) in Eq. (7.58), we get

$$u(x,y) = \sum_{n=1}^{\infty}\frac{2\alpha}{n\pi\left(1-e^{\frac{2bn\pi}{a}}\right)}\left[-(-1)^n+1\right]\sin\left(\frac{n\pi}{a}\right)x\left(e^{\frac{n\pi}{a}y}-e^{\frac{n\pi(2b-y)}{a}}\right),$$

which is the required solution.

7.2.2 The Neumann Problem in Rectangle

In a rectangle, the Neumann problem is defined as follows:

Laplace Equation: $\frac{\partial^2 u}{\partial x^2}+\frac{\partial^2 u}{\partial y^2}=0,\ 0\le x\le a,\ 0\le x\le b.$ (7.61)

Boundary conditions: $u_x(0,y)=0,\ 0\le y\le b,$ (7.61a)

$$u_x(a,y)=0,\ 0\le y\le b, \tag{7.61b}$$

$$u_y(x,0)=0,\ 0\le x\le a, \tag{7.61c}$$

$$u_y(x,b)=f(x), 0\le x\le a. \tag{7.61d}$$

Now we evaluate the solution of the Eq. (7.61) with boundary conditions Eq. (7.61a) - (7.61d).

Assume that $u(x,y)=X(x)Y(y)$ or $u=XY$ (7.62)

be a solution of (7.61) where X is a function of *x* alone and Y is a function of *y* alone.

Differentiating twice Eq. (7.62) partially w.r.t *x* and *y*, we get

$$\frac{\partial^2 u}{\partial x^2}=X''Y,\ \frac{\partial^2 u}{\partial y^2}=XY''. \tag{7.63}$$

Substituting Eqs. (7.62) and (7.63) in (7.61), we get

$$\therefore X''Y+XY''=0 \Rightarrow \frac{X''}{X}=-\frac{Y''}{Y}. \tag{7.64}$$

Now in Eq. (7.64) variables are separable since x and y are independent variables, this equation can hold only when both sides reduce to a constant say k.

$$\therefore X''Y + XY'' = 0 \Rightarrow \frac{X''}{X} = -\frac{Y''}{Y} = k, \tag{7.65}$$

$$\therefore \frac{X''}{X} = k \Rightarrow X'' - kX = 0, \tag{7.66}$$

and $-\frac{Y''}{Y} = k \Rightarrow Y'' + kY = 0.$ (7.67)

Eqs. (7.66) and (7.67) is second order ordinary differential equation. Now let us look at the boundary condition

BC1: $u_x(0, y) = 0,$ **BC2**: $u_x(a, y) = 0.$

The solution of Eq. (7.61) is given by Eq. (7.62) i.e., $u(x,t) = X(x)T(t)$. Using Eq. (7.62) and **BC1**, we get $u_x(0, y) = X'(0)Y(y) \Rightarrow X'(0)Y(y) = 0 \Rightarrow X'(0) = 0.$

(7.68)

($\because Y(y) \neq 0$ as we are look for non –zero solution)

Similarly, using Eq. (7.22) and **BC2**, we get

$$u_x(a,t) = X'(a)T(t) \Rightarrow X'(a)T(t) = 0 \Rightarrow X'(a) = 0.$$

(7.69)

($\because Y(y) \neq 0$ as we are look for non –zero solution)

Now we solve Eq. $X'' - kX = 0,$ (7.70)

with boundary conditions $X'(0) = 0$ and $X'(a) = 0.$ (7.71)

Case (i) when k is positive. Let $k = p^2$,for some $p \in R - \{0\}$.

The Eq. (7.70) reduces to $X'' - p^2X = 0$. The auxiliary equation associate with $X'' - p^2X = 0$ is $m^2 - p^2 = 0 \Rightarrow m = p, -p$.

Since the roots of the auxiliary equation are real and distinct, therefore

$$X(x) = Ae^{px} + Be^{-px}, \tag{7.72}$$

$$X'(x) = Ape^{px} - pBe^{-px}. \tag{7.73}$$

where *A* and *B* are arbitrary constant. Now we apply the boundary condition given by Eq. (7.71) to find the value of *A* and *B*. Put $x = 0$ in Eq. (7.73), we get $X'(0) = Ap - pB$. (7.74)

Using Eq. (7.71) in Eq. (7.74), we get $A = B$. (7.75)

Put $x = a$ in Eq. (7.73), we get

$$X'(a) = pAe^{pa} - poBe^{-pa} \Rightarrow X(a) = pAe^{pa} - pBe^{-pa}. \tag{7.76}$$

Using Eqs. (7.71) and (7.74) in Eq. (7.76), we get
$X(a) = Ae^{-pa}\left(e^{2pa} - 1\right) = 0$.

$$\Rightarrow A = 0 \quad \left(\because e^{pa}\left(e^{2pa} - 1\right) \neq 0\right). \tag{7.77}$$

Since $A = 0$, therefore from Eq. (7.74), we get $B = 0$. So, Eq. (7.72) reduces to $X(x) = 0$.

Substituting $X(x)=0$ in Eq. (7.62), the required solution is $u(x,t)=0$. So, when k is positive, there is no Eigen value, and we get a trivial solution.

Case (ii) when k is zero.

Then auxiliary equation of Eqs. (7.70) is $m^2=0 \Rightarrow m=0,0$. Since, the roots are real and repeated, therefore solution of Eq. (7.70) is

$$X(x)=(A+Bx) \tag{7.78}$$

Differentiating Eq. (7.77), we get $X'(x)=B$. (7.79)

Now, we use boundary condition $X'(0)=0$ and $X'(l)=0$ to find the value of arbitrary constant.

Put $x=0$ and $x=a$ in Eq. (7.78), we get $X'(0)=B=0$,

(7.80)

Since $B=0$, therefore Eq. (7.77) gives $X(x)=A$.

Case (iii) when k is negative. Let $k=-p^2$, where $p\in R-\{0\}$.

Then auxiliary equation of Eqs. (7.70) is $m^2+p^2=0 \Rightarrow m=\pm ip$. Since, the roots are complex, therefore solution of Eq. (7.70) is

$$X(x)=(A\cos px+B\sin px) \tag{7.81}$$

where *A* and *B* are arbitrary constant. Differentiating Eq. (7.81) w.r.t x, we get

$$X'(x)=(-Ap\sin px+Bp\cos px). \tag{7.82}$$

Now, we use boundary condition $X'(0)=0$ and $X'(a)=0$ to find the value of arbitrary constant.

Put $x=0$ in Eq. (7.82), we get $X'(0)=(-Ap\sin 0+Bp\cos 0)$.

$$\Rightarrow Bp=0 \Rightarrow B=0 \ (\because X'(0)=0) \tag{7.83}$$

Put $x=a$ in Eq. (7.82), we get $X'(a)=(-Ap\sin pa+Bp\cos pa)$,

$$\Rightarrow X'(a)=-Ap\sin pa \Rightarrow -Ap\sin pa=0,$$

$$\Rightarrow A=0 \text{ or } \sin pa=0,$$

$$\Rightarrow \sin pa=0, (\because B\neq 0 \text{ as } X(x)=0)$$

$$\Rightarrow pa=\sin^{-1}(0)$$

$$\Rightarrow pa=n\pi,$$

$$\Rightarrow p=\frac{n\pi}{a},\ n=1,2,3,..... \tag{7.84}$$

For $n=1,2,3,.....$, we get different value of p, so we denote p by p_n. Therefore $p=p_n=\frac{n\pi}{a}$. Since $p_n=\frac{n\pi}{a}$, therefore from Eq. (7.81) $X(x)=A\cos\left(\frac{n\pi}{a}\right)x$.

For $n=1,2,3,.....$, we get different value of $X(x)$ using Eq. (7.43), so we denote $X(x)$ by $X_n(x)$. Therefore $X_n(x)=A_n\cos\left(\frac{n\pi}{a}\right)x$.

(7.85)

Similarly, when k is negative i.e., $k=-p^2$ then Eq. (7.67) changes to $Y''-\frac{n^2\pi^2}{a^2}Y=0$.

The auxiliary equation associate with $Y''-\frac{n^2\pi^2}{a^2}Y=0$ is

$$m^2-\frac{n^2\pi^2}{a^2}=0$$

$$\Rightarrow m = \frac{n\pi}{a}, -\frac{n\pi}{a}.$$

Since the roots of the auxiliary equation are real and distinct, therefore

$$Y(y) = Ce^{\frac{n\pi}{a}y} + De^{-\frac{n\pi}{a}y}, \tag{7.86}$$

where *A* and *B* are arbitrary constant. Using boundary condition given by Eq. (7.61c) in Eq. (7.62), we get

$$u_y(x,0) = X(x)Y'(0) = 0 \Rightarrow X(x) = 0 \text{ or } Y'(0) = 0$$

$$\Rightarrow Y'(0) = 0. \tag{7.87}$$

($\because$ $X(x) \neq 0$, as $X(x) = 0$ gives $u(x,y) = 0$, which is a trivial solution and we are looking for non-trivial solution).

Differentiating Eq. (7.86) w.r.t y, we get

$$Y'(y) = \frac{n\pi}{a}Ce^{\frac{n\pi}{a}y} - \frac{n\pi}{a}De^{-\frac{n\pi}{a}y}. \tag{7.88}$$

Substituting $y = 0$ in Eq. (7.88), we get

$$Y'(y) = \frac{n\pi}{a}C - \frac{n\pi}{a}D \Rightarrow C = D. \text{ (From (7.88)))} \tag{7.89}$$

Substituting the value of $C = D$ in Eq. (7.86), we get

$$Y(y) = Ce^{\frac{n\pi}{a}y} + Ce^{-\frac{n\pi}{a}y},$$

$$\Rightarrow Y(y) = C\left(e^{\frac{n\pi}{a}y} + e^{-\frac{n\pi}{a}y}\right). \tag{7.90}$$

For $n = 1, 2, 3, \ldots\ldots$, we get different value of $Y(y)$ using Eq. (7.90), so we denote $Y(y)$ by $Y_n(y)$. Therefore

$$Y_n(y) = C_n\left(e^{\frac{n\pi}{a}y} + e^{-\frac{n\pi}{a}y}\right).$$

(7.91)

Therefore, from Eqs. (7.85) and (7.91) and using the principle of superposition, we get

$$u(x,y) = \sum_{n=1}^{\infty} X_n(x) Y_n(x),$$

$$u(x,y) = \sum_{n=1}^{\infty} A_n \cos\left(\frac{n\pi}{a}\right) x\, C_n\left(e^{\frac{n\pi}{a}y} + e^{-\frac{n\pi}{a}y}\right),$$

$$u(x,y) = \sum_{n=1}^{\infty} B_n \cos\left(\frac{n\pi}{a}\right) x\left(e^{\frac{n\pi}{a}y} + e^{-\frac{n\pi}{a}y}\right).$$

(7.92)

where $B_n = A_n C_n$.

Eq. (7.92) is the solution satisfying the boundary conditions Eq. (7.61a) -(7.61c). Now to find B_n, we use the boundary condition (7.61 d) i.e. $u_x(x,b) = f(x)$. Substituting $y = b$ in Eq. (7.92), we get $u(x,b) = \sum_{n=1}^{\infty} B_n \cos\left(\frac{n\pi}{a}\right) x\left(e^{\frac{n\pi b}{a}} + e^{-\frac{n\pi b}{a}}\right)$.

From (7.61d), the above equation changes to $f(x) = \sum_{n=1}^{\infty} B_n \cos\left(\frac{n\pi}{a}\right) x\left(e^{\frac{n\pi b}{a}} + e^{-\frac{n\pi b}{a}}\right)$, which represent the half range Fourier cosine series of $f(x)$ over the interval $(0\ b)$.Hence , we get $B_n\left(e^{\frac{n\pi b}{a}}+e^{-\frac{n\pi b}{a}}\right)=\frac{2}{b}\int_0^a f(x)\cos\left(\frac{n\pi x}{a}\right)dx$,

$$B_n=\frac{2}{b\left(e^{\frac{n\pi b}{a}}+e^{-\frac{n\pi b}{a}}\right)}\int_0^a f(x)\cos\left(\frac{n\pi x}{a}\right)dx.$$

Therefore, the required solution of the Laplace equation given by Eq. (7.61) with the boundary conditions (7.61a) -(7.61d) is

$$u(x,y)=\sum_{n=1}^{\infty}B_n\cos\left(\frac{n\pi}{a}\right)x\left(e^{\frac{n\pi}{a}y}+e^{-\frac{n\pi}{a}y}\right), \tag{7.93}$$

where $B_n=\frac{2}{b\left(e^{\frac{n\pi b}{a}}+e^{-\frac{n\pi b}{a}}\right)}\int_0^a f(x)\cos\left(\frac{n\pi x}{a}\right)dx.$ (7.94)

7.2.3 Mixed Problem in Rectangle

Example: Evaluate the steady temperature in a rectangular plate of length a and width b, the sides of which are kept at temperature zero, the lower end is kept at temperature f(x) and the upper edge is kept insulated.

Solution: In steady state temperature $u(x,t)$of two dimensional plate is given by Laplace ‘s equation $\frac{\partial^2 u}{\partial x^2}+\frac{\partial^2 u}{\partial y^2}=0,$ (7.95)

Subject to boundary condition **BC1:** $u(0,y)=0, u(a,y)=0,$ (7.96)

BC2: $u_y(x,b)=0$ i.e., $\left(\frac{\partial u}{\partial y}\right)_{y=b}=0$, (7.97)

BC3: $u(x,0)=f(x).$ (7.98)

Eq. (7.98) can be physical interpreted as the upper edge $y = b$ is kept insulated. Assume that

$$u(x, y) = X(x)Y(y) \text{ or } u = XY. \tag{7.99}$$

be a solution of (7.95), where X is a function of x alone and Y is a function of y alone.

Differentiating twice Eq. (7.99) partially w.r.t x and y, we get

$$\frac{\partial^2 u}{\partial x^2} = X''Y, \quad \frac{\partial^2 u}{\partial y^2} = XY''. \tag{7.100}$$

Substituting the values from Eqs. (7.99) and (7.100) in (7.95), we get

$$\therefore X''Y + XY'' = 0 \Rightarrow \frac{X''}{X} = -\frac{Y''}{Y}. \tag{7.101}$$

Now in Eq. (7.101) variables are separable since x and y are independent variables, this equation can hold only when both sides reduce to a constant say k.

$$\therefore X''Y + XY'' = 0 \Rightarrow \frac{X''}{X} = -\frac{Y''}{Y} = k, \tag{7.102}$$

$$\therefore \frac{X''}{X} = k \Rightarrow X'' - kX = 0, \tag{7.103}$$

$$\text{and } -\frac{Y''}{Y} = k \Rightarrow Y'' + kY = 0. \tag{7.104}$$

Eqs. (7.103) and (7.104) are second order ordinary differential equations. Now let us look at the boundary condition **BC1**:

$$u(0, y) = 0, u(a, y) = 0. \tag{7.105}$$

The solution of Eq. (7.97) is given by Eq. (7.100) i.e., $u(x,t) = X(x)Y(y)$. Using Eq. (7.99) and **BC1**, we get

$$u(0,y) = X(0)Y(y) = 0 \Rightarrow X(0) = 0, \tag{7.106}$$

and

$$u(a,y) = X(a)Y(y) = 0 \Rightarrow X(a) = 0. \tag{7.107}$$

($\because Y(y) \neq 0$ as we are look for non –zero solution)

Now we solve Eq. $X'' - kX = 0$, (7.108)

with boundary conditions $X(0) = 0$ and $X(a) = 0$. (7.109)

Case (i) when k is positive. Let $k = p^2$,for some $p \in R - \{0\}$.

The Eq. (7.70) reduces to $X'' - p^2 X = 0$. The auxiliary equation associate with $X'' - p^2 X = 0$ is $m^2 - p^2 = 0 \Rightarrow m = p, -p$.

Since the roots of the auxiliary equation are real and distinct, therefore

$$X(x) = Ae^{px} + Be^{-px}, \tag{7.110}$$

where *A* and *B* are arbitrary constant. Now we apply the boundary condition given by Eq. (7.109) to find the value of *A* and *B*. Put $x = 0$ in Eq. (7.110), we get $X(0) = A + B$. (7.111)

Using Eq. (7.109) in Eq. (7.111), we get $A = -B$. (7.112)

Put $x = a$ in Eq. (7.110), we get $X(a) = Ae^{pa} + Be^{-pa}$. (7.113)

Using Eqs. (7.109) and (7.112) in Eq. (7.113), we get $X(a) = Ae^{-pa}\left(e^{2pa} - 1\right) = 0$.

$\Rightarrow A = 0 \left(\because e^{pa}\left(e^{2pa} - 1\right) \neq 0\right)$.

Since $A = 0$, therefore from Eq. (7.112), we get $B = 0$. So, Eq. (7.110) reduces to $X(x) = 0$.

Substituting $X(x) = 0$ in Eq. (7.99), the required solution is $u(x, y) = 0$. So, when k is positive, there is no Eigen value and we get a trivial solution.

Case (ii) when k is zero.

Then auxiliary equation of Eqs. (7.103) is $m^2 = 0 \Rightarrow m = 0, 0$. Since, the roots are real and repeated, therefore solution of Eq. (7.108) is

$$X(x) = (A + Bx) \tag{7.114}$$

Now, we use boundary condition $X(0) = 0$ and $X(a) = 0$ to find the value of arbitrary constant.

Put $x = 0$ and $x = a$ in Eq. (7.114), we get $X(0) = A = 0$,

(7.115)

and $X(a) = A + Ba = 0$. Since $A = 0$, therefore

$$X(a) = A + Ba = 0 \text{ gives } B = 0. \tag{7.116}$$

From Eqs. (7.115) and (7.116), Eq. (7.114) gives $X(x) = 0$. Substituting $X(x) = 0$ in Eq. (7.99), the required solution is $u(x, y) = 0$. So, when k is zero, there is no Eigen value and we get a trivial solution.

Case (iii) when k is negative. Let $k=-p^2$, where $p \in R-\{0\}$.

Then auxiliary equation of Eqs. (7.108) is $m^2+p^2=0 \Rightarrow m=\pm ip$.

Since, the roots are complex, therefore solution of Eq. (7.103) is

$$X(x)=(A\cos px+B\sin px) \tag{7.117}$$

where *A* and *B* are arbitrary constant. Now, we use boundary condition $X(0)=0$ and $X(a)=0$ to find the value of arbitrary constant.

Put $x=0$ in Eq. (7.117), we get $X(0)=(A\cos p0+B\sin p0)$.

$$\Rightarrow A=0 \quad (\because X(0)=0) \tag{7.118}$$

Put $x=a$ in Eq. (7.117), we get $X(a)=(A\cos pa+B\sin pa)$,

$\Rightarrow X(a)=B\sin pa=0$, (From (7.118))

$\Rightarrow B=0$ or $\sin pa=0$,

$\Rightarrow \sin pa=0$, ($\because B\neq 0$ as $X(x)=0$)

$\Rightarrow pa=\sin^{-1}(0)$

$\Rightarrow pa=n\pi$,

$$\Rightarrow p=\frac{n\pi}{a},\ n=1,2,3,\ldots.. \tag{7.119}$$

For $n=1,2,3,\ldots..$, we get different value of p, so we denote p by p_n.Therefore $p=p_n=\frac{n\pi}{a}$. Since $p_n=\frac{n\pi}{a}$, therefore from Eq. (7.117)

$$X(x)=B\sin\left(\frac{n\pi}{a}\right)x. \tag{7.120}$$

For $n = 1, 2, 3, ...$, we get different value of $X(x)$ using Eq. (7.120), so we denote $X(x)$ by $X_n(x)$. Therefore $X_n(x) = B_n \sin\left(\frac{n\pi}{a}\right)x$. (7.121)

Similarly, when k is negative i.e., $k = -p^2$ then Eq. (7.105) changes to $Y'' - \frac{n^2\pi^2}{a^2}Y = 0$.

The auxiliary equation associate with $Y'' - \frac{n^2\pi^2}{a^2}Y = 0$ is

$$m^2 - \frac{n^2\pi^2}{a^2} = 0 .$$

$$\Rightarrow m = \frac{n\pi}{a}, -\frac{n\pi}{a} .$$

Since the roots of the auxiliary equation are real and distinct, therefore

$$Y(y) = Ce^{\frac{n\pi}{a}y} + De^{-\frac{n\pi}{a}y}, \tag{7.122}$$

where *C* and *D* are arbitrary constant. Using boundary condition BC2 given by Eq. (7.97), we get

$$u_y(x,b) = X(x)Y'(b) = 0 \Rightarrow X(x) = 0 \text{ or } Y'(b) = 0$$

$$\Rightarrow Y'(b) = 0. \tag{7.123}$$

($\because$ $X(x) \neq 0$, as $X(x) = 0$ gives $u(x, y) = 0$, which is a trivial solution and we are looking for non-trivial solution).

Differentiating Eq. (7.87) w.r.t y, we get

$$Y'(y)=\frac{n\pi}{a}Ce^{\frac{n\pi}{a}y}-\frac{n\pi}{a}De^{-\frac{n\pi}{a}y}. \tag{7.124}$$

Substituting $y=b$ in Eq. (7.124), we get

$$Y'(b)=\frac{n\pi}{a}Ce^{\frac{n\pi}{a}b}-\frac{n\pi}{a}De^{-\frac{n\pi}{a}b}. \tag{7.125}$$

$\frac{n\pi}{a}Ce^{\frac{n\pi}{a}b}-\frac{n\pi}{a}De^{-\frac{n\pi}{a}b}=0$, (From (7.123))

$$D=\frac{Ce^{\frac{n\pi b}{a}}}{e^{-\frac{n\pi b}{a}}}. \tag{7.126}$$

Substituting the value of D from Eq. (7.126) in Eq. (7.122), we get

$$Y(y)=Ce^{\frac{n\pi}{a}y}+\frac{Ce^{\frac{n\pi b}{a}}}{e^{-\frac{n\pi b}{a}}}e^{-\frac{n\pi}{a}y},$$

$$Y(y)=\frac{Ce^{\frac{n\pi}{a}y}e^{-\frac{n\pi b}{a}}+Ce^{\frac{n\pi b}{a}}e^{-\frac{n\pi}{a}y}}{e^{-\frac{n\pi b}{a}}},$$

$$Y(y)=e^{\frac{n\pi b}{a}}\left(Ce^{\frac{n\pi}{a}(b-y)}+Ce^{-\frac{n\pi}{a}(b-y)}\right). \tag{7.127}$$

For $n=1,2,3,.....$, we get different value of $Y(y)$ using Eq. (7.127), so we denote $Y(y)$ by $Y_n(y)$. Therefore Eq. (7.127) can be expressed as $Y_n(y)=C_n e^{\frac{n\pi b}{a}}\left(e^{\frac{n\pi}{a}(b-y)}+e^{-\frac{n\pi}{a}(b-y)}\right)$.

$$Y_n(y)=2C_ne^{\frac{n\pi b}{a}}\cosh\left\{\frac{n\pi(b-y)}{a}\right\}$$

(7.128)

as $e^{\theta}+e^{-\theta}=2\cos h\theta$. Therefore, from Eqs. (7.121) and (7.128) and using the principle of superposition, we get

$$u(x,y)=\sum_{n=1}^{\infty}X_n(x)Y_n(x),$$

$$u(x,y)=\sum_{n=1}^{\infty}B_n\sin\left(\frac{n\pi}{a}\right)x\,2C_ne^{\frac{n\pi b}{a}}\cosh\left\{\frac{n\pi(b-y)}{a}\right\},$$

$$u(x,y)=\sum_{n=1}^{\infty}A_ne^{\frac{n\pi b}{a}}\sin\left(\frac{n\pi}{a}\right)x\cosh\left\{\frac{n\pi(b-y)}{a}\right\}.$$

(7.129)

where $A_n=2B_nC_n$. Eq. (7.129) is the solution satisfying the boundary conditions Eq. (7.96) -(7.97). Now to find A_n, we use the boundary condition (7.98) i.e. $u(x,0)=f(x)$. Substituting $y=0$ in Eq. (7.129), we get $u(x,0)=\sum_{n=1}^{\infty}A_ne^{\frac{n\pi b}{a}}\sin\left(\frac{n\pi}{a}\right)x\cosh\left(\frac{n\pi b}{a}\right)$.

From (7.99), the above equation changes to

$$f(x)=\sum_{n=1}^{\infty}A_ne^{\frac{n\pi b}{a}}\sin\left(\frac{n\pi}{a}\right)x\cosh\left(\frac{n\pi b}{a}\right),$$

which represent the half range Fourier sine series of $f(x)$ over the interval $(0\ a)$.Hence , we get

$$A_n\,e^{\frac{n\pi b}{a}}\cosh\left(\frac{n\pi b}{a}\right)=\frac{2}{a}\int_0^a f(x)\sin\left(\frac{n\pi x}{a}\right)dx,$$

$$A_n = \frac{2}{ae^{\frac{n\pi b}{a}} \cosh\left(\frac{n\pi b}{a}\right)} \int_0^a f(x) \sin\left(\frac{n\pi x}{a}\right) dx.$$

Therefore, the required solution of the Laplace equation given by Eq. (7.95) with the boundary conditions (7.96) -(7.98) is

$$u(x,y) = \sum_{n=1}^{\infty} A_n e^{\frac{n\pi b}{a}} \sin\left(\frac{n\pi}{a}\right) x \cosh\left\{\frac{n\pi(b-y)}{a}\right\}, \tag{7.130}$$

where $A_n = \frac{2}{ae^{\frac{n\pi b}{a}} \cosh\left(\frac{n\pi b}{a}\right)} \int_0^a f(x) \sin\left(\frac{n\pi x}{a}\right) dx.$ (7.131)

Example: Find a steady state temperature distribution in a rectangular plate bounded by the lines $x=0, x=a, y=0$ and $y=b$ if the edge $y=0$ is insulated , the edge $x=0$ and $x=a$ are kept at $0^\circ C$ and edge $y=b$ is kept at temperature $f(x)$.

Solution: In 2-dimensional plate the required temperature $u(x,t)$ in the steady state is given by Laplace Equation $\frac{\partial^2 u}{\partial x^2} + \frac{\partial^2 u}{\partial y^2} = 0.$

(7.132)

According to given problem, the boundary condition are as follows:

BC1: $u(0,y)=0, u(a,y)=0,$ (7.133)

BC2: $u_y(x,0)=0$ i.e., $\left(\frac{\partial u}{\partial y}\right)_{y=0} = 0,$ (7.134)

as the edges $y=0$ is insulated for $0 \le x \le a$.

BC3: $u(x,b)=f(x).$ (7.135)

for $0 \le x \le a$

Assume that $u(x,y)=X(x)Y(y)$ or $u=XY$.

(7.136)

be a solution of (7.132) where X is a function of *x* alone and Y is a function of *y* alone.

Differentiating twice Eq. (7.136) partially w.r.t *x* and *y*, we get

$$\frac{\partial^2 u}{\partial x^2}=X''Y,\ \frac{\partial^2 u}{\partial y^2}=XY''. \tag{7.137}$$

Substituting the values from Eqs. (7.136) and (7.137) in (7.132), we get

$$\therefore X''Y+XY''=0 \Rightarrow \frac{X''}{X}=-\frac{Y''}{Y}. \tag{7.138}$$

Now in Eq. (7.138) variables are separable since x and y are independent variables, this equation can hold only when both sides reduce to a constant say k.

$$\therefore X''Y+XY''=0 \Rightarrow \frac{X''}{X}=-\frac{Y''}{Y}=k, \tag{7.139}$$

$$\therefore \frac{X''}{X}=k \Rightarrow X''-kX=0, \tag{7.140}$$

and $-\dfrac{Y''}{Y}=k \Rightarrow Y''+kY=0.$ (7.141)

Eqs. (7.140) and (7.141) is second order ordinary differential equation. Now let us look at the boundary condition **BC1**: $u(0,y)=0, u(a,y)=0$.

The solution of Eq. (7.132) is given by Eq. (7.136) i.e., $u(x,t)=X(x)Y(y)$. Using Eq. (7.136) and **BC1**, we get

$$u(0,y)=X(0)Y(y)=0 \Rightarrow X(0)=0, \tag{7.142}$$

and

$$u(a,y)=X(a)Y(y)=0 \Rightarrow X(a)=0. \tag{7.143}$$

($\because Y(y)\neq 0$ as we are look for non –zero solution)

Now we solve Eq. $X'' - kX = 0$, (7.144)

with boundary conditions $X(0)=0$ and $X(a)=0$. (7.145)

Case (i) when k is positive. Let $k=p^2$,for some $p \in R-\{0\}$. The Eq. (7.144) reduces to $X''-p^2X=0$. The auxiliary equation associate with $X''-p^2X=0$ is $m^2-p^2=0 \Rightarrow m=p,-p$.

Since the roots of the auxiliary equation are real and distinct, therefore

$$X(x)=Ae^{px}+Be^{-px}, \tag{7.146}$$

where *A* and *B* are arbitrary constant. Now we apply the boundary condition given by Eq. (7.145) to find the value of *A* and *B*. Put $x=0$ in Eq. (7.146), we get

$$X(0)=A+B. \tag{7.147}$$

Using Eq. (7.145) in Eq. (7.147), we get $A=-B$. (7.148)
Put $x=a$ in Eq. (7.146), we get $X(a)=Ae^{pa}+Be^{-pa}$. (7.149)

Using Eqs. (7.145) and (7.148) in Eq. (7.149), we get $X(a)=Ae^{-pa}\left(e^{2pa}-1\right)=0$.

$$\Rightarrow A=0 \ \left(\because e^{pa}\left(e^{2pa}-1\right)\neq 0\right).$$

Since $A=0$, therefore from Eq. (7.148), we get $B=0$. So, Eq. (7.146) reduces to $X(x)=0$.

Substituting $X(x)=0$ in Eq. (7.136), the required solution is $u(x,y)=0$. So, when k is positive, there is no Eigen value and we get a trivial solution.

Case (ii) when k is zero.

The auxiliary equation of Eqs. (7.144) is $m^2=0 \Rightarrow m=0,0$. Since, the roots are real and repeated, therefore solution of Eq. (7.144) is

$$X(x)=(A+Bx) \tag{7.150}$$

Now, we use boundary condition $X(0)=0$ and $X(a)=0$ to find the value of arbitrary constant.

Put $x=0$ and $x=a$ in Eq. (7.150), we get $X(0)=A=0$,

(7.151)

and $X(a)=A+Ba=0$. Since $A=0$, therefore

$X(a)=A+Ba=0$ gives $B=0$. (7.152)

From Eqs. (7.151) and (7.152) , Eq. (7.150) gives $X(x)=0$. Substituting $X(x)=0$ in Eq. (7.136), the required solution is $u(x,y)=0$. So, when k is zero, there is no Eigen value and we get a trivial solution.

Case (iii) when k is negative. Let $k=-p^2$, where $p \in R-\{0\}$.

Then auxiliary equation of Eqs. (7.144) is $m^2+p^2=0 \Rightarrow m=\pm ip$.

Since, the roots are complex, therefore solution of Eq. (7.144) is

$$X(x)=(A\cos px+B\sin px) \tag{7.153}$$

where *A* and *B* are arbitrary constant. Now, we use boundary condition $X(0)=0$ and $X(a)=0$ to find the value of arbitrary constant.

Put $x=0$ in Eq. (7.153), we get $X(0)=(A\cos p0+B\sin p0)$.

$\Rightarrow A=0$ ($\because X(0)=0$) (7.154)

Put $x=a$ in Eq. (7.153), we get $X(a)=(A\cos pa+B\sin pa)$,

$\Rightarrow X(a)=B\sin pa=0$, (From (7.118))

$\Rightarrow B=0$ or $\sin pa=0$,

$\Rightarrow \sin pa=0$, ($\because B\neq 0$ as $X(x)=0$)

$\Rightarrow pa=\sin^{-1}(0)$

$\Rightarrow pa=n\pi$,

$\Rightarrow p=\dfrac{n\pi}{a}$, $n=1,2,3,.....$ (7.155)

For $n=1,2,3,.....$, we get different value of p, so we denote p by p_n. Therefore $p=p_n=\dfrac{n\pi}{a}$. Since $p_n=\dfrac{n\pi}{a}$, therefore from Eq. (7.153), we get $X(x)=B\sin\left(\dfrac{n\pi}{a}\right)x$. (7.156)

For $n=1,2,3,.....$, we get different value of $X(x)$ using Eq. (7.156), so we denote $X(x)$ by $X_n(x)$. Therefore $X_n(x)=B_n\sin\left(\dfrac{n\pi}{a}\right)x$. (7.157)

Similarly, when k is negative i.e., $k=-p^2$ then Eq. (7.141) changes to $Y''-\dfrac{n^2\pi^2}{a^2}Y=0$.

The auxiliary equation associate with $Y'' - \frac{n^2\pi^2}{a^2}Y = 0$ is

$$m^2 - \frac{n^2\pi^2}{a^2} = 0.$$

$$\Rightarrow m = \frac{n\pi}{a}, -\frac{n\pi}{a}.$$

Since the roots of the auxiliary equation are real and distinct, therefore

$$Y(y) = Ce^{\frac{n\pi}{a}y} + De^{-\frac{n\pi}{a}y}, \tag{7.158}$$

where *C* and *D* are arbitrary constant. Using boundary condition BC2 given by Eq. (7.134) in Eq. (7.158), we get

$$u_y(x,0) = X(x)Y'(0) = 0 \Rightarrow X(x) = 0 \text{ or } Y'(0) = 0$$

$$\Rightarrow Y'(0) = 0. \tag{7.159}$$

($\because$ $X(x) \neq 0$, as $X(x) = 0$ gives $u(x,y) = 0$, which is a trivial solution and we are looking for non-trivial solution).

Differentiating Eq. (7.158) w.r.t *y*, we get

$$Y'(y) = \frac{n\pi}{a}Ce^{\frac{n\pi}{a}y} - \frac{n\pi}{a}De^{-\frac{n\pi}{a}y}. \tag{7.160}$$

Substituting $y = 0$ in Eq. (7.160), we get $Y'(0) = \frac{n\pi}{a}C - \frac{n\pi}{a}D$.

(7.161)

$\frac{n\pi}{a}C - \frac{n\pi}{a}D = 0$, (From (7.159))

$$D = C. \tag{7.162}$$

Substituting the value of D from Eq. (7.162) in Eq. (7.158), we get

$$Y(y) = Ce^{\frac{n\pi}{a}y} + Ce^{-\frac{n\pi}{a}y},$$

$$Y(y) = C\left(e^{\frac{n\pi}{a}y} + e^{-\frac{n\pi}{a}y}\right),$$

$$Y(y) = 2C\cosh\left(\frac{n\pi y}{a}\right). \quad (7.163)$$

as $e^{\theta} + e^{-\theta} = 2\cosh\theta$

For $n = 1, 2, 3, \ldots\ldots$, we get different value of $Y(y)$ using Eq. (7.163), so we denote $Y(y)$ by $Y_n(y)$. Therefore Eq. (7.163) can be expressed as $Y_n(y) = 2C_n \cosh\left(\frac{n\pi y}{a}\right)$. (7.164)

Therefore, from Eqs. (7.157) and (7.164) and using the principle of superposition, we get

$$u(x, y) = \sum_{n=1}^{\infty} X_n(x) Y_n(x),$$

$$u(x, y) = \sum_{n=1}^{\infty} B_n \sin\left(\frac{n\pi}{a}\right) x\, 2C_n \cosh\left(\frac{n\pi y}{a}\right),$$

$$u(x, y) = \sum_{n=1}^{\infty} A_n \sin\left(\frac{n\pi}{a}\right) x \cosh\left(\frac{n\pi y}{a}\right). \quad (7.165)$$

where $A_n = 2B_nC_n$. Eq. (7.165) is the solution satisfying the boundary conditions Eq. (7.133) -(7.134). Now to find A_n, we use the boundary condition (7.135) i.e. $u(x,b)=f(x)$. Substituting $y=b$ in Eq.(7.165), we get $u(x,b)=\sum_{n=1}^{\infty} A_n \sin\left(\frac{n\pi}{a}\right) x \cosh\left(\frac{n\pi b}{a}\right)$.

From (7.135), the above equation changes to $f(x)=\sum_{n=1}^{\infty} A_n \sin\left(\frac{n\pi}{a}\right) x \cosh\left(\frac{n\pi b}{a}\right)$,

which represent the half range Fourier sine series of $f(x)$ over the interval $(0\ a)$.Hence , we get

$$A_n \cosh\left(\frac{n\pi b}{a}\right)=\frac{2}{a}\int_0^a f(x)\sin\left(\frac{n\pi x}{a}\right)dx,$$

$$A_n=\frac{2}{a\cosh\left(\frac{n\pi b}{a}\right)}\int_0^a f(x)\sin\left(\frac{n\pi x}{a}\right)dx.$$

Therefore, the required solution of the Laplace equation given by Eq. (7.132) with the boundary conditions (7.133) -(7.135) is

$$u(x,y)=\sum_{n=1}^{\infty} A_n \sin\left(\frac{n\pi}{a}\right) x \cosh\left(\frac{n\pi y}{a}\right),$$

where $A_n=\frac{2}{a\cosh\left(\frac{n\pi b}{a}\right)}\int_0^a f(x)\sin\left(\frac{n\pi x}{a}\right)dx$.

Example: A rectangular metal plate is bounded by the lines $x=0, x=a, y=0$ and $y=b$. The three sides $x=0, x=a$ and $y=b$ are insulated and the side $y=0$ is kept at temperature

$u_0 \cos\left(\frac{\pi x}{a}\right)$. Find the steady state temperature at any point of the plate.

Solution: In 2-dimensional plate the required temperature $u(x,t)$ in the steady state is given by Laplace Equation $\frac{\partial^2 u}{\partial x^2}+\frac{\partial^2 u}{\partial y^2}=0$.

(7.166)

According to given problem three sides of rectangular plate $x=0, x=a$ and $y=b$ are insulated, therefore the boundary condition are as follows:

BC1: $\left(\frac{\partial u}{\partial x}\right)_{x=0}=0, \left(\frac{\partial u}{\partial x}\right)_{x=a}=0$,

(7.167)

BC2: $u_y(x,b)=0$ i.e., $\left(\frac{\partial u}{\partial y}\right)_{y=b}=0$,

(7.168)

BC3: $u(x,a)=u_0 \cos\left(\frac{\pi x}{a}\right)$. (7.169)

Assume that $u(x,y)=X(x)Y(y)$ or $u=XY$.

(7.170)

be a solution of (7.166) where X is a function of *x* alone and Y is a function of *y* alone.

Differentiating twice Eq. (7.170) partially w.r.t *x* and *y*, we get

$$\frac{\partial^2 u}{\partial x^2}=X''Y,\ \frac{\partial^2 u}{\partial y^2}=XY''. \quad (7.171)$$

Substituting the values from Eqs. (7.171) in (7.166), we get

$$\therefore X''Y + XY'' = 0 \Rightarrow \frac{X''}{X} = -\frac{Y''}{Y}. \tag{7.172}$$

Now in Eq. (7.172) variables are separable since x and y are independent variables, this equation can hold only when both sides reduce to a constant say k.

$$\therefore X''Y + XY'' = 0 \Rightarrow \frac{X''}{X} = -\frac{Y''}{Y} = k, \tag{7.173}$$

$$\therefore \frac{X''}{X} = k \Rightarrow X'' - kX = 0, \tag{7.174}$$

and $-\frac{Y''}{Y} = k \Rightarrow Y'' + kY = 0.$ (7.175)

Eqs. (7.174) and (7.175) is second order ordinary differential equation. Now let us look at the boundary condition **BC1**: $\left(\frac{\partial u}{\partial x}\right)_{x=0} = 0, \left(\frac{\partial u}{\partial x}\right)_{x=a} = 0.$

The solution of Eq. (7.166) is given by Eq. (7.170) i.e., $u(x,t) = X(x)Y(y)$. Using Eq. (7.170) and **BC1**, we get $u'(0,y) = X'(0)Y(y) = 0 \Rightarrow X'(0) = 0,$

(7.176)

and $u'(a,y) = X'(a)Y(y) = 0 \Rightarrow X'(a) = 0.$

(7.177)

($\because Y(y) \neq 0$ as we are look for non –zero solution)

Now we solve Eq. $X'' - kX = 0,$ (7.178)

with boundary conditions

$$X'(0)=0 \text{ and } X'(a)=0. \tag{7.179}$$

Case (i) when k is positive. Let $k=p^2$,for some $p\in R-\{0\}$.

The Eq. (7.178) reduces to $X''-p^2X=0$. The auxiliary equation associate with $X''-p^2X=0$ is $m^2-p^2=0 \Rightarrow m=p,-p$ .Since the roots of the auxiliary equation are real and distinct, therefore $X(x)=Ae^{px}+Be^{-px}$,

$$\Rightarrow X'(x)=Ape^{px}-Bpe^{-px} \tag{7.180}$$

where *A* and *B* are arbitrary constant. Now we apply the boundary condition given by Eq. (7.179) to find the value of *A* and *B*. Put $x=0$ in Eq. (7.180), we get

$$X'(0)=Ap-Bp. \tag{7.181}$$

Using Eq. (7.179) in Eq. (7.181), we get $A=B$. (7.182)

Put $x=a$ in Eq. (7.180), we get $X'(a)=pAe^{pa}-pBe^{-pa}$.

(7.183)

Using Eqs. (7.179) and (7.182) in Eq. (7.183), we get $X'(a)=Ape^{pa}-Ape^{-pa}=0$.

$Ap\left(e^{pa}-e^{-pa}\right)=0$

$\Rightarrow A=0$, as $\left(e^{pa}-e^{-pa}\right)\neq 0$.

Since $A=0$, therefore from Eq. (7.182), we get $B=0$. So, $X(x)=Ae^{px}+Be^{-px}$ reduces to $X(x)=0$. Substituting

$X(x)=0$ in Eq. (7.170) , the required solution is $u(x,y)=0$. So, when k is positive, there is no Eigen value and we get a trivial solution.

Case (ii) when k is zero.

Then auxiliary equation of Eqs. (7.178) is $m^2=0 \Rightarrow m=0,0$. Since, the roots are real and repeated, therefore solution of Eq. (7.178) is

$$X(x)=(A+Bx), \tag{7.184}$$

$$\Rightarrow X'(x)=B. \tag{7.185}$$

Now, we use boundary condition $X'(0)=0$ and $X'(a)=0$ to find the value of arbitrary constant.

Put $x=0$ and $x=a$ in Eq. (7.185), we get $X'(0)=B=0$,

(7.186)

and $X'(a)=B=0$. Since $A=0$, therefore $X(a)=A+Ba=0$ gives $B=0$. (7.187)

From Eqs. (7.186) and (7.187) , Eq. (7.184) gives $X(x)=A$.

Case (iii) when k is negative. Let $k=-p^2$, where $p \in R-\{0\}$.

Then auxiliary equation of Eqs. (7.178) is $m^2+p^2=0 \Rightarrow m=\pm ip$. Since, the roots are complex, therefore solution of Eq. (7.178) is

$$X(x)=(A\cos px+B\sin px) \tag{7.188}$$

where *A* and *B* are arbitrary constant. Differentiating Eq. (7.188) w.r.t *x*, we get

$$X'(x) = -pA\sin px + Bp\cos px. \tag{7.189}$$

Now, we use boundary condition $X'(0)=0$ and $X'(a)=0$ to find the value of arbitrary constant.

Put $x=0$ in Eq. (7.189), we get $X'(0) = -pA\sin 0 + Bp\cos 0$.

$\Rightarrow Bp = 0$, ($\because X'(0)=0$)

$$\Rightarrow B = 0. \tag{7.190}$$

Put $x=a$ in Eq. (7.189), we get $X'(a) = -pA\sin pa + Bp\cos pa$,

$\Rightarrow X'(a) = -pA\sin pa$, (From (7.190))

$\Rightarrow -pA\sin pa = 0$,

$\Rightarrow A = 0$ or $\sin pa = 0$,

$\Rightarrow \sin pa = 0$, ($\because A=0$ as $X(x)=0$) and we look for non-trivial solution.

$\Rightarrow pa = \sin^{-1}(0)$

$\Rightarrow pa = n\pi$,

$$\Rightarrow p = \frac{n\pi}{a},\ n = 1, 2, 3, \ldots\ldots \tag{7.191}$$

For $n = 1, 2, 3, \ldots\ldots$, we get different value of p, so we denote p by p_n. Therefore $p = p_n = \frac{n\pi}{a}$. Since $p_n = \frac{n\pi}{a}$, therefore from Eq. (7.191), we get

$$X(x) = A\cos\left(\frac{n\pi}{a}\right)x. \tag{7.192}$$

For $n = 1, 2, 3, ...$, we get different value of $X(x)$ using Eq. (7.192), so we denote $X(x)$ by $X_n(x)$. Therefore

$$X_n(x) = A_n \cos\left(\frac{n\pi}{a}\right)x. \tag{7.193}$$

Similarly, when k is negative i.e., $k = -p^2$ then Eq. (7.175) changes to $Y'' - \frac{n^2\pi^2}{a^2}Y = 0$.

The auxiliary equation associate with $Y'' - \frac{n^2\pi^2}{a^2}Y = 0$ is $m^2 - \frac{n^2\pi^2}{a^2} = 0$.

$$\Rightarrow m = \frac{n\pi}{a}, -\frac{n\pi}{a}.$$

Since the roots of the auxiliary equation are real and distinct, therefore

$$Y(y) = Ce^{\frac{n\pi}{a}y} + De^{-\frac{n\pi}{a}y}, \tag{7.194}$$

where *C* and *D* are arbitrary constant. Using boundary condition BC2 given by Eq. (7.168), we get

$$u_y(x,b) = X(x)Y'(b) = 0 \Rightarrow X(x) = 0 \text{ or } Y'(b) = 0$$

$$\Rightarrow Y'(b) = 0. \tag{7.195}$$

($\because$ $X(x) \neq 0$, as $X(x) = 0$ gives $u(x,y) = 0$, which is a trivial solution and we are looking for non-trivial solution).

Differentiating Eq. (7.194) w.r.t *y*, we get

$$Y'(y) = \frac{n\pi}{a} Ce^{\frac{n\pi}{a}y} - \frac{n\pi}{a} De^{-\frac{n\pi}{a}y}. \tag{7.196}$$

Substituting $y = b$ in Eq. (7.196), we get

$$Y'(b) = \frac{n\pi}{a} Ce^{\frac{n\pi b}{a}} - \frac{n\pi}{a} De^{-\frac{n\pi b}{a}}. \tag{7.197}$$

$$\frac{n\pi}{a} Ce^{\frac{n\pi b}{a}} - \frac{n\pi}{a} De^{-\frac{n\pi b}{a}} = 0, \text{ (From (7.195))}$$

$$D = \frac{Ce^{\frac{n\pi b}{a}}}{e^{-\frac{n\pi b}{a}}}. \tag{7.198}$$

Substituting the value of D from Eq. (7.198) in Eq. (7.194), we get

$$Y(y) = Ce^{\frac{n\pi}{a}y} + \frac{Ce^{\frac{n\pi b}{a}}}{e^{-\frac{n\pi b}{a}}} e^{-\frac{n\pi}{a}y},$$

$$Y(y) = \frac{Ce^{\frac{n\pi}{a}y} e^{-\frac{n\pi b}{a}} + Ce^{\frac{n\pi b}{a}} e^{-\frac{n\pi}{a}y}}{e^{-\frac{n\pi b}{a}}},$$

$$Y(y)=e^{\frac{n\pi b}{a}}\left(Ce^{\frac{n\pi}{a}(b-y)}+Ce^{-\frac{n\pi}{a}(b-y)}\right). \tag{7.199}$$

For $n=1,2,3,.....$, we get different value of $Y(y)$ using Eq. (7.199), so we denote $Y(y)$ by $Y_n(y)$. Therefore Eq. (7.199) can be expressed as $Y_n(y)=C_n e^{\frac{n\pi b}{a}}\left(e^{\frac{n\pi}{a}(b-y)}+e^{-\frac{n\pi}{a}(b-y)}\right)$.

$$Y_n(y)=2C_n e^{\frac{n\pi b}{a}}\cosh\left\{\frac{n\pi(b-y)}{a}\right\} \tag{7.200}$$

as $e^{\theta}+e^{-\theta}=2\cos h\theta$. Therefore, from Eqs. (7.193) and (7.200) and using the principle of superposition, we get

$$u(x,y)=\sum_{n=1}^{\infty}X_n(x)Y_n(x),$$

$$u(x,y)=\sum_{n=1}^{\infty}A_n\cos\left(\frac{n\pi}{a}\right)x\,2C_n e^{\frac{n\pi b}{a}}\cosh\left\{\frac{n\pi(b-y)}{a}\right\},$$

$$u(x,y)=\sum_{n=1}^{\infty}B_n\cos\left(\frac{n\pi}{a}\right)x\cosh\left\{\frac{n\pi(b-y)}{a}\right\}. \tag{7.201}$$

where $B_n=2A_nC_n e^{\frac{n\pi b}{a}}$. Eq. (7.201) is the solution satisfying the boundary conditions Eq. (7.167) -(7.168). Now to find B_n, we use the boundary condition (7.169) i.e. $u(x,a)=u_0\cos\left(\frac{\pi x}{a}\right)$. Substituting $y=a$ in Eq. (7.201), we get

$$u(x,a)=\sum_{n=1}^{\infty}B_n\cos\left(\frac{n\pi}{a}\right)x\cosh\left\{\frac{n\pi(b-a)}{a}\right\}.$$

From (7.169) , the above equation changes to

$$u_0\cos\left(\frac{\pi x}{a}\right)=\sum_{n=1}^{\infty}B_n\cos\left(\frac{n\pi}{a}\right)x\cosh\left\{\frac{n\pi(b-a)}{a}\right\},$$

$$u_0\cos\left(\frac{\pi x}{a}\right)=B_0+B_1\cos\left(\frac{\pi x}{a}\right)\cosh\left\{\frac{\pi(b-a)}{a}\right\}+B_2\cos\left(\frac{2\pi x}{a}\right)\cosh\left\{\frac{2\pi(b-a)}{a}\right\}$$
$$+B_3\cos\left(\frac{3\pi x}{a}\right)\cosh\left\{\frac{3\pi(b-a)}{a}\right\}+\ldots,$$

Equating the coefficient on both the side of the above equation, we get

$$B_0=0,\ B_1\cosh\left(\frac{\pi(b-a)}{a}\right)=u_0,\ B_n=0, n\geq 2.$$

Substituting the value of $B_i, i=0,1,2,3,\ldots.$ in Eq. (201), we get

$$u(x,y)=u_0\operatorname{sech}\left\{\frac{(b-a)\pi}{a}\right\}\cosh\left\{\frac{(b-y)\pi}{a}\right\}.$$

Exercise

1. Solve the following Laplace equation $\frac{\partial^2 u}{\partial x^2}+\frac{\partial^2 u}{\partial y^2}=0$ in a rectangle with $u(0,y)=0, u(\mathrm{a},y)=0, u(x,b)=0$ and $u(x,0)=f(x)$ along x axis.

Ans:

$$u(x,y)=\sum_{n=1}^{\infty} b_n \sin\frac{n\pi x}{a}\left\{e^{\frac{n\pi y}{a}}-e^{-\frac{n\pi(y-2b)}{a}}\right\}$$

2. Find the steady state temperature distribution in a rectangular plate of sides $20cm$ and $10cm$ insulated on the lateral surface and maintained at zero temperature along three of its edges and at temperature $100^\circ C$ along the remaining longer edge.

Ans:

$$u(x,y)=\sum_{n=odd}^{\infty}\frac{400}{n\pi}\frac{1}{\sin\frac{n\pi}{a}}\sin\frac{n\pi x}{a}\sin\frac{n\pi y}{a}=\frac{400}{\pi}\sum\frac{\sin\left(\frac{2n-1}{20}\right)\pi x}{(2n-1)\sinh\left(\frac{2n-1}{2}\right)\pi}\sinh\left(\frac{2n-1}{20}\right)\pi y$$

3. Find the deflection $u(x,y,t)$ of a rectangular membrane, $0\le x\le a, 0\le y\le b$, given that is entire its entire boundary is fixed, initial velocity is zero (starts from rest) and initial deflection $f(x,y)=kxy(a-x)(b-y)$.

Ans:

$$u(x,y,t)=\left[\frac{64a^2bk}{\pi}\sum_{m=1}^{\infty}\sum_{n=1}^{\infty}\frac{1}{m^3}\sin\frac{m\pi x}{a}\right]\times\frac{1}{n^3}\sin\frac{n\pi y}{b}$$

4. Solve $u_{xx}+u_{yy}=0$, subject to $u(x,0)=x-x^2, u(x,1)=0$ $u(0,\mathrm{y})=0, u(1,\mathrm{y})=0$.

Ans:

$$u(x,y)=\frac{4}{\pi^3}\sum_{n=1}^{\infty}\frac{1-(-1)^n}{n^3}\frac{\sinh n\pi(1-x)}{\sinh n\pi}\sin n\pi y$$

5. Solve $u_{xx}+u_{yy}=0 \quad , 0<x<L_x, 0<y<L_y,$

Subjcct to $u(x,0)=0, u(x,1)=0,$
$u(0,\mathrm{y})=g(y), u(1,\mathrm{y})=0.$

Ans: $$u=\sum_{n=1}^{\infty} b_n \frac{\sin\frac{n\pi(1-x)}{L_y}}{\sinh\frac{n\pi}{L_y}} \sin\frac{n\pi y}{L_y}$$ where

$$b_n=\frac{2}{L_y}\int_0^{L_y} g(y)\sin\frac{n\pi y}{L_y}\,dy.$$

7.3 Solution of Laplace Equation Using Finite Difference Method

The Laplace 's equation is given by $\frac{\partial^2 u}{\partial x^2}+\frac{\partial^2 u}{\partial y^2}=0,$ (7.202)

Consider a rectangular region R as shown in Fig. 7.1, for which $u(x,y)$ is known at the boundary. Divide region R in to a network of square mesh of side h, (assuming that an exact sub-division of R is possible). Replacing the derivatives in Eq. (7.202) by their difference approximations, we have

$$\tfrac{1}{h^2}\left[u_{i-1,j}-2u_{i,j}+u_{i+1,j}\right]+\tfrac{1}{h^2}\left[u_{i,j-1}-2u_{i,j}+u_{i,j+1}\right]=0, \tag{7.203}$$

or

$$u_{i,j}=\tfrac{1}{4}\left[u_{i-1,j}+u_{i+1,j}++u_{i,j+1}+u_{i,j-1}\right]. \tag{7.204}$$

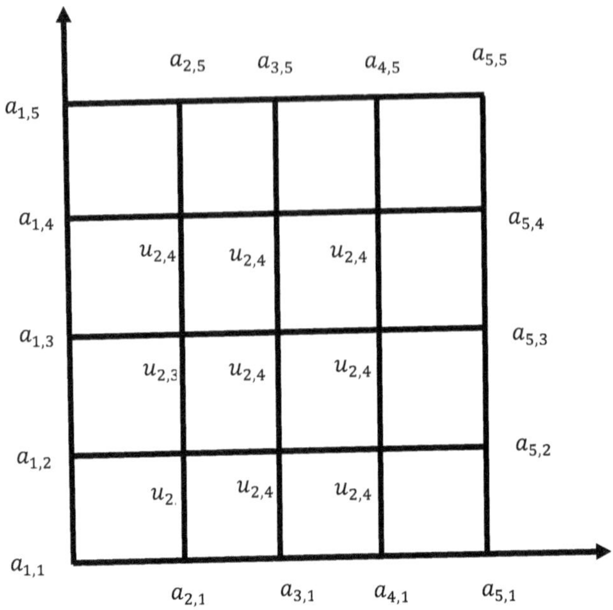

Figure 7.1: Mesh grid

The value at every internal mesh point is equal to the mean of the values at the four neighbouring that are closest to it on the left, right, above, and below. The typical five-point formula, or Eq. (7.204), is shown in Fig. 5.2. Occasionally, a formula equivalent to Eq. (7.204), which is provided by

$$u_{i,j} = \tfrac{1}{4}\left[u_{i-1,j+1} + u_{i+1,j-1} + u_{i+1,j+1} + u_{i-1,j-1}\right] \tag{7.205}$$

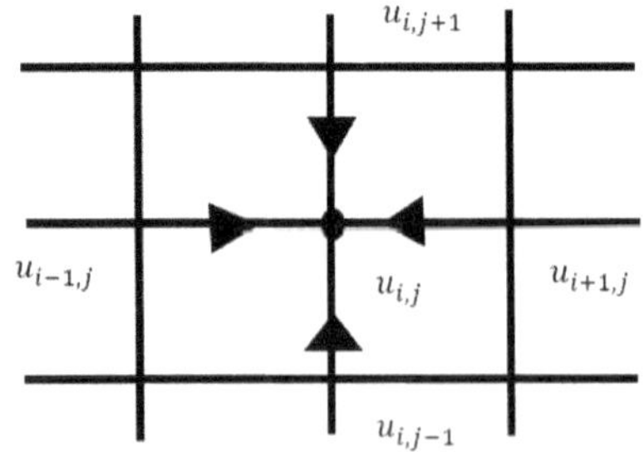

Figure 7.2: Standard five-point formula

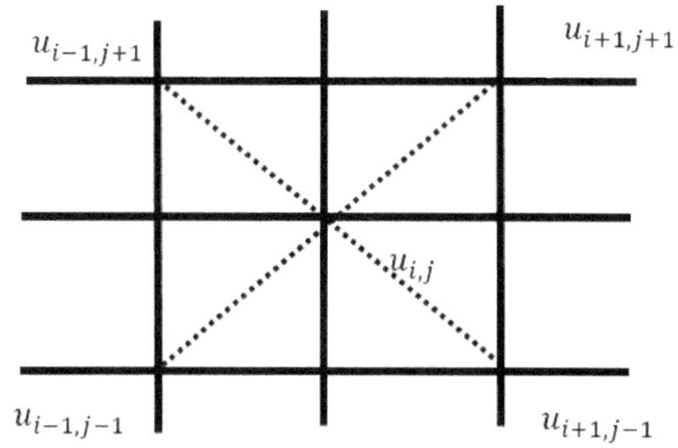

Figure 7.3: Diagonal five-point formula

This shows that the value of $u_{i,j}$ is the average of its values at the four neighbouring diagonal mesh points. The formula given by Eq. (7.205) is called the diagonal 5-point formula which is represented in Fig. 7.3. Although Eq. (7.205) is less accurate than Eq. (7.204), yet it serves as a reasonably good approximation for obtaining the starting values at the mesh points. Now to find the initial values of u at the interior mesh points, we first use diagonal five point formula Eq. (7.205) and compute

$u_{3,3}$, $u_{2,4}$, $u_{4,4}$, $u_{4,2}$ and $u_{2,2}$ in this order. Thus we get,

$$u_{3,3} = \tfrac{1}{4}\left[b_{1,5} + b_{5,1} + b_{5,5} + b_{1,1}\right]; \quad u_{2,4} = \tfrac{1}{4}\left[b_{1,5} + u_{3,3} + b_{3,5} + b_{1,3}\right]$$

$$u_{4,4} = \tfrac{1}{4}\left[b_{3,5} + b_{5,3} + b_{5,5} + u_{3,3}\right]; u_{4,2} = \tfrac{1}{4}\left[u_{3,3} + b_{5,1} + b_{3,1} + b_{5,3}\right]$$

$$u_{2,2} = \tfrac{1}{4}\left[b_{1,3} + b_{3,1} + u_{3,3} + b_{1,1}\right].$$

The values at the remaining interior points i.e. $u_{3,4}$, $u_{2,3}$, $u_{4,3}$, and $u_{3,2}$ are computed by the standard five- point formula Eq.(7.204).

Thus, we obtain

$$u_{2,3}=\frac{1}{4}\left[b_{1,3}+u_{3,3}+u_{2,4}+u_{2,2}\right]; u_{3,4}=\frac{1}{4}\left[u_{2,4}+u_{4,4}+b_{3,5}+u_{3,3}\right],$$

$$u_{4,3}=\frac{1}{4}\left[u_{3,3}+b_{5,3}+u_{4,4}+u_{4,2}\right]; u_{3,2}=\frac{1}{4}\left[u_{2,2}+u_{4,2}+u_{3,3}+u_{3,1}\right].$$

Having found all the nine values of $u_{i,j}$ once, their accuracy is improved by repeated application of Eq.(7.204) in the form

$$u_{i,j}^{(n+1)}=\frac{1}{4}\left[u_{i-1,j}^{(n+1)}+u_{i+1,j}^{(n)}+u_{i,j+1}^{(n+1)}+u_{i,j-1}^{(n)}\right]. \quad (7.206)$$

This formula utilizes the mesh points symmetrically from left to right along successive rows, using the most recent iterative value available. This procedure is continued until there is little to no difference in the values between one round and the next and this process is called iterations Liebmann's.

Example: Solve the elliptic equation $\frac{\partial^2 u}{\partial x^2}+\frac{\partial^2 u}{\partial y^2}=0$ for the square mash of Fig. 7.4 with boundary values as shown.

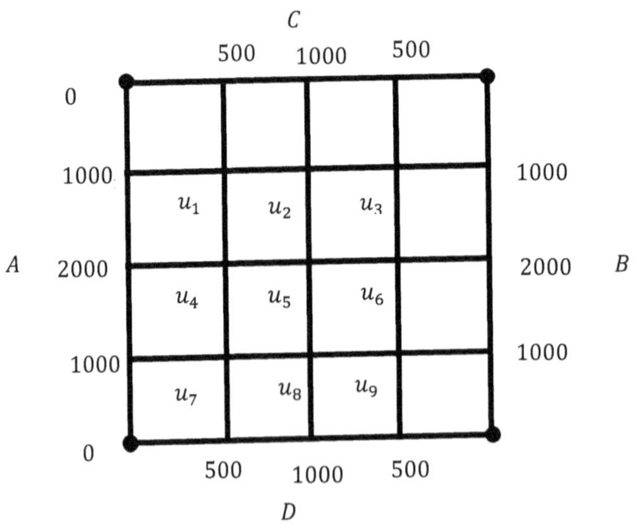

Figure 7.4: Mesh –points

Solution. Let $u_1, u_2, \ldots\ldots. u_9$ be the values of u at the interior mesh-points. Since the boundary values of u are symmetrical about AB.
$\therefore u_7 = u_1, u_8 = u_2, u_9 = u_3$

Also, the values of u being symmetrical about CD, $u_3 = u_1, u_6 = u_4, u_9 = u_7$.

Thus, it is sufficient to find the values u_1, u_2, u_4 and u_5. Now we find their initial value in the following order:

$$u_5 = \frac{1}{4}(2000 + 2000 + 1000 + 1000) = 1500 (\text{Std. formula})$$

$$u_1 = \frac{1}{4}(0 + 1500 + 1000 + 2000) = 1125 (\text{Diag. formula})$$

$$u_2 = \frac{1}{4}(1125 + 1125 + 1000 + 1500) = 1128 (\text{Std. formula})$$

$$u_4 = \frac{1}{4}(2000 + 1500 + 1125 + 1125) \approx 1438 (\text{Std. formula})$$

We carry out the iteration process using the formulae:

$$u_1^{(n+1)} = \tfrac{1}{4}\left[1000 + u_2^{(n)} + 500 + u_4^{(n)}\right]$$

$$u_2^{(n+1)} = \tfrac{1}{4}\left[u_1^{(n+1)} + u_1^{(n)} + 1000 + u_5^{(n)}\right]$$

$$u_4^{(n+1)} = \tfrac{1}{4}\left[2000 + u_5^{(n)} + u_1^{(n+1)} + u_1^{(n)}\right]$$

$$u_5^{(n+1)} = \tfrac{1}{4}\left[u_4^{(n+1)} + u_4^{(n)} + u_2^{(n+1)} + u_2^{(n)}\right]$$

First iteration: Put $n = 0$

$$u_1^{(1)} = \frac{1}{4}(1000 + 1188 + 500 + 1438) \approx 1032$$

$$u_2^{(1)} = \frac{1}{4}(1032 + 1032 + 1000 + 1500) = 1141$$

$$u_4^{(1)} = \frac{1}{4}(2000 + 1500 + 1032 + 1032) = 1391$$

$$u_5^{(1)} = \frac{1}{4}(1391 + 1391 + 1141 + 1141) = 1266$$

Second iteration: Put $n = 1$

$$u_1^{(2)} = \frac{1}{4}(1000 + 1141 + 500 + 1391) = 1008$$

$$u_2^{(2)} = \frac{1}{4}(1008 + 1008 + 1000 + 1266) = 1069$$

$$u_4^{(2)} = \frac{1}{4}(2000 + 1266 + 1008 + 1008) = 1321$$

$$u_5^{(2)} = \frac{1}{4}(1321 + 1321 + 1069 + 1069) = 1195$$

Third iteration: Put $n = 2$

$$u_1^{(3)} = \frac{1}{4}(1000 + 1069 + 500 + 1321) = 973$$

$$u_2^{(3)} = \frac{1}{4}(973 + 973 + 1000 + 1195) = 1035$$

$$u_4^{(3)} = \frac{1}{4}(2000 + 1195 + 973 + 973) = 1288$$

$$u_5^{(3)} = \frac{1}{4}(1288 + 1288 + 1035 + 1035) = 1162$$

Fourth iteration: Put $n = 3$

$$u_1^{(4)} = \frac{1}{4}(1000 + 1035 + 500 + 1288) = 956$$

$$u_2^{(4)} = \frac{1}{4}(956 + 956 + 1000 + 1162) = 1019$$

$$u_4^{(4)} = \frac{1}{4}(2000 + 1162 + 956 + 956) = 1269$$

$$u_5^{(4)} = \frac{1}{4}(1269 + 1269 + 1019 + 1019) = 1144$$

Fifth iteration: Put $n = 4$

$$u_1^{(5)} = \frac{1}{4}(1000 + 1019 + 500 + 1269) = 947$$

$$u_2^{(5)} = \frac{1}{4}(947 + 947 + 1000 + 1144) = 1010$$

$$u_4^{(5)} = \frac{1}{4}(2000 + 1144 + 947 + 947) = 1260$$

$$u_5^{(5)} = \frac{1}{4}(1260 + 1260 + 1010 + 1010) = 1135$$

Similarly,

$$u_1^{(6)} = 942, u_2^{(6)} = 1005, u_4^{(6)} = 1255, u_5^{(6)} = 1130$$

$$u_1^{(7)} = 940, u_2^{(7)} = 1003, u_4^{(7)} = 1253, u_5^{(7)} = 1128$$

$$u_1^{(8)} = 939, u_2^{(8)} = 1002, u_4^{(8)} = 1252, u_5^{(8)} = 1127$$

$$u_1^{(9)} = 939, u_2^{(9)} = 1001, u_4^{(9)} = 1251, u_5^{(9)} = 1126$$

Thus, there is a negligible difference between the values obtained in the eighth and ninth iterations.

Hence $u_1 = 939, u_2 = 1001, u_4 = 1251 \text{ and } u_5 = 1126$

OR

Solution: Let the interior values of u at the 9 grid points be $u_1, u_2, u_3, \ldots\ldots\ldots, u_9$.

Take the central horizontal and vertical line AB and CD. The values of u on the boundary are symmetrical w. r. t the lines AB and CD. Hence the value of u inside the mesh will also be symmetrical about AB and CD.

$u_1 = u_3 = u_7 = u_9; u_2 = u_8; u_4 = u_6$

Therefore, it is enough to find u_1, u_2, u_4, u_5.

Find the initial value at the grid points using standard five-point formula and diagonal five-point formula.

$u_5 = \frac{1}{4}(2000 + 2000 + 1000 + 1000) = 1500$ (Standard Five Point Formula (SFPF))

$u_1 = \frac{1}{4}(0+1500+1000+2000) = 1125$ (Apply Diagonal Five Point Formula (DFPF))

$u_2 = \frac{1}{4}(1000+1500+1125+1125) = 1187.5$ (Apply Standard Five Point Formula (SFPF))

$u_4 = \frac{1}{4}(1125+1125+2000+1500) = 1437.5$ (Apply Standard Five Point Formula (SFPF))

Find the iterative formula using standard five-point formula using Gauss Seidel Method

$$u_1 = \frac{1}{4}[500+u_4+1000+u_2]$$

$$u_2 = \frac{1}{4}[1000+u_5+u_1+u_3]$$

$$u_4 = \frac{1}{4}[2000+u_5+u_1+u_7]$$

$u_5 = \frac{1}{4}[u_2+u_8+u_4+u_6]$ But $u_2 = u_8; u_4 = u_6$

So $u_5 = \frac{1}{4}[2u_2+2u_4]$

Using the above iterative formula to improve the values at the grid points using Gauss Seidel points

Iteration	u_1	u_2	u_4	u_5
Initial value	1125	1187.5	1437.5	1500
1	1031.25	1093.75	1343.75	1218.75
2	984.38	1046.88	1296.88	1171.88
3	960.94	1023.44	1273.44	1148.44
4	949.22	1011.72	1261.72	1136.72
5	943.36	1005.86	1255.86	1130.86
6	940.43	1002.93	1252.93	1127.93
7	939.1	1001.6	1250.8	1126.6
8	938.3	1000.4	1250.2	1125.8

9	937.7	1000.2	1250.1	1125.2
10	937.6	1000.1	1250.1	1125.1
11	937.6	1000.1	1250.1	1125.1

Since the value in last two rows are same.

Since $u_1 = u_3 = u_7 = u_9; u_2 = u_8; u_4 = u_6$ therefore

$u_1 = u_3 = u_7 = u_9 = 937.6$,

$u_2 = u_8 = 1000.1; u_4 = u_6 = 1250.1; u_5 = 1125.1$.

Example: Given the values of $u(x, y)$ on the boundary of the square in the Figure 5.5, evaluate the function $u(x, y)$ satisfying the Laplace equation $\nabla^2 u = 0$ at the pivotal points of this figure.

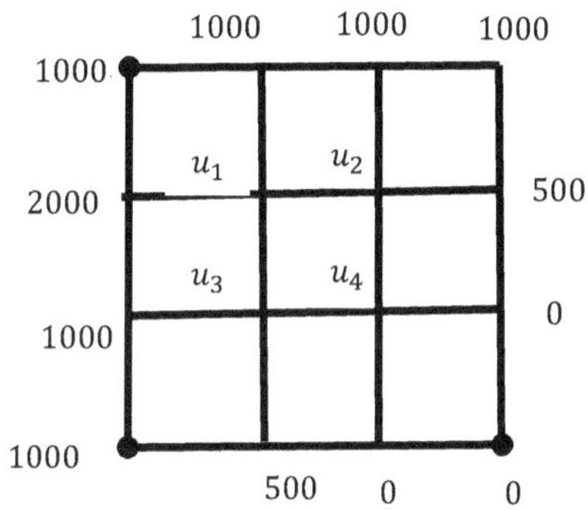

Figure 7.5: Grid points or mesh points

Solution: To get the initial values of u_1, u_2, u_3, u_4 we assume $u_4 = 0$.Then

$$u_1 = \frac{1}{4}(1000 + 0 + 1000 + 2000) = 1000 \left(\text{Diag. formula}\right)$$

$$u_2 = \frac{1}{4}(1000 + 500 + 1000 + 0) = 625 \left(\text{Std. formula}\right)$$

$$u_3 = \frac{1}{4}(2000 + 0 + 1000 + 500) = 875 \left(\text{Std. formula}\right)$$

$$u_4 = \frac{1}{4}(875 + 0 + 625 + 0) = 375 \left(\text{Std. formula}\right)$$

We carry out the successive iterations, using the formulae

$$u_1^{(n+1)} = \tfrac{1}{4}\left[2000 + u_2^{(n)} + 1000 + u_3^{(n)}\right]$$

$$u_2^{(n+1)} = \tfrac{1}{4}\left[u_1^{(n+1)} + 500 + 1000 + u_4^{(n)}\right]$$

$$u_3^{(n+1)} = \tfrac{1}{4}\left[2000 + u_4^{(n)} + u_1^{(n+1)} + 500\right]$$

$$u_4^{(n+1)} = \tfrac{1}{4}\left[u_3^{(n+1)} + 0 + u_2^{(n+1)} + 0\right]$$

First iteration: Put $n = 0$

$$u_1^{(1)} = \frac{1}{4}(2000 + 625 + 1000 + 875) = 1125$$

$$u_2^{(1)} = \frac{1}{4}(1125 + 500 + 1000 + 375) = 750$$

$$u_3^{(1)} = \frac{1}{4}(2000 + 375 + 1125 + 500) = 1000$$

$$u_4^{(1)} = \frac{1}{4}(1000 + 0 + 750 + 0) = 438$$

Second iteration: Put $n = 1$

$$u_1^{(2)} = \frac{1}{4}(2000 + 750 + 1000 + 1000) = 1188$$

$$u_2^{(2)} = \frac{1}{4}(1188 + 500 + 1000 + 438) = 782$$

$$u_3^{(2)} = \frac{1}{4}(2000 + 438 + 1188 + 500) = 1032$$

$$u_4^{(2)} = \frac{1}{4}(1032 + 0 + 782 + 0) = 454$$

Third iteration: Put $n = 2$

$$u_1^{(3)} = \frac{1}{4}(2000 + 782 + 1000 + 1032) = 1204$$

$$u_2^{(3)} = \frac{1}{4}(1204 + 500 + 1000 + 454) = 789$$

$$u_3^{(3)} = \frac{1}{4}(2000 + 454 + 1204 + 500) = 1040$$

$$u_4^{(3)} = \frac{1}{4}(1040 + 0 + 789 + 0) = 458$$

Similarly, $u_1^{(4)} = 1207, u_2^{(4)} = 791, u_3^{(4)} = 1041, u_4^{(4)} = 458,$

$$u_1^{(5)} = 1208, u_2^{(5)} = 791.5, u_3^{(5)} = 1041.5, u_4^{(5)} = 458.25.$$

Thus there is no significant difference between the fourth and fifth iteration values. Hence,

$u_1 = 1208, u_2 = 792, u_3 = 1042$ and $u_4 = 458.$

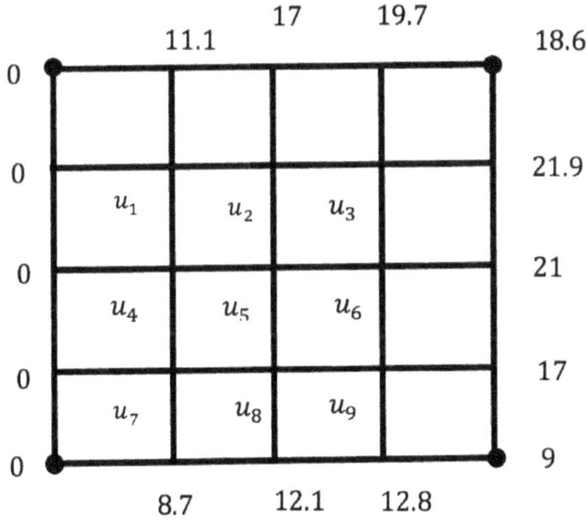

Figure 7.6: Grid points or mesh points

Example: Solve the Laplace equation $\frac{\partial^2 u}{\partial x^2} + \frac{\partial^2 u}{\partial y^2} = 0$given that Figure 7.6.

Solution: We first find the initial values in the following order:

$$u_5 = \frac{1}{4}(0 + 17 + 21 + 12.1) = 12.5 \left(\text{Std. formula}\right)$$

$$u_1 = \frac{1}{4}(0 + 12.5 + 0 + 17) = 7.4 \left(\text{Diag. formula}\right)$$

$$u_3 = \frac{1}{4}(12.5 + 18.6 + 17 + 21) = 17.28 \left(\text{Diag. formula}\right)$$

$$u_7 = \frac{1}{4}(12.5 + 0 + 0 + 12.1) = 6.15 \left(\text{Diag. formula}\right)$$

$$u_9 = \frac{1}{4}(12.5 + 9 + 21 + 12.1) = 13.65 \left(\text{Diag. formula}\right)$$

$$u_2 = \frac{1}{4}(17 + 12.5 + 7.4 + 17.3) = 13.55 \left(\text{Std. formula}\right)$$

$$u_4 = \frac{1}{4}(7.4 + 6.2 + 0 + 12.5) = 6.52 \left(\text{Std. formula}\right)$$

$$u_6 = \frac{1}{4}(17.3 + 13.7 + 12.5 + 21) = 16.12 \left(\text{Std. formula}\right)$$

$$u_8 = \frac{1}{4}(12.5 + 12.1 + 6.2 + 13.7) = 11.12 \left(\text{Std. formula}\right)$$

Now we carry out the iteration process using the standard formula:

$$u_1^{(n+1)} = \tfrac{1}{4}\left[0 + 11.1 + u_4^{(n)} + u_2^{(n)}\right]$$

$$u_2^{(n+1)} = \tfrac{1}{4}\left[u_1^{(n+1)} + 17 + u_5^{(n)} + u_3^{(n)}\right]$$

$$u_3^{(n+1)} = \tfrac{1}{4}\left[u_2^{(n+1)} + 19.7 + u_6^{(n)} + 21.9\right]$$

$$u_4^{(n+1)} = \tfrac{1}{4}\left[u_1^{(n+1)} + 19.7 + u_7^{(n)} + u_5^{(n)}\right]$$

$$u_5^{(n+1)} = \tfrac{1}{4}\left[u_4^{(n+1)} + u_2^{(n+1)} + u_8^{(n)} + u_6^{(n)}\right]$$

$$u_6^{(n+1)} = \tfrac{1}{4}\left[u_5^{(n+1)} + u_3^{(n+1)} + u_9^{(n)} + 21\right]$$

$$u_7^{(n+1)} = \tfrac{1}{4}\left[0 + u_4^{(n+1)} + 8.7 + u_8^{(n)}\right]$$

$$u_8^{(n+1)} = \tfrac{1}{4}\left[u_7^{(n+1)} + u_5^{(n+1)} + 12.1 + u_9^{(n)}\right]$$

$$u_9^{(n+1)} = \tfrac{1}{4}\left[u_8^{(n+1)} + u_6^{(n)} + 12.8 + 17\right]$$

First iteration: Put $n = 0$

$$u_1^{(1)} = \tfrac{1}{4}\left[0 + 11.1 + u_4^{(0)} + u_2^{(0)}\right] = \tfrac{1}{4}[0 + 11.1 + 6.52 + 13.55] = 7.79$$

$$u_2^{(1)} = \tfrac{1}{4}[7.79 + 17 + 12.5 + 17.28] = 13.64$$

$$u_3^{(1)} = \tfrac{1}{4}[13.64 + 19.7 + 16.12 + 21.9] = 12.84$$

$$u_4^{(1)} = \tfrac{1}{4}[0 + 7.79 + 6.15 + 12.5] = 6.61$$

$$u_5^{(1)} = \tfrac{1}{4}[6.61 + 13.64 + 11.12 + 16.12] = 11.88$$

$$u_6^{(1)} = \tfrac{1}{4}[11.88 + 17.84 + 13.65 + 21] = 16.09$$

$$u_7^{(1)} = \tfrac{1}{4}[0 + 6.61 + 8.7 + 11.12] = 6.61$$

$$u_8^{(1)} = \tfrac{1}{4}[6.61 + 11.88 + 12.1 + 13.65] = 11.06$$

$$u_9^{(1)} = \tfrac{1}{4}[11.06 + 16.09 + 12.8 + 17] = 12.238$$

Second iteration: Put $n = 1$

$$u_1^{(2)} = \tfrac{1}{4}[0 + 11.1 + 6.61 + 13.64] = 7.84$$

$$u_2^{(2)} = \tfrac{1}{4}[7.84 + 17 + 11.88 + 17.84] = 13.64$$

$$u_3^{(2)} = \tfrac{1}{4}[13.64 + 19.7 + 16.09 + 21.9] = 17.83$$

$$u_4^{(2)} = \tfrac{1}{4}[0 + 7.84 + 6.61 + 11.88] = 6.58$$

$$u_5^{(2)} = \tfrac{1}{4}[6.58 + 13.64 + 11.06 + 16.09] = 11.84$$

$$u_6^{(2)} = \tfrac{1}{4}[11.84 + 17.83 + 14.24 + 21] = 16.23$$

$$u_7^{(2)} = \tfrac{1}{4}[0 + 6.58 + 8.7 + 11.06] = 6.58$$

$$u_8^{(2)} = \tfrac{1}{4}[6.58 + 11.84 + 12.1 + 14.24] = 11.19$$

$$u_9^{(2)} = \tfrac{1}{4}[11.19 + 16.23 + 12.8 + 17] = 14.30$$

Third iteration: Put $n = 2$

$$u_1^{(3)}=\tfrac{1}{4}[0+11.1+6.58+13.64]=7.83$$
$$u_2^{(3)}=\tfrac{1}{4}[7.83+17+11.84+17.83]=13.637$$
$$u_3^{(3)}=\tfrac{1}{4}[13.63+19.7+16.23+21.9]=17.86$$
$$u_4^{(3)}=\tfrac{1}{4}[0+7.83+6.58+11.84]=6.56$$
$$u_5^{(3)}=\tfrac{1}{4}[6.56+13.63+11.19+16.23]=11.90$$
$$u_6^{(3)}=\tfrac{1}{4}[11.90+17.86+14.30+21]=16.27$$
$$u_7^{(3)}=\tfrac{1}{4}[0+6.56+8.7+11.19]=6.61$$
$$u_8^{(3)}=\tfrac{1}{4}[6.61+11.90+12.1+14.30]=11.23$$
$$u_9^{(3)}=\tfrac{1}{4}[11.23+16.27+12.8+17]=14.32$$

Similarly,

$$u_1^{(4)}=7.82,\ u_2^{(4)}=13.65,\ u_3^{(4)}=17.88,\ u_4^{(4)}=6.58,\ u_5^{(4)}=11.94,$$
$$u_6^{(4)}=16.28,\ u_7^{(4)}=6.63,\ u_8^{(4)}=11.25,\ u_9^{(4)}=14.33$$
$$u_1^{(5)}=7.83,\ u_2^{(5)}=13.66,\ u_3^{(5)}=17.89,\ u_4^{(5)}=6.50,\ u_5^{(5)}=11.95, u_6^{(5)}=16.29$$
$$u_7^{(5)}=6.64,\ u_8^{(5)}=11.25,\ u_9^{(5)}=14.34$$

7.4 Solution of Poisson's Equation

$$\frac{\partial^2 u}{\partial x^2}+\frac{\partial^2 u}{\partial y^2}=f(x,y), \tag{7.207}$$

is an elliptic equation which can be solved numerically at the interior mesh points of a square network when the boundary values are known. The standard five-point formula for Eq. (7.207) takes the form

$$u_{i-1,j}+u_{i+1,j}+u_{i,j+1}+u_{i,j-1}-4u_{i,j}=h^2 f(ih, jh). \tag{7.208}$$

By applying Eq. (7.208) at each mesh-point, we arrive at linear equations in the pivotal values i, j. These equations can be solved by Gauss-Seidal iteration method.

Example: Solve the Poisson equation

$$\frac{\partial^2 u}{\partial x^2}+\frac{\partial^2 u}{\partial y^2}=-81xy, 0<x<1, 0<y<1$$ given that

$$u(0, y)=0, u(x,0)=0,\ u(1,y)=100,\ u(x,1)=100 \text{ and } h=\frac{1}{3}.$$

Solution:

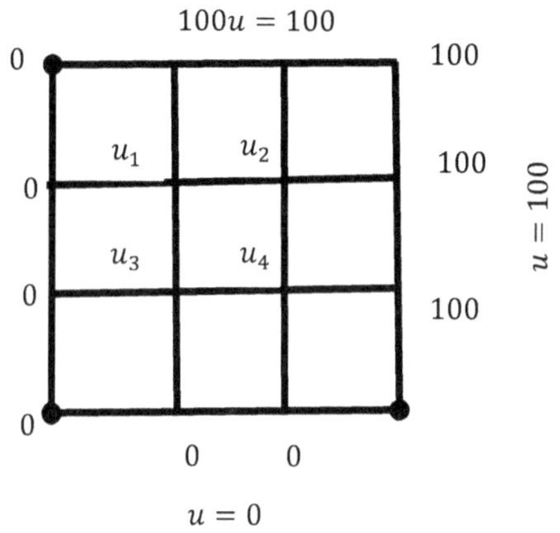

Figure 7.7: Grid points or mesh points

Here $h=\frac{1}{3}u_{i,j-1}u_{i,j}$. The standard 5-point formula for the given equation is

$$u_{i-1,j}+u_{i+1,j}+u_{i,j+1}+u_{i,j-1}-4u_{i,j}=h^2 f(ih, jh)=h^2\left[-81(ih.jh)\right]=h^4(-81)ij=-ij$$

. (7.209)

Put $i=1, j=2$ for u_1 , Eq. (7.209) gives $,0+u_2+u_3+100-4u_1=-2$,
i.e. $u_2+u_3-4u_1=-102$. (7.210)

Put $i=2, j=2$ for u_2 , Eq. (7.210) gives $u_1+100+u_4+100-4u_2=-4$,
i.e. $u_1-4u_2+u_4=-204$. (7.211)

Put $i=1, j=1$ for u_3 , Eq. (7.210) gives $0+u_4+0+u_1-4u_3=-1$,
i.e. $u_1-4u_3+u_4=-1$. (7.212)

Put $i = 2, j = 1$ for u_4, Eq.(7.211) gives $u_3 + 100 + u_2 - 4u_4 = -2$, i.e.
$u_2 + u_3 - 4u_4 = -102$ (7.213)

Subtracting Eq. (7.213) from Eq.(7.210), $-4u_1 + 4u_4 = 0$.
$\therefore u_1 = u_4$

Then Eq. (7.211) becomes $2u_1 - 4u_2 = -204$. (7.214)

and Eq. (7.212) becomes $2u_1 - 4u_3 = -1$. (7.215)

Now (4) x (7.210) + (7.214) gives $-14u_1 + 4u_3 = -612$. (7.216)

Eq. (7.215) + (7.216) gives $-12u_1 = -613$

Thus $u_1 = \dfrac{613}{12} = 51.0833 = u_4$.

From Eq. (7.214), $u_2 = \dfrac{1}{2}(u_1 + 102) = 76.5477$.

From Eq. (7.215), $u_3 = \dfrac{1}{2}\left(u_1 + \dfrac{1}{2}\right) = 25.7916$.

Example: Solve the partial differential equation $\nabla^2 u = -10(x^2 + y^2 + 10)$ over the square with sides $x = 0 = y$, $x = 3 = y$ with $u = 0$ on the boundary and mesh length $= 1$.

Solution: Here h = 1.

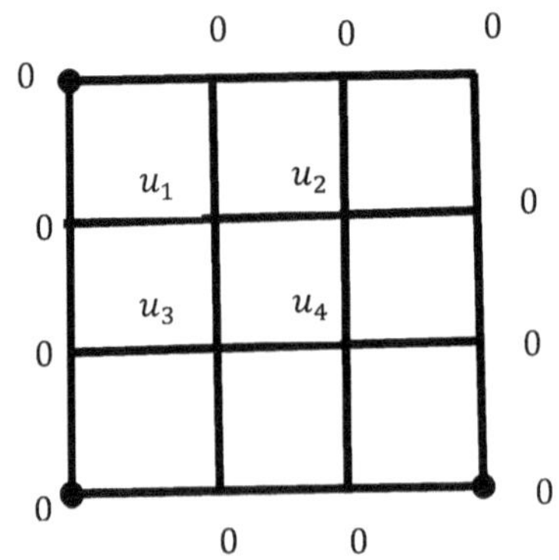

Figure 7.8: Grid points or mesh points

The standard 5-point formula for the given equation is

$$u_{i-1,j}+u_{i+1,j}+u_{i,j+1}+u_{i,j-1}-4u_{i,j}=-10(i^2+j^2+10). \qquad (7.217)$$

Put $i=1, j=2$ for u_1, Eq. (7.217) gives

$0+u_2+0+u_3-4u_1=-10(1+4+10)$

i.e. $u_1=\frac{1}{4}(u_2+u_3+150)$. (7.218)

Put $i=2, j=2$ for u_2, Eq. (7.217) gives $u_2=\frac{1}{4}(u_1+u_4+180)$.

(7.219)

Put $i=1, j=1$ for u_1, Eq. (7.217) gives $u_3=\frac{1}{4}(u_1+u_4+120)$. (7.220)

Put $i=2, j=1$ for u_2, Eq. (7.217) gives $u_4=\frac{1}{4}(u_2+u_3+150)$.

(7.221)

Eqs. (7.218) and (7.221) show that $u_4=u_1$. Thus the above equations reduce to

$$u_1=\frac{1}{4}(u_2+u_3+150). \qquad (7.222)$$

$$u_2=\frac{1}{2}(u_1+90). \qquad (7.223)$$

$$u_3=\frac{1}{2}(u_1+60) \qquad (7.224)$$

Solving system of linear equations given by (7.222) -(7.224) by Gauss-Seidal iteration method.

First iteration: Starting from the approximations $u_2=0, u_3=0,$, we obtain

$$u_1^{(1)}=37.5.\text{Then } u_2^{(1)}=\tfrac{1}{2}\left[37.5+90\right]\approx 64; u_3^{(1)}=\tfrac{1}{2}\left[37.5+60\right]\approx 49$$

Second iteration:

$$u_1^{(2)}=\tfrac{1}{4}\left[64+49+150\right]\approx 66;\ u_2^{(2)}=\tfrac{1}{2}\left[66+90\right]=78; u_3^{(2)}=\tfrac{1}{2}\left[66+60\right]=63$$

Third iteration:

$$u_1^{(3)} = \tfrac{1}{4}[78+63+150]=73;\ u_2^{(3)} = \tfrac{1}{2}[73+90] \approx 82; u_3^{(3)} = \tfrac{1}{2}[73+60] \approx 67$$

Fourth iteration:

$$u_1^{(4)} = \tfrac{1}{4}[82+67+150]=75;\ u_2^{(4)} = \tfrac{1}{2}[75+90] = 82.5; u_3^{(4)} = \tfrac{1}{2}[75+60] = 67.5$$

Fifth iteration:

$$u_1^{(5)} = \tfrac{1}{4}[82.5+67.5+150]=75;\ u_2^{(5)} = \tfrac{1}{2}[75+90] = 82.5; u_3^{(5)} = \tfrac{1}{2}[75+60] = 67.5$$

Since these values are the same as those of fourth iteration we have

$$u_1 = 75;\ u_2 = 82.5; u_3 = 67.5, u_4 = 75.$$

Example: Solve $u_{xx} + u_{yy} = 0$ for the following square mesh with boundary values as shown in the figure below.

Solution: Let u_1, u_2, u_3, u_4 be the values at the interior grid points. The boundary values are symmetrical about the diagonal AC but not about BD.

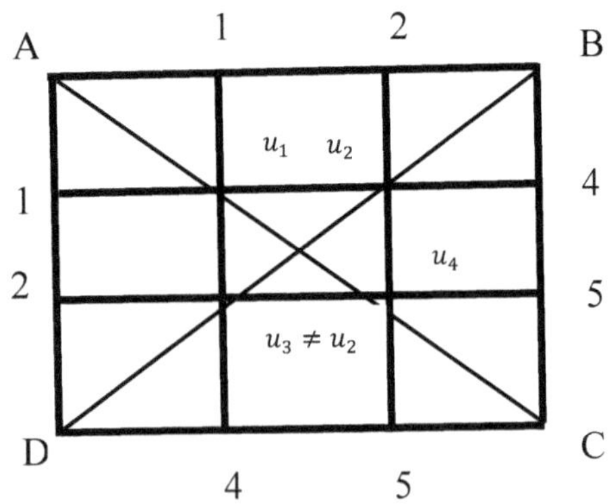

Figure 7.9: Grid points or mesh points

By symmetry, $u_2 = u_3; u_1 \neq u_4$. We need to find only u_1, u_2, u_4. Now we evaluate the initial values using standard five point formula by assuming $u_2 = u_3 = 0$. The initial values are

$$u_1 = \frac{1}{4}[0+0+1+1] = 0.5,$$

$$u_4 = \frac{1}{4}[0+5+5+0] = 2.5.$$

Using standard five-point formula, the iterative formulas are given by

$$u_1 = \frac{1}{4}[u_2 + u_2 + 1 + 1], \quad (7.225)$$

$$u_2 = \frac{1}{4}[u_4 + u_1 + 2 + 4], \quad (7.226)$$

$$u_4 = \frac{1}{4}[u_2 + u_2 + 5 + 5]. \quad (7.227)$$

Solving system of linear equations given by (7.225) -(7.226) by Gauss-Seidal iteration method.

Itera-tion	u_1	u_2	u_4
0	0.5	0.5625	2.5
1	0.78125	2.3203125	3.66015625
2	1.6515625	2.82729	3.913964844
3	1.913645	2.956902461	3.978451231
4	1.978451231	2.989225615	3.994612808
5	1.994612808	2.997306404	3.998653202
6	1.998653202	2.999326601	3.999663301
7	1.999663301	2.99983165	3.999915825

Since the value of u_1, u_2, u_4 are same up to three decimal ,therefore $u_1 = 1.999 \approx 2$, $u_2 = 2.999 \approx 3$, $u_3 = 3.999 \approx 4$. Also $u_2 = u_3$ therefore $u_3 = 2.999 \approx 3$.

Example: Solve $u_{xx} + u_{yy} = 0$ over the square mesh of side 4 units satisfying the following boundary conditions correct up to two decimal places:

$(i) u(0, y) = 0 \text{ for } 0 \leq y \leq 4$

$(ii) u(4, y) = 12 + y \text{ for } 0 \leq y \leq 4$

$(iii) u(x, 0) = 3x \text{ for } 0 \leq x \leq 4$

$(iv) u(x, 4) = x^2 \text{ for } 0 \leq x \leq 4$

Solution: We divide the square mesh into 16 sub-square of side 1 unit and calculate the numerical values of u on the boundary using the given boundary conditions.

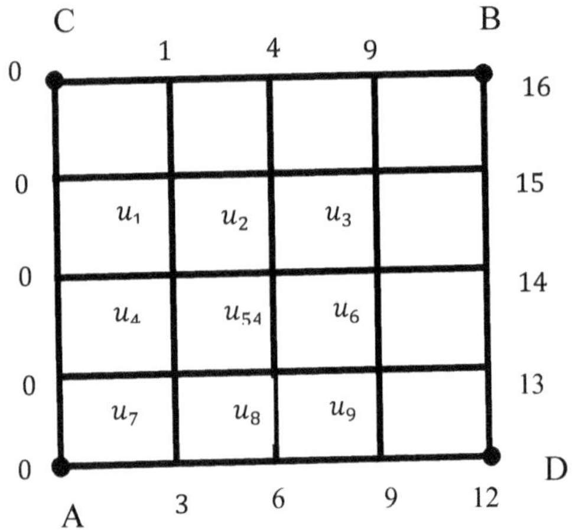

Figure 7.10: Grid points or mesh points

The boundary values are calculated using given boundary condition:

$(i) u(0, y) = 0 \text{ for } 0 \leq y \leq 4$

When x=0 and y=0,1,2,3,4 the value of function is 0 i.e.,

$x=0, y=0 \Rightarrow u(0,0)=0$

$x=0, y=1 \Rightarrow u(0,1)=0$

$x=0, y=2 \Rightarrow u(0,0)=0$

$x=0, y=3 \Rightarrow u(0,0)=0$

$x=0, y=4 \Rightarrow u(0,0)=0$

$(ii) u(4,y)=12+y \; for \; 0 \le y \le 4$

When x=4 and y=0,1,2,3,4 the value of function is 12+y i.e.,

$x=4, y=0 \Rightarrow u(4,0)=12+0=12$

$x=4, y=1 \Rightarrow u(4,1)=12+1=13$

$x=4, y=2 \Rightarrow u(4,0)=12+2=14$ (Give the value on line CB)

$x=4y=3 \Rightarrow u(4,0)=12+3=15$

$x=4y=4 \Rightarrow u(4,0)=12+4=16$

$(iii) u(x,0)=3x \; for \; 0 \le x \le 4$

When x=0,1,2,3,4 and y=0 the value of function is 3x i.e.,

$x=0, y=0 \Rightarrow u(0,0)=3 \times 0=0$

$x=1, y=0 \Rightarrow u(1,0)=3 \times 1=3$

$x=2, y=0 \Rightarrow u(2,0)=3 \times 2=6$ (Give the value on line D C)

$x=3, y=0 \Rightarrow u(3,0)=3 \times 3=9$

$x=4y=0 \Rightarrow u(4,0)=3 \times 4=12$

$(iv) u(x,4)=x^2 \; for \; 0 \le x \le 4$

When x=0,1,2,3,4 and y=4 the value of function is $\mathbf{x}^2$ i.e.,

$x=0, y=4 \Rightarrow u(0,4)=0^2=0$

$x=1, y=4 \Rightarrow u(1,4)=1^2=1$

$x=2, y=4 \Rightarrow u(2,4)=2^2=4$ (Give the value on line D C)

$x=3, y=4 \Rightarrow u(3,4)=3^2=9$

$x=4, y=4 \Rightarrow u(4,4)=4^2=16$

Let u_1, u_2, u_3, u_4 be the values at the interior grid points. The initial values for the problem are computed using standard five point formula and diagonal five point formula.

$$u_5 = \frac{1}{4}(4+6+0+14) = 6 \text{ (SFPF)}$$

$$u_1 = \frac{1}{4}(0+6+4+0) = 2.5 \text{ (DFPF)}$$

$$u_3 = \frac{1}{4}(16+6+14+4) = 17.3 \text{ (DFPF)}$$

$$u_7 = \frac{1}{4}(0+6+6+0) = 3 \text{ (DFPF)}$$

$$u_9 = \frac{1}{4}(6+14+6+12.1) = 13.7 \text{ (DFPF)}$$

$$u_2 = \frac{1}{4}(4+6+2.5+10) = 5.625 \text{ (SFPF)}$$

$$u_4 = \frac{1}{4}(0+6+2.5+3) = 3.125 \text{ (SFPF)}$$

$$u_6 = \frac{1}{4}(6+14+10+9.5) = 9.875 \text{ (SFPF)}$$

$$u_8 = \frac{1}{4}(6+6+3+9.5) = 13.7 \text{ (SFPF)}$$

Now, wc cvaluate the iterative formula using standard five-point formula using Gauss Seidel Method

$$u_1 = \frac{1}{4}[1+u_4+0+u_2]$$

$$u_2 = \frac{1}{4}[4+u_5+u_1+u_3]$$

$$u_3 = \frac{1}{4}[9 + u_2 + u_6 + 15]$$

$$u_4 = \frac{1}{4}[0 + u_5 + u_1 + u_7]$$

$$u_5 = \frac{1}{4}[u_2 + u_8 + u_4 + u_6]$$

$$u_6 = \frac{1}{4}[u_3 + u_9 + u_5 + 14]$$

$$u_7 = \frac{1}{4}[u_4 + u_8 + 3 + 0]$$

$$u_8 = \frac{1}{4}[u_5 + u_7 + u_9 + 6]$$

$$u_9 = \frac{1}{4}[u_6 + u_8 + 9 + 13]$$

Using the above iterative formula to improve the values at the grid points using Gauss Seidel points

Itera-tion	u_1	u_2	u_3	u_4	u_5	u_6	u_7	u_8	u_9
Initial value	6	2.5	5.625	10	6	9.875	3	6.125	9.5
1	2.4375	5.6094	9.8711	2.8594	6.1172	9.8721	2.9948	6.153	9.5063
2	2.3672	5.5888	9.8652	2.8698	6.1209	9.8731	3.0057	6.1582	9.5078
3	2.37	5.59	9.87	2.88	6.13	9.883.01	3.01	6.16	9.51

Since all the 9 values of u in third iteration up to two decimals are same as the corresponding values of the second iteration. Hence we stop the procedure, therefore

$$u_1 = 2.37, u_2 = 5.59, u_3 = 9.87, u_4 = 2.88, u_5 = 6.13, u_6 = 9.88, u_7 = 3.01, u_8 = 6.16, u_9 = 9.51.$$

Here we discuss the MATLAB code for the Laplace Equation $u_{xx} + u_{yy} = 0$ with the Dirichlet boundary condition given by

- $u(0, y) = 0$ for all *y* in the domain.
- $u(x, 0) = 0$ for all *x* in the domain.
- $u(1, y) = 0$ for all *y* in the domain.
- $u(x, 1) = 0$ for all *x* in the domain.

These conditions explicitly specify the values of the solution ϕ at the boundaries of the rectangular domain.

Matlab code for Laplace Equation

```
% Parameters
Lx = 1;       % Length of the domain in x-direction
Ly = 1;       % Length of the domain in y-direction

Nx = 50;      % Number of grid points in x-direction
Ny = 50;      % Number of grid points in y-direction
dx = Lx / (Nx-1); % Grid spacing in x-direction
dy = Ly / (Ny-1); % Grid spacing in y-direction
max_iter = 1000; % Maximum number of iterations
tol = 1e-5;    % Tolerance for convergence
% Initialize the solution
phi = zeros(Nx, Ny);
% Boundary conditions
phi(:,1) = 0;     % Bottom boundary
phi(:,Ny) = 1;    % Top boundary
phi(1,:) = 0;     % Left boundary
phi(Nx,:) = 0;    % Right boundary

% Main loop for iterative solution
for iter = 1:max_iter
  phi_old = phi;
   % Update the solution at interior points
  for i = 2:Nx-1
    for j = 2:Ny-1
      phi(i,j) = 0.25 * (phi_old(i+1,j) + phi_old(i-1,j) + phi_old(i,j+1) +
phi_old(i,j-1));
    end
  end
    % Check for convergence
  if max(abs(phi - phi_old), [], 'all') < tol
    fprintf('Converged after %d iterations.\n', iter);
    break;
  end
end
% Plot the solution
[X, Y] = meshgrid(0:dx:Lx, 0:dy:Ly);
```

```
figure;
surf(X, Y, phi');
xlabel('x');
ylabel('y');
zlabel('phi');
title('Solution of Laplace Equation');
```

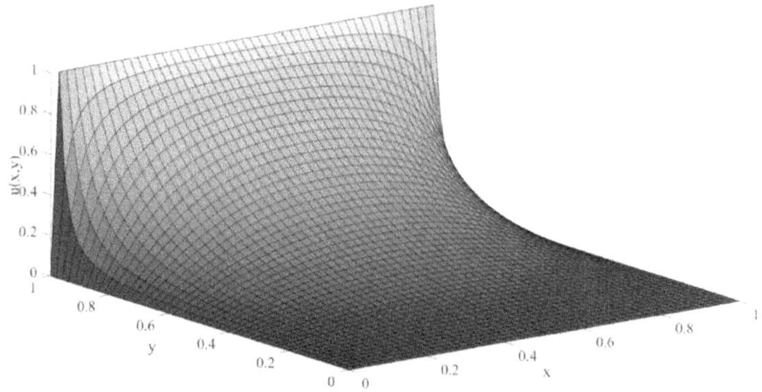

Figure 7.11: Solution of Laplace equation

www.ingramcontent.com/pod-product-compliance
Lightning Source LLC
LaVergne TN
LVHW012345041125
825065LV00014B/889
* 9 7 8 1 9 2 2 6 1 7 6 9 9 *